Physical Chemistry of Semiconductor Materials and Processes

Physical Chemistry of Semiconductor Materials and Processes

SERGIO PIZZINI

WILEY

This edition first published 2015
© 2015 John Wiley & Sons, Ltd

Registered office
John Wiley & Sons Ltd, The Atrium, Southern Gate, Chichester, West Sussex, PO19 8SQ, United Kingdom

For details of our global editorial offices, for customer services and for information about how to apply for permission to reuse the copyright material in this book please see our website at www.wiley.com.

The right of the author to be identified as the author of this work has been asserted in accordance with the Copyright, Designs and Patents Act 1988.

Wiley also publishes its books in a variety of electronic formats. Some content that appears in print may not be available in electronic books.

Designations used by companies to distinguish their products are often claimed as trademarks. All brand names and product names used in this book are trade names, service marks, trademarks or registered trademarks of their respective owners. The publisher is not associated with any product or vendor mentioned in this book.

Limit of Liability/Disclaimer of Warranty: While the publisher and author have used their best efforts in preparing this book, they make no representations or warranties with respect to the accuracy or completeness of the contents of this book and specifically disclaim any implied warranties of merchantability or fitness for a particular purpose. It is sold on the understanding that the publisher is not engaged in rendering professional services and neither the publisher nor the author shall be liable for damages arising herefrom. If professional advice or other expert assistance is required, the services of a competent professional should be sought

The advice and strategies contained herein may not be suitable for every situation. In view of ongoing research, equipment modifications, changes in governmental regulations, and the constant flow of information relating to the use of experimental reagents, equipment, and devices, the reader is urged to review and evaluate the information provided in the package insert or instructions for each chemical, piece of equipment, reagent, or device for, among other things, any changes in the instructions or indication of usage and for added warnings and precautions. The fact that an organization or Website is referred to in this work as a citation and/or a potential source of further information does not mean that the author or the publisher endorses the information the organization or Website may provide or recommendations it may make. Further, readers should be aware that Internet Websites listed in this work may have changed or disappeared between when this work was written and when it is read. No warranty may be created or extended by any promotional statements for this work. Neither the publisher nor the author shall be liable for any damages arising herefrom.

Library of Congress Cataloging-in-Publication Data applied for.

A catalogue record for this book is available from the British Library.

ISBN: 9781118514573

Typeset in 10/12pt TimesLTStd by SPi Global, Chennai, India

1   2015

Contents

Preface

Sunt igitur solida ac sine inani corpora prima. Preaterea quoniam genities in rebus inanest, materiam circum solidam constare necessest, nec res ulla potest vera ratione probari corpore inane suo celare atque intus habere, si non, quod cohibet, solidum constare reliquas.

Lucretius (56-95 BEC) 'De rerum natura'.

Physical chemistry today plays a critical role in our basic understanding, modelling, diagnostics and theoretical forecast of semiconductor materials' properties, albeit some of Lucretius's concepts remain alive, such as the concept of vacuum/vacancies (materia inane) in solid materials.

Their mechanical, electrical and optical properties depend not only on their structure and composition, but also on their defects and impurities content and on their deviations from stoichiometry, when compound semiconductors are considered. We have learned how to manage their properties by doping and defect engineering processes.

Physical chemistry is behind most of these processes and is crucial in the growth, purification and post-growth treatments of semiconductors, such as impurity gettering and passivation. In this respect, this book represents the first attempt to treat semiconductor materials and processes from a purely physico-chemical viewpoint.

This subject is treated at a tutorial level, for students and specialists having a background knowledge in solid-state physics. For this reason the book starts with some elementary thermodynamic concepts, then continues by dealing with issues concerning point and extended defects in elemental and compound semiconductors, having in mind the thermodynamics and kinetics at the base of their behaviour. The physico-chemical aspects of growth-and post-growth processes of semiconductor materials are the final issues considered, giving substantial attention to the majority of semiconductors of industrial interest, but also to semiconductor nanowires and thin film semiconductors for photovoltaic and optoelectronic applications.

This book is dedicated to Professor Giovanni (Nanni) Giacometti, Academician, who stimulated in me a strong interest for the physico-chemical aspects of material science when he was my teacher, and who became a colleague and a close friend in the subsequent years.

This book is dedicated also to the hundreds of masters and doctoral degree students whom I taught, while not having the time to write a book for them.

I express my gratitude to a number of friends and colleagues worldwide who have supported my work during the preparation of this book with advice and delivery of material. Among them, my particular gratitude goes to Stefan Estreicher, Ichiro Yonenaga, Koichi Kakimoto, Chris Van de Walle, Arthur Pelton, Andrew R. Barron, Peter Weinberger, Michael Stavola, Nicola Marzari, Sandro Scandolo, Annalisa Fasolino, Arul Kumar, Margit Zacharias, Gudrun Kissinger, Harmut Bracht, Leonid Zhigilei, Kai Tang, Yong Du, Biao Hu, Erik Mazur, Eugene Yakimov, Alexander Thorsten Blumenau, Otto Sankey, David Holec, Jonathan S. Barnard, Anrew Barron, Caterina Summonte, Naoki Fukata, Eike Weber, Giovanni Isella, Alessia Irrera, Tzanimir Arguirov, Pierre Ruterana, Wladek Walukiewicz, and to my former coworkers Simona Binetti and Maurizio Acciarri, as well as to my son Michele, his wife Elena, and Silvia Mazzon for their help in improving, when possible, the figures of this book.

1

Thermodynamics of Homogeneous and Heterogeneous Semiconductor Systems

1.1 Introduction

Elemental and compound semiconductors represent a vast family of materials of strategic interest for a variety of mature and advanced applications in micro- and opto-electronics, solid state lighting (SSL), solid state physical and chemical sensors, high efficiency solar cells and nanodevices. The materials themselves have always been technology enablers and their role today is even more significant in view of the increasing demand for sustainable development applications and high temperature, high pressure technologies.

The semiconductors family includes elemental solids such as silicon, the material of choice for the microelectronic and photovoltaic industry, binary alloys such as the Si-Ge alloys used for their elevated carrier mobilities, and compound semiconductors, of which SiC is used for high power, high frequency devices and phosphides, arsenides and nitrides for the most advanced optoelectronic applications.

Their preparation under defined limits of stoichiometry (in the case of compounds) and purity requires a deep knowledge of the chemistry and physics of liquid-, solid-, vapour- and plasma-growth and post-growth processes.

Semiconductors, like other inorganic and organic solids, may be stable in different structural configurations, depending on the composition, temperature, hydrostatic pressure and strain. Their chemical, physical and mechanical properties under different environmental conditions (temperature, pressure, strain) depend on their elemental composition, stoichiometry, impurity contamination, and also on their point- and extended-defects content. In fact, although solids are a typical class of materials characterized by microscopic order, most of their electronic and optoelectronic properties depend on or are influenced by impurities and point and extended defects.

Physical Chemistry of Semiconductor Materials and Processes, First Edition. Sergio Pizzini.
© 2015 John Wiley & Sons, Ltd. Published 2015 by John Wiley & Sons, Ltd.

Knowledge of their macroscopic features, such as their structural, thermodynamic, chemical, electrical and mechanical properties over a broad range of temperatures and pressures, is critical for their practical use. These properties, when not already available, should be experimentally or computationally determined.

This objective, addressed at metals, metal alloys and non-metallic solids, has been in the last few decades the traditional goal of physical metallurgy and physical chemistry. It is also the subject of several excellent textbooks and monographs [1–4], where emphasis is mainly given to structure–property relationships, solution- and defect-theories and nonstoichiometry of non-metallic solids [5], devoting, until very recently [6], only limited attention to elemental and compound semiconductors.

The aim of this chapter, and of the entire book, is to fill this gap and present in the most concise and critical manner possible the application of thermodynamics and physical chemistry to elemental and compound semiconductors, assuming knowledge of the fundamental laws of thermodynamics [7] and the basic principles of solid state and semiconductor physics [8, 9].

The intent is also to show that physical chemistry applied to semiconductors has been, and still is, of unique value for the practical and theoretical understanding of their environmental compliance and for the optimization of their growth and post-growth processes, all having a strong impact on the final properties of the material.

As impurities have a significant role in the optical and electronic properties of semiconductors, their thermodynamic behaviour will be considered in terms of their solubility and distribution among neighbouring phases as well as in terms of formation of complex species with other impurities and point defects.

For elemental semiconductors the main interest will be devoted to Group IV and VI elements (carbon, germanium, silicon, selenium and tellurium), the first of which being characterized by a number of stable phases, some of these of extreme scientific and technological interest, as is the case for diamond and graphene. For compound semiconductors we will consider the II–VI and the III–V compounds, such as the arsenides, phosphides, selenides, sulfides, tellurides and nitrides, all of which are of crucial interest for optoelectronic applications, SSL and radiation detection.

The most up to date physical and structural data of the different systems will be used: the reader interested in thermodynamic databases and phase diagram computation is referred to the Scientific Group Thermodata Europe (SGTE) Solution database, NSM Archive, www.ioffe.rssi.ru/SVA/NSM/Semicond/ and to Gibbs [10].

1.2 Basic Principles

A semiconductor is a *thermodynamic system* for which one has to define the equilibrium state and the nature of the transformations which occur when it is subjected to external thermal, mechanical, chemical, magnetic or electromagnetic forces during its preparation and further processing.

This system may consist of a homogeneous elemental or multicomponent *phase* or a heterogeneous mixture of several phases, depending on the temperature, pressure and composition.

A phase is conventionally defined as a portion of matter, having the property of being chemically and physically homogeneous at the microscopic level and of being confined within a surface which embeds it entirely.

The surface itself may be an external surface if it separates a phase from vacuum or from a gaseous environment. It is an internal surface, or an interface, when it separates a phase from another identical or different phase. According to Gibbs, the surface itself may be considered a phase of reduced (2D) dimensionality.

When one is concerned with microscopic or nanoscopic phases, such as nanodots, nanowires and nanotubes, the surface area to volume ratio, A_s/V increases considerably, as does the ratio R of the number of atoms at the surface to those in the bulk (see Table 1.1), with reduction in size of the crystallite phase. This has a significant impact on the physical and chemical properties of the phase itself and of its surface, enhancing in particular its chemical reactivity, but also other properties of relevant importance in semiconductor physics, such as the distribution and electrical activity of dopant impurities.

A phase may be gaseous, liquid or solid. In extreme conditions it could be stable in a plasma configuration, consisting of a mixture of electrons and ionized atoms/molecules. A phase is *condensed* when its aggregation state is that of a liquid or a solid material.

The thermodynamic state of a system is defined by specifying the minimum set of measurable properties needed for all the remaining properties to be fully determined. Properties which do not depend on mass (e.g. P, T) are called intensive. Those depending on mass (i.e. on composition) are called extensive.

A critical thermodynamic state of a system is its equilibrium state. It represents the condition where the system sits in a state of minimum energy and there are no spontaneous changes in any of its properties.

For a system consisting of a single, homogeneous multicomponent phase it is possible to define its thermodynamic state using thermodynamic functions (e.g. the internal energy U, the Helmholtz free energy F, the Gibbs free energy G, the entropy S and the chemical potential μ), whose values depend on macroscopic parameters, such as the hydrostatic pressure P, the absolute temperature T and the composition, this last given conventionally in terms of the atomic fraction of the components $x_i = \dfrac{n_i}{\sum_i n_i}$, n_i being the number of atoms of i-type.

Table 1.1 *Cell size dependence of the surface to volume ratio (A_s/V) and of the ratio R of atoms sitting at the surface vs those sitting in the volume, for a cubic crystal having an atomic density of 10^{21} cm^{-3}*

Cell edge length (nm)	Volume (nm^3)	Surface area (nm^2)	A_s/V (nm^{-1})	Atoms in the volume (N_V)	Atoms at the surface (N_s)	$R = N_s/N_v$
10^7	10^{21}	$6 \cdot 10^{14}$	$6 \cdot 10^{-7}$	10^{21}	$6 \cdot 10^{14}$	$6 \cdot 10^{-7}$
10^5	10^{15}	$6 \cdot 10^{10}$	$6 \cdot 10^{-5}$	10^{15}	$6 \cdot 10^{10}$	$6 \cdot 10^{-5}$
10^4	10^{12}	$6 \cdot 10^8$	$6 \cdot 10^{-4}$	10^{12}	$6 \cdot 10^8$	$6 \cdot 10^{-4}$
10^3	10^9	$6 \cdot 10^6$	$6 \cdot 10^{-3}$	10^9	$6 \cdot 10^6$	$6 \cdot 10^{-3}$
10^2	10^6	$6 \cdot 10^4$	$6 \cdot 10^{-2}$	10^6	$6 \cdot 10^4$	$6 \cdot 10^{-2}$

For Ge and Si the actual values of atomic densities are $4.42 \cdot 10^{22}$ and $5 \cdot 10^{22}$ (cm^{-3}).

A system is said to be in *mechanical equilibrium* when there are no unbalanced mechanical forces within the system and between the system and its surrounding. The system is also said to be in mechanical equilibrium when the pressure P throughout the system and between the system and the environment is the same. This condition is typical of the liquid state but not of the solid state unless internal mechanical stresses are fully relaxed.

Two systems are said to be in mechanical equilibrium with each other when their pressures are the same.

A system is said to be in chemical equilibrium when there are no chemical reactions going on within the system or they are fully balanced, such that there is no transfer of matter from one part of the system to another due to a composition gradient. Two systems are said to be in chemical equilibrium with each other when the chemical potentials of their components are the same. A definition of the chemical potential will be given below.

When the temperature T of the system is uniform and not changing inside the system, the system is said to be in thermal equilibrium. Two systems are said to be in thermal equilibrium with each other when their temperatures are the same. By convention, the temperature is expressed in degrees Kelvin.

Mechanical equilibrium conditions in a solid phase may be modified by an imbalance of mechanical forces arising from the presence of thermal or composition gradients and of lattice misfits. This imbalance is the driving force for the migration of point defects and for the formation of extended defects during crystallization processes from a liquid phase or a vapour phase and during heteroepitaxial depositions, as will be shown in Chapter 4.

We call *transformation* any change of the structural, physical and chemical properties of a system and, therefore, of its thermodynamic state, induced by the work w carried out on it by external mechanical, thermal, chemical, magnetic or electromagnetic forces, that is as the result of interaction of the system with the ambient.

To deal properly with the properties of a transformation it is necessary to know whether it is uniquely driven by mechanical forces, or by thermal, chemical, magnetic or electromagnetic ones.

The definition of a *thermally insulated system*, originally associated to a system impervious to any exchange of heat, should be extended to the case of systems impervious to exchanges of any kind of energy of non-mechanical nature with an external system (its environment), which behaves as the source or the sink of non-mechanical energy.

In the rest of this chapter heat is identified with the electromagnetic energy emitted by a specific radiation source (the flame of a combustion process, the hot filament of a light bulb, the surface of a star, a light emitting diode (LED) or a laser), which emits a broad- or line-like spectrum of radiations, depending on the physics of the emission process. Thermal sources behave as black bodies and the property of their spectrum is described by Planck's radiation law

$$E_\lambda = \frac{8\pi hc}{\lambda^5 (e^{(hc/\lambda kT)} - 1)} \tag{1.1}$$

where E_λ is the emitted energy per unit volume, h is the Planck constant (6.626×10^{-34} J s), λ is the wavelength and c is the speed of light (3×10^8 m s^{-1}).

The black body radiation spectra for different black body temperatures are displayed in Figure 1.1.

As conventional hot filament light bulbs operate with a maximum colour temperature around 3500 K, their emission occurs almost entirely in the infrared, with a

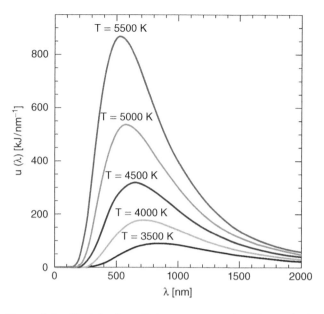

Figure 1.1   *Black body radiation curves. earthguide.ucsd.edu*

fraction not larger than 5% in the visible. This condition grants huge advantages to semiconductor-based SSL devices, whose monochromatic light emission depends on the energy gap E_g of the semiconductor used as the active substrate for quantum device applications which will be discussed later in this section.

By definition, a transformation which occurs without any input or output of heat is defined as *adiabatic.*

In this context, the work carried out by the system or supplied to the system is directly related to the variation in its internal energy U

$$\Delta U = \pm w \tag{1.2}$$

An *isothermal transformation,* however, occurs when the temperature of the system remains constant during the transformation.

Important examples of isothermal transformations are the changes of phase associated with solidification, melting, vaporization and sublimation or phase transitions of systems crystallizing in phases of different structure, as we will see later in this chapter.

In this last case, the system consists of two phases in thermodynamic equilibrium at the transformation temperature, any heat exchange from/with an external source/sink merely serves to modify the mass ratio of the two phases.

For a generic process involving a system that adsorbs heat from an external source and delivers work, the following equation holds for the variation of the internal energy U of the system

$$\Delta U = -w + q \tag{1.3}$$

where q is the energy absorbed as heat, w is the supplied work and $\Delta U = U_2 - U_1$ is the difference in the internal energy of the system.

Heat can be converted to mechanical work using a transfer fluid (liquid, gas or vapour) and operating a thermodynamic cycle (e.g. a Carnot cycle), whose maximum efficiency η is given by the second law of thermodynamics

$$\eta = \frac{W_{out}}{W_{in}} = \frac{T_s - T_a}{T_s} \tag{1.4}$$

where W_{out} and W_{in} are the output and input power, respectively, T_s is the temperature of the source and T_a is the temperature of a sink which dissipates the excess heat not converted to work.

As semiconductor-based devices play a crucial role in harvesting light and transforming it into electrical energy or emitting light as a result of a supply of electrical work, it is interesting to examine here these processes from the viewpoint of thermodynamics.

The electromagnetic energy of a radiation can be directly transformed into energy/work without making use of a transfer fluid using, for example, a semiconductor-based single- or multiple-junction device, where light is converted into electrical work qV, where q is the charge and V is the tension.[1]

Conversely, the electrical work qV supplied by polarizing an n–p junction fabricated on a direct gap semiconductor can be converted into the energy of a beam of monochromatic light. In fact, upon excitation, a fraction $\varepsilon = \frac{hv}{kT}\eta_{EQE}$ of the generated free electrons and holes, where η_{EQE} is the quantum efficiency of the process, relaxes by radiative recombination and generates a beam of photons with energy $\hbar v = E_g$, where E_g is the energy gap of the semiconductor. In principle, this process could occur with a theoretical 100% efficiency, as it is submitted only to the restrictions of the first principle of thermodynamics [11]. The second law of thermodynamics provides, instead, an upper limit to the theoretical efficiency of a direct transformation of the light adsorbed by a semiconducting material into electrical work qV, occurring through the recombination of the generated free electrons and holes in an external circuit.

In this case, if the radiation is that of a black body at a temperature T, as is the case of the light emitted by the sun, the efficiency of a solar cell is again ruled by the temperature of the source T_s and that of the substrate T_a, which behaves as a selective light absorber for light photons with energy $hv \geq E_g$ and dissipates the light not converted into energy as heat.

This upper theoretical limit cannot be reached by a single junction solar cell because only the photons of the incident light beam having an energy $hv \geq E_g$ are absorbed and excite electrons from the valence to the conduction band.

According to Shockley and Queisser [12], the variables on which the efficiency of a single junction semiconductor-based light converter depends are: the temperature of the sun T_s

$$kT_s = qV_s \tag{1.5}$$

the temperature of the solar cell T_c

$$kT_c = qV_c \tag{1.6}$$

---

[1] It is not the author's intention to deal here with the physics of solar cells or of solid state light emitters, but only to discuss the behaviour of these devices with respect to the second law of thermodynamics.

and the energy gap of the absorber

$$E_g = \hbar v_g = qV_g \tag{1.7}$$

so that the efficiency of the device involves only the two ratios

$$x_g = \frac{E_g}{kT_s} \tag{1.8}$$

$$\text{and } x_c = \frac{T_c}{T_s} \tag{1.9}$$

but depends also on the probability t_s that a photon of energy $\hbar v \geq E_g$ would produce an electron–hole pair and on a geometrical factor ζ related to the angle of incidence of the light on the surface of the device.

The ultimate efficiency is given by the equation

$$\eta(x_g, x_a, t_s, \xi) = \frac{qV}{W_{in}} \tag{1.10}$$

where the product qV is the maximum electrical power output of the device and W_{in} is the power (in watts) of the incident radiation. If the light source emits, instead, a beam of monochromatic light of energy $\hbar v'$ and the absorber has an energy gap $E_g = \hbar v'$, the system obeys only the first law and the maximum theoretical conversion efficiency should be 100%, as the electrical work qV corresponds to the variation of the internal energy of the system.

Other important kinds of transformation involve chemical forces, which drive matter fluxes among neighbouring phases of different structure and chemical composition. These transformations may by reversible or irreversible, and the latter ones are the concern of the thermodynamics of irreversible processes [13]. Most of the modern growth processes, such as physical vapour deposition (PVD), the chemical vapour deposition (CVD), atomic layer deposition (ALD) and metal-organic chemical vapour deposition (MO-CVD), which will be discussed in Chapter 4, belong to this kind of chemical processes. The systems involved in these processes are *open* systems.

1.3 Phases and Their Properties

1.3.1 Structural Order of a Phase

The properties of a phase are determined by its short- and long-range order and by its macroscopic and microscopic composition. While only short-range order is present in liquid phases, short- and long-range order is present in crystalline phases. In liquid and solid phases the short-range order can be investigated by extended X-ray absorption fine structure (EXAFS) spectroscopy [14–19].

Long-range order is investigated by diffraction experiments using X-rays, electrons or neutrons of adequate energy hv or wavelength λ, while high resolution transmission electron microscopy (HRTEM) directly reveals short- and long-range order with sub-nanometric resolution.

Today, knowledge of local composition is an additional, crucial requirement in meso-scopic and nanoscopic applications of semiconductors. Depending on the detection range and on the chemical sensitivity required, X-ray fluorescence (XRF), infrared spectroscopy (IRS), Scanning Ion mass spectroscopy (SIMS, both time of flight (TOF)-SIMS and dynamic SIMS), Raman spectroscopy, energy dispersive X-ray spectroscopy (EDS), field-emission Auger electron spectroscopy (FE-AES), electron energy loss spectroscopy (EELS) and atom probe tomography (APT) may be used [20–24], with local resolutions ranging from a few micrometres for Fourier transform infrared (FTIR) to sub-nanometres for APT and a detection sensitivity down to 100 ppt for dynamic SIMS.

To get information simultaneously on local structure and composition, APT is partic-ularly suited, as it allows both atomic range structural resolution and excellent chemical selectivity for the study of local order [25]. Figure 1.2 shows an interesting example of the application of APT to Ge-Sn alloys which will be discussed in Section 1.7.

A phase may consist of a single crystal or a polycrystalline aggregate. The crystalline materials found in nature, with several exceptions, consist mainly of polycrystalline aggregates, where the single crystallite size may vary from a few micrometres to several centimetres. The size- and orientation-distribution of the crystallites in a natural polycrys-talline matrix is generally disordered. The interfaces between crystallites are called grain boundaries (GBs).

Synthetic crystals can be grown as single-crystalline or polycrystalline ingots, or as thick or thin polycrystalline or amorphous films, as will be shown in Chapter 4. Synthetic poly-crystalline ingots can present a disordered or an ordered microstructure, depending on the growth process adopted.

GBs in natural and synthetic polycrystalline semiconductor materials (see Figure 1.3) are characterized by conditions of local disorder and electrical activity, associated with the

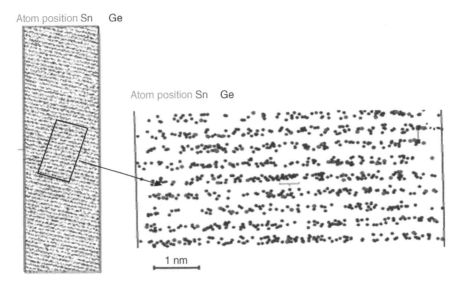

Figure 1.2   *Atom probe tomography of a Ge-Sn alloy sample: lattice planes lie in the <111> direction. Distance between the <111> planes for this sample is 0.377 nm. Kumar et al., 2012, [24]. Reproduced with permission from John Wiley & Sons, Ltd*

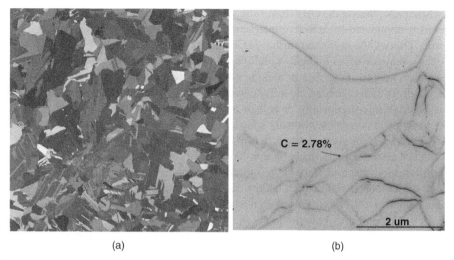

(a)   (b)

Figure 1.3   *(a) Scanning electron microscopy (SEM) micrography of a multicrystalline silicon sample. (b) Light beam induced current (LBIC) image of the electrical activity of GBs associated with recombination of minority carriers minority carriers recombination*

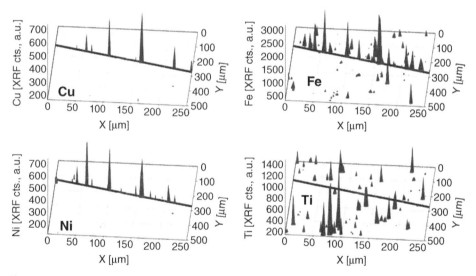

Figure 1.4   *Impurity segregation patterns in multicrystalline silicon, localized by synchrotron X-ray measurements. Buonassisi et al., 2006, [28] Reproduced with permission from John Wiley & Sons, Ltd*

presence of dislocations, distorted or broken (dangling) bonds and impurities segregated on them as individual species, precipitates or microprecipitates. SIMS, APT [26], X-ray absorption microspectroscopy (μ-XAS) [27] and X-rays microfluorescence (Figure 1.4) [28–31] are the main tools for direct impurity localization on GBs in semiconductors.

One can see in Figure 1.4 that Ni and Cu impurities segregate in correspondence to a GB (the solid line in the figure), Fe segregates both at the GB and in the bulk and Ti segregates elsewhere rather than at the GB. This is already an indication that segregation of impurities at a GB is dominated by selective interactions depending on the chemical nature of the impurity.This will be discussed further in Chapters 2 and 3.

A phase may consist of a single or multiple components. Homogeneous multicomponent solid phases may be stable as a *solution* or consist of stoichiometric or non-stoichiometric compounds. In both cases, the chemical composition of a phase in thermodynamic equilibrium should be microscopically, or at least mesoscopically, homogeneous. Solid solutions of semiconductors are conventionally called *alloys,* as in the case of metallic alloys.

The *solvent* and the *solutes,* respectively, are the components which are present in larger and smaller amount, respectively, in a homogeneous solution. Solutions are discussed in terms of ideality or non-ideality of their thermodynamic behaviour.

Ideal solution behaviour, which will be discussed in Section 1.6 and occurs very rarely in semiconductor alloys, is accomplished when the components do not interact chemically with each other and distribute randomly in the condensed phase, giving rise to a compositional disorder. Deviations from the ideal behaviour are also discussed in Section 1.6 in the frame of the most recent applications of the regular solution approach [32] and of the generalized quasichemical approximations (QCAs) [33, 34].

The solution is of the substitutional-type when the solute atoms replace the solvent atoms in their stable lattice positions, as is the case for the solution of Ge, Al, B, Ga, P in Si.

The solution is of the interstitial-type when the solutes enter in interstitial positions of the lattice. This is the case for transition metals (TMs), carbon and oxygen in Si and Ge.

Several impurities in elemental or compound semiconductors, however, may enter in both lattice and interstitial positions, behaving as substitutional and interstitial species. An additional possibility is present in compound semiconductors, where impurities may share occupancy in both sublattices.

Solutes which, added in trace amounts, as is the case of B, P, Sb, Ga, As in Si, modify the carrier concentration in the semiconductor are conventionally called dopants. The impurity/dopant content is generally reported in terms of atoms cm^{-3}, parts per million by weight (ppmw) or in parts per million atoms (ppma). As the concentration of dopants and impurities in semiconductors is directly correlated with the concentration of the corresponding shallow and deep levels, the best concentration notation is that expressed in atoms cm^{-3}.

Solutions of two different compounds, such as the solutions of alkali halides, alkali earth oxides or the III–V and II–VI compounds, could be considered pseudo-binary when the content of one of the components is invariant towards the composition changes of the other components.

A solution may be stable and homogeneous in a limited range of concentration or in a continuous range of compositions. A solution of different elemental or compound semiconductors is called a multicomponent alloy.

In the case of ionic solids forming a continuous series of substitutional solid solutions, Vegard [35] observed that often the effective lattice constants change linearly with the solute concentration, expressed in terms of atomic fraction, with a law which took his name

$$a^{o}_{A_x B_{(1-x)}} = x a^{o}_{A} + (1-x) a^{o}_{B} \tag{1.11}$$

(where $a^{o}_{A_x B_{(1-x)}}$ is the lattice constant of the solution for a certain value of x and a^{o}_{A} and a^{o}_{B} are the lattice constants of the pure components). In spite of many experimental and theoretical proofs which show that this empirical law is very rarely followed, ideal solid solution formation in semiconductors has often been correlated in the literature on their accomplishment with Vegard's law.

As will be discussed in detail in Chapter 2, solid solution formation may involve the additional presence of point defects [36–39].

This is the case for ionic and compound semiconductors, when aliovalent dopants are added to favour the ionic conductivity or to induce a p-type or n-type carrier excess, as occurs when AlN is doped with Mg_2N_3

$$MgN_{1.5} \rightarrow Mg_{Al} + 1.5\ N_N + 0.5\ V_{Al} \tag{1.12}$$

where Mg_{Al} is a Mg atom in an Al-lattice site, behaving as an acceptor, N_N is a nitrogen atom in a regular lattice position and V_{Al} is an aluminium vacancy.[2]

Amorphous solids (and glasses) exhibit the absence of long-range order and the presence of distorted and broken bonds: they might be conceived as metastable phases because they transform spontaneously in crystalline phases under suitable activation of the transformation.

1.4 Equations of State of Thermodynamic Systems

1.4.1 Thermodynamic Transformations and Functions of State

The properties of a thermodynamic system are described using equations of state that account for the dependence of a specific property of the system, such as its internal energy U, on the external variables and take the general form

$$f(P, V, \dots T) = 0 \tag{1.13}$$

A well known equation of state is that of ideal gases, written as

$$PV = RT \tag{1.14}$$

where R is the gas constant

Any transformation of a thermodynamic system from a generic state 1 to a generic state 2 is quantitatively accounted for by the corresponding changes in the function of state X which better describes the thermodynamic state of that system in the state 1 and in the state 2.

A specific character of a function of state X is that its change ΔX_{1-2} when the system passes from a state 1 to a state 2 must be independent of the path followed passing from 1 to 2.

Figure 1.5 shows such a change of state in a PV diagram, where $X_1 = P_1 V_1$ and $X_2 = P_2 V_2$.

A is well known, the functions which have this property are the internal energy U, the enthalpy H, the entropy S, the Helmoltz free energy F and the Gibbs free energy G.

---

[2] The Kröger and Vink notation [40] for defects in solids will be used throughout this book.

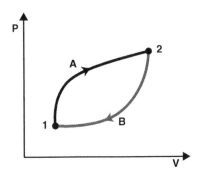

Figure 1.5   *Two possible paths for the transformation of the system from state 1 to state 2 and vice versa*

As it is experimentally impossible, although theoretically feasible, to define the absolute value of a function of state, it is common practice to define an arbitrary standard reference state, to which all changes associated with a transformation refer.

Thermodynamic transformations may be carried out reversibly or irreversibly. Reversible transformations are carried out along a continuous, formally infinite series of intermediate equilibrium states, each of them presenting a negligible energy difference from the neighbouring ones. Such types of transformations are invariant with respect to time and are fully deterministic. This condition allows one to bring a system from its initial state 1 to the final state 2, and to bring it back to the initial state 1 without hysteresis effects.

It should be noted that the transformations of real systems seldom follow full reversibility conditions, for various chemical, physical and structural reasons. The value of this procedure, however, rests in the fact that it allows calculation of the maximum work done by a transformation or the minimum work needed to induce the transformation.

1.4.2   Work Associated with a Transformation, Entropy and Free Energy

Every change in the thermodynamic state of a system is associated with an exchange of energy, and in particular of heat q and work w with the environment or with a neighbouring phase.

Like work, heat is not a function of state, because the amount of heat exchanged or supplied in any kind of transformation depends on the path followed in the transformation.

The infinitesimal change of internal energy dU associated with a reversible stage of a transformation is the sum of a mechanical work term and a heat term

$$dU = \delta w + \delta q \tag{1.15}$$

where the partial derivative notation for the heat and work terms shows that the values of δq and δw in the transformation depend on the route followed along the transformation.

In a closed cycle $\oint dU = 0$ and thus $\delta q = -\delta w$.

The work, here, is formally associated with a volumetric expansion or contraction of the system and is given by the integral

$$w = \int_{1}^{2} pdV \tag{1.16}$$

Thus, the total variation of internal energy of the system is given by the sum

$$\Delta U = \int_1^2 dq \pm \int_1^2 p dV = T \int_1^2 \frac{dq}{T} \pm \int_1^2 p dV \qquad (1.17)$$

where T is the absolute temperature.

The value of the integral $\int_1^2 p dV$, for an isothermal expansion/compression of an ideal gas

$$\int_1^2 p dV = RT \int_1^2 \frac{dV}{V} = RT \ln \frac{V_2}{V_1} \qquad (1.18)$$

depends only on the values V_2 and V_1 of the final and initial states and represents the maximum work done by the system or, changing the sign, the minimum work needed to carry out the transformation.

Furthermore, the value of the integral $\int_1^2 \frac{dq}{T}$ relative to a reversible transformation from state 1 to 2 at constant temperature T, is independent of the path followed, depending only on the states 1 and 2 and allows definition of a new function of state S

$$\int_1^2 \frac{dq}{T} = \frac{q_2}{T} - \frac{q_1}{T} = S_2 - S_1 = \Delta S \qquad (1.19)$$

which is the entropy.

According to Eq. (1.17)

$$\Delta U = T\Delta S + RT \ln \frac{V_2}{V_1} \qquad (1.20)$$

It must be emphasized here that only the variation ΔU of the internal energy associated with a transformation is thermodynamically well defined and measurable. The complete definition of U for a specific system would require, in fact, an arbitrary definition of its state of zero energy. Once this reference state is fixed, any other state is univocally determined.

It is then useful to define another function of state, the Helmholtz free energy F, whose change $\Delta F = RT \ln \frac{V_2}{V_1}$ is a measure of the maximum (here mechanical) work done by the system in isothermal conditions.

We have, therefore, for a reversible transformation, that

$$\Delta F = \Delta U - T\Delta S \qquad (1.21)$$

It should be mentioned that U, S and F are extensive properties of the system: it is, therefore, common practice to account for them in terms of unit of mass or mole.

Another thermodynamic function is G, the Gibbs free energy, whose variation is defined as the sum

$$\Delta G = \Delta U - T\Delta S + P\Delta V \qquad (1.22)$$

where the $P\Delta V$ term is a measure of the work carried out on the system to pass from an initial volume V_o to the final volume V during a composition change or an isobaric and isothermal transformation. Typical examples of processes involving a volume change are chemical reactions, the sublimation of solids and a number of solid state phase transformations, as is the case of silicon dioxide (i.e. silica) (see Table 1.2)

Table 1.2 *Density and crystal structure of silica polymorphs*

Phase	Structure	Density (g cm^{-3})	Temperature (°C)	Pressure (MPa)
β-Cristobalite	Cubic	2.33	>1470	—
β-Tridymite	Hexagonal	2.28	>870	—
β-Quartz	Hexagonal	2.53	>570	—
α-Quartz	Rhombohedral	2.65	<570	—
Coesithe	Monoclinic	2.93	—	>2000
Stishovite	Tetragonal	4.30	—	>8000

The definition of G allows also the definition of enthalpy H, by means of the equation

$$\Delta H = \Delta U + P\Delta V \tag{1.23}$$

Because in semiconductor alloys the volume dependence on composition is often negligible, the Helmoltz free energy F is interchangeable with the Gibbs free energy [33].

The standard thermodynamic properties of inorganic and organic solids[3] are tabulated in the CRC Handbook of Chemistry and Physics [41]. An entry value of 0.00 for ΔH or ΔG for an element indicates the reference state of the element.

1.4.3   Chemical Potentials

The transformations discussed so far occur in the absence of matter exchanges. Of special relevance are, however, the transformations, or processes, which involve mass transfer within neighbouring phases of different structure and composition, and which occur in the vast majority of chemical and physical processes used in the technology of materials of current use.

To quantify the variation of the internal energy of a homogeneous phase associated with a defined amount of mass exchange, it is convenient to use a new function, called the chemical potential μ_i

$$\mu_i = \left(\frac{dU}{dn_i} \right)_{S,V,n_i} \tag{1.24}$$

which accounts for the variation of the internal energy of a phase consisting of a solution of j components, when only its i-content changes by an infinitesimal amount dn_i.

Therefore, the total change in the internal energy of a homogeneous phase associated with an infinitesimal composition change dn of all its components is given by:

$$dU = TdS - PdV + \sum_i \mu_i dn_i \tag{1.25}$$

Given the definition of the Gibbs free energy (see Eq. (1.22)), we have also

$$dG = \sum_i \mu_i dn_i \tag{1.26}$$

---

[3] The temperature dependence of thermodynamic functions may be found in the NIST-JANAF Thermochemical Tables, making reference to their standard conditions, corresponding to an absolute temperature of 298.15 K and a pressure of 1 atm or 101.3 kPa.

In a system consisting of only one component, Eq. (1.26) reads

$$\mu_i = \left(\frac{dG}{dn_i}\right)_{T,P} \tag{1.27}$$

and

$$G_i^o = \mu_i \tag{1.28}$$

where the chemical potential μ_i takes the value of the molar free energy $G^o{}_i$ of the species i.

Let us now denote by $\left(\frac{dG}{dn_1}\right)_{\alpha\to\beta}$ the work needed to transfer an infinitesimal amount dn_i of the species i from the phase α to the phase β and by $\left(\frac{dG}{dn_1}\right)_{\beta\to\alpha}$ the work needed to transfer the same infinitesimal amount of mass dn_i from the phase β to the phase α, at the equilibrium coexistence temperature $T_{\alpha\beta}$ in the absence of any change of the concentration of the other species in both α and β.

To maintain the system in full equilibrium conditions during and after the transformation $(dG = 0)$

$$\left(\frac{dG}{dn_i}\right)_{a\to\beta} = \left(\frac{dG}{dn_i}\right)_{\beta\to\alpha} \tag{1.29}$$

and

$$\mu_i^\alpha = \mu_i^\beta \tag{1.30}$$

The equilibrium between phases of different composition is, therefore, given by a set of equations (1.30) for each j component of the solution. The relationship between μ_i and the composition will be given in Section 1.5.

1.4.4 Free Energy and Entropy of Spontaneous Processes

Let us now consider a thermally isolated system consisting, as before, of an intimate mixture of two phases α and β of identical composition in mutual equilibrium.

It holds, therefore, in equilibrium conditions

$$T_\alpha = T_\beta; G_\alpha = G_\beta \tag{1.31}$$

and

$$S = S_\alpha + S_\beta \tag{1.32}$$

if the two phases maintain their original integrity during and after the mass transfer process.

Let us suppose, instead, that the two phases are at different temperatures

$$T_\alpha > T_\beta \tag{1.33}$$

In this case a heat flux[4] $\Phi_q = \frac{dq}{dt}$ (where t is the time) would spontaneously flow from α to β until $T_\alpha = T_\beta$. Consequently, we will have a change of entropy of each of the two phases

---

[4] Heat fluxes may involve conduction, convection and radiation.

$dS_\alpha = -\left(\frac{dq}{T_\alpha}\right)$ and $dS_\beta = \left(\frac{dq}{T_\beta}\right)$ and a variation of entropy dS of the system

$$dS = -\left(\frac{dq}{T_\alpha}\right) + \left(\frac{dq}{T_\beta}\right) = dq\left(\frac{1}{T_\beta} - \frac{1}{T_\alpha}\right) > 0 \qquad (1.34)$$

Therefore, $dS > 0$ for a spontaneous process. We expect, similarly, that $dF < 0$ and $dG < 0$ for a spontaneous process.

1.4.5 Effect of Pressure on Phase Transformations, Polymorphs/Polytypes Formation and Their Thermodynamic Stability

Knowledge of the effects of an applied gas or hydrostatic pressure on the properties and the stability of a phase has important technological outcomes, as it may allow, as a not exclusive example, the synthesis of materials outside conventional temperature and pressure conditions, where their preparation would be challenging or impossible.

As an example, the equilibrium vapour pressure or the decomposition pressure of phosphide, arsenide and nitride alloys may be so high at their growth temperatures (even well below the melting point) as to induce partial or full decomposition of the material or/and the onset of non-stoichiometry. This would make their growth as bulk single crystal or single crystalline- or polycrystalline-thin films very challenging, unless the applied gas pressure is sufficiently high to prevent the decomposition.

Meanwhile, an understanding of the effects of a mechanical strain, arising from external or internal sources of stress, on the properties of a material, is important in order to optimize its growth or deposition processes and to accomplish for the presence of extended defects. A typical, but not exclusive example is the dislocation generation induced by lattice mismatch, occurring in the heteroepitaxial deposition processes which will be discussed in Chapter 4.

As a further example, the combined effect of temperature and applied pressure may induce a sequence of phase transformations in elemental or compound semiconductors. In this section we will deal with the effects of an externally applied hydrostatic pressure on the equilibrium transformation temperatures of liquid/solid systems and of solid/solid systems, leaving the analysis of the effects of internal mechanical stresses on defect generation to later chapters.

The effect of the applied pressure on the melting temperature T_f and the molar volume of a number of model systems, including Ge and Si, is reported in Table 1.3 [42, 43], while Table 1.4 reports the value of the critical pressure at which typical phase transformations of alkali halides and oxides occur.

Pressure-induced phase transformations in silicon and germanium were studied by Bundy [44], who showed a transition from semiconductor to metallic character in both cases, although the silicon transition was more sluggish. In the case of silicon this transition occurs at 10–13 GPa and is associated with a volume decrease of 22%. Diamond anvil cell and micro- and nano-indentation experiments showed the formation of at least 12 different silicon polymorphs following the application of increasing pressures [45]. Other examples can be found within compound semiconductors (see Section 1.7), as is the case of GaAs, which crystallizes in the cubic zinc blende structure at ambient pressures but undergoes a transition to at least two orthorhombic structures, the first of which occurs at 12 ± 1.5 GPa at 300 K [46]. CdS, CdSe and ZnS, eventually, show unique behaviour

Table 1.3 *Effect of the applied hydrostatic pressure on the melting temperature of model systems*

Material	T_f(K)	$\Delta_f H$ (kJ mol^{-1})	ΔV_m(cm^3 mol^{-1})	ΔT_{calc} (0.9 GPa)	ΔT_{exp} (0.9 GPa)
H$_2$O	273.2	6.012	−1.6308	−7.5	−7.4
Sn	505	7.03 (white tin)	+0.4617	+3.4	3.28
Bi	544.7	11.3	−0.7147	−3.56	−3.10
Si [42]	1687	50.21	−6.767	−20.46	900 (10 GPa)
Ge	1211.4	34.7	−20.121	−78.048	—
Al [43]	933.47	10.71	+7.473	+58.62	1500 (10 GPa)

Adapated from Soma and Matsuo, 1982, [42] and from Hänström and Lazor, 2000, [43].

Table 1.4 *Critical phase transformation pressures for a number of model systems*

Material	Phase transformation	P_c (GPa)	$\Delta V_{m\alpha\rightarrow\beta}$(cm^3 mol^{-1})	$\Delta H_{\alpha\rightarrow\beta}$(kJ mol^{-1})
KCl	NaCl → CsCl	1.96	−4.11	8.03
KBr	NaCl → CsCl	1.80	−4.17	7.65
RbCl	NaCl → CsCl	0.57	−6.95	3.39
ZnO	Wurtzite → NaCl	8.86	−2.55	19.23
SiO$_2$	Quartz → coesite	1.88	−2.0	2.93
SiO$_2$	Coesite → stishovite	9.31	−6.6	52.27

in view of their wurtzite–zinc blende transformations which resemble quasi-equilibrium conditions.

An even more important case of polymorphism for its relevance to material applications is that exhibited by carbon and silicon carbide, which will be discussed in Section 1.7. The polymorphism of silicon carbide, with its more than 200 different polymorphs, falls into the category of polytypism, as each different polymorph can be regarded as built up by stacking layers of (nearly) identical structure and composition, and the difference lies only in their stacking sequence. Polytypism is, therefore, a special case of polymorphism where the two-dimensional translations within the layers are essentially preserved [47].

While the evaluation of the critical pressure at which a structural phase transformation occurs may be obtained by *ab initio* or molecular dynamics (MD) computations, the effect of the pressure on the phase transformation temperature $T_{\alpha\rightarrow\beta}$ of a generic material may be deduced by the Clausius–Clapeyron equation

$$\frac{dP}{dT} = \frac{\Delta H_{\alpha\rightarrow\beta}}{T_{\alpha\rightarrow\beta}\Delta V_{\alpha\rightarrow\beta}} \tag{1.35}$$

where $\Delta H_{\alpha-\beta}$ is the enthalpy of the $\alpha \rightarrow \beta$ transformation, $T_{\alpha-\beta}$ is the equilibrium transformation temperature and $\Delta V_{\alpha-\beta}$ is the volume change associated with the $\alpha \rightarrow \beta$ transformation.

Equation (1.35) could also be written

$$dT_{\alpha\rightarrow\beta} = \frac{T_{\alpha\rightarrow\beta}\Delta V_{\alpha\rightarrow\beta}}{\Delta H_{\alpha\rightarrow\beta}}dP \tag{1.36}$$

for the change in the transformation temperature as a function of pressure.

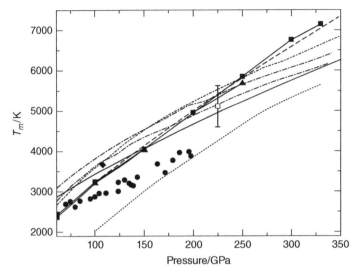

Figure 1.6 *Effect of the pressure on the melting point of iron. Luo et al., 2011, [48]. Repro-duced with permisssion from American Chemical Society*

If both the $\Delta H_{\alpha-\beta}$ and $\Delta V_{\alpha-\beta}$ terms are taken as independent of both temperature and pressure, a linear increase or decrease of the melting temperature is expected with increase in the applied pressure, depending on whether there is a volume contraction or expansion on melting, as shown in Table 1.3.

The melting temperature, therefore, increases with the applied pressure in the case of metals, see Figures 1.6 and 1.7, and decreases in semiconductors, see Figure 1.8 for the typical case of silicon, because, almost systematically, there is, on freezing, a volume con-traction in metals and a volume expansion in elemental semiconductors [43, 48–50]. As the Clapeyron equation has been deduced for conditions very close to the equilibrium melt-ing temperature $T^{\circ}{}_{\alpha-\beta}$, we might expect sensible deviations from the linearity at elevated pressures, as is, in fact, shown in Figures 1.6–1.8.

The validity of the Clapeyron equation may be, however, extended to a wider pressure and temperature range once the pressure and temperature dependence of the $\Delta H_{\alpha-\beta}$ and $\Delta V_{\alpha-\beta}$ terms are known or could be calculated.

A critical role is expected to be played by the $\Delta V_{\alpha-\beta}$ term in Eqs. (1.35) and (1.36), which is related to the effect of the applied pressure on the density of the two phases in equilibrium. Density should decrease in absolute value with increase in the applied pressure, that is with increase in the mechanical deformation. Figure 1.9 shows, as an example, the calculated strong decrease of the atomic volume of hexagonal close packed (HCP) solid iron with increase in pressure, which is associated with a decrease in the $\Delta V_{\alpha-\beta}$ term value from $0.16 \, \text{cm}^3 \, \text{mol}^{-1}$ at atmospheric pressure to $0.10 \, \text{cm}^3 \, \text{mol}^{-1}$ at 250 GPa [48].

The effect of volume changes on the melting temperature is phenomenologically taken into account by the Lindeman equation [51] in its integrated form [43]

$$T_m = T_m^{\circ} \left(\frac{V_{T_m}}{V_{T_m}^{\circ}} \right)^{\frac{2}{3}} \exp \left[\frac{2\gamma^{\circ}}{q} \left(1 - \frac{V_{T_m}}{V_{T_m}^{\circ}} \right)^q \right] \tag{1.37}$$

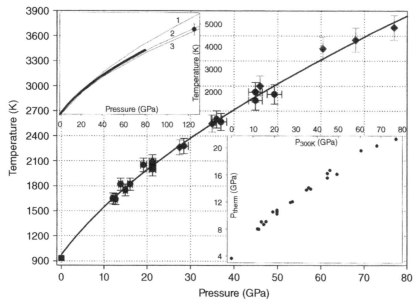

Figure 1.7 *Experimental and calculated melting temperatures of aluminium under high pressure. Hänström & Lazor, 2000, [43]. Reproduced with permisssion from Elsevier*

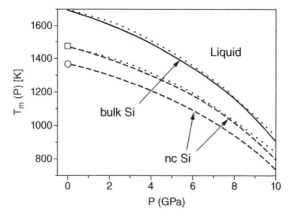

Figure 1.8 *Effect of the pressure on the melting point of silicon (the effect of crystallite size is also shown). Yang et al., 2003, [49]. Reproduced with permisssion from Institute of Physics and Q. Jiang*

where $\gamma = \gamma^0 \left(\dfrac{V_{T_m}}{V_{T_m}^0} \right)^q$, $V_{T_m}^0$ is the molar volume at the onset of linearity deviations, V_{T_m} is the molar volume at the experimental melting temperature, T_m^0 is the melting temperature in correspondence to $V_{T_m}^0$ and q is the volume dependence for γ, often taken equal to 1.

The Lindeman equation has been successfully applied to fit the experimental points of Figure 1.7 for the case of Al, where $V_{T_m}^0$ is the experimental molar volume of Al at 100 kPa,

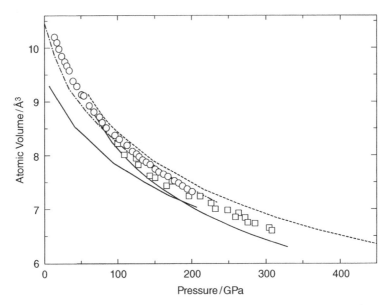

Figure 1.9   *Atomic volume of iron as a function of the applied pressure. Luo et al., 2011, [48]. Reproduced with permisssion from American Chemical Society*

V_{T_m} is the experimental molar volume of Al at the melting temperature under pressure, T_m^o is the melting temperature of Al at 933 K and 100 kPa and q is taken equal to 1 [43].

Alternatively, when the data concerning the pressure dependence of the melting temperature of a material are known experimentally they could be satisfactorily fitted, as was done for the case of Al, see again Figure 1.7, using the empirical Simon's melting equation:

$$T_m = T_m^o \left(\frac{P}{A} + 1 \right) B \tag{1.38}$$

where T_m^o is the experimental value of the melting temperature of Al at 100 kPa and A and B are fitting parameters, treating aluminium as an harmonic Debye solid.

It is important to remark here that, different from the past, there is a growing research interest today concerning the behaviour of metals and semiconductors under elevated pressures. Independently of the advances in the understanding of their basic physico-chemical behaviour, knowledge of the effect of the hydrostatic pressure on the melting temperature is important for the optimization of growth and further thermal processes of high temperature melting or highly corrosive semiconductor materials. In several cases a decrease in the melting temperature might permit the use of cheaper crucibles or reduce corrosion effects which are more detrimental as the operational temperatures increase, as well as reducing energy costs. This would be the case for silicon, for which the growth of single crystalline or multicrystalline ingots could be carried out at about 1000 K at 10 GPa instead of at 1685 K at ambient pressure. Diamond is a limiting case, as will be seen in Section 1.7.6, where temperatures of several thousand degrees K and pressures of 1000 GPa might be obtained by shock waves [52].

The growing interest in pressure effects on the melting temperatures of metals and semiconductors is demonstrated by a number of recent theoretical studies [42, 49, 53–56] dedicated to the melting behaviour of face centred cubic (FCC) metals, semiconductors and semiconductor alloys under pressure.

All these studies are based on the application of a variant of the Lindeman's melting law [51] which postulates that the melting process in FCC solids occurs when the root-mean-square displacement, $\langle u^2 \rangle^{1/2}$ of the lattice vibrations reaches a critical fraction x_m of the nearest-neighbour distance and that this critical fraction is assumed to be the same for all crystalline solids. It was, however, shown that in various cubic metals and alkali halides this fraction was actually not constant. A better account with the experimental data is obtained [54] by assuming as the critical fraction x_m for melting the ratio of twice the root-mean-square displacement $\langle u^2 \rangle^{1/2}$, at the melting temperature, T_m, to the nearest-neighbour distance, $R_1 = (2^{1/2} a)/2$, given by

$$x_m = \frac{\langle u^2 \rangle^{1/2}}{R_1} \tag{1.39}$$

Considering, further, the pressure dependence of the mean square displacement, $\langle u^2 \rangle$ and of the nearest-neighbour distance, R_1, and assuming as before x_m as the melting criterion, it is possible to estimate the pressure effect on the melting temperature T_m when both the following relations are simultaneously satisfied

$$x_m = \frac{\langle u^2 \rangle^{1/2}_{T_m,P}}{R_1(P)} \tag{1.40}$$

$$R_1(P) = \left[\frac{\Omega(P)}{\Omega^\circ} \right] R_1(\Omega^\circ) \tag{1.41}$$

where Ω° and $\Omega(P)$ are the crystal volumes under atmospheric pressure and under a pressure P, respectively, and $R_1(\Omega^\circ)$ is the nearest-neighbour distance at atmospheric pressure.

The comparison between the experimental and calculated values is reported in Figure 1.10, which shows that the calculated values deviate slightly from the experimental ones, still exhibiting a linear dependence on the applied pressure.

Among issues not considered here, those concerning the pressure-induced widening of the solid solubility domains of impurities in silicon and germanium has relevant technological consequences and will be discussed in Section 1.7.5.

1.4.6   Electrochemical Equilibria and Electrochemical Potentials of Charged Species

It is a common practice in physical chemistry and solid state physics [7, 9] to discuss the equilibrium properties of charged species, such as ions, charged point defects or electrons, between phases of equal or different chemical compositions, like two differently doped semiconductors or two solutions of electrolytes, using the electrochemical potential η_i formally defined as

$$\eta_i^\alpha = \mu_i^\alpha + ze\varphi_\alpha \tag{1.42}$$

where μ_i is the chemical potential of the i species, ze is its charge and φ is the internal potential of the phase.

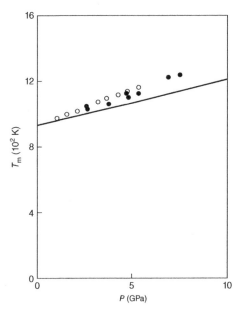

Figure 1.10 *Calculated dependence of the melting point of Al as a function of the pressure (solid line):* O *and* • *are experimental points. Kagaya et al., [54]. Reproduced with permisssion from Chapman and Hall)*

The condition of equilibrium of a charged species i^{+z} between two phases α and β requires that

$$\eta_i^\alpha = \eta_i^\beta \tag{1.43}$$

and that $(\varphi_\alpha - \varphi_\beta) = 0$ \hfill (1.44)

when the composition of the two neighbouring phases is equal, that is $\mu_i^\alpha = \mu_i^\beta$.

The equilibrium condition for a charged species i present in two neighbouring phases α and β of different chemical compositions still requires that

$$\eta_i^\alpha = \eta_i^\beta \tag{1.43}$$

but that a potential drop $\Delta\varphi$ occurs at the α/β interface

$$\Delta\varphi = (\varphi_\beta - \varphi_\alpha) = \frac{\mu_i^\alpha - \mu_i^\beta}{zF} \tag{1.45}$$

It is very important to note here that the electrochemical potentials refer to defined charged species (point defects and electrons in metals and semiconductors, ions in electrolytes) while the internal potential is a property of the phase. It should be additionally noted that a difference $\Delta\varphi = (\varphi_\beta - \varphi_\alpha)$ in the internal potential of two phases is physically defined and directly measurable only if the two phases have identical chemical composition.

This means that a system where a difference of internal potentials exists should be properly connected to two electrodes of equal chemical composition to make the difference $\Delta\varphi = (\varphi_\beta - \varphi_\alpha)$ directly measurable.

We will discuss in the Appendix a direct application of these concepts when dealing with the equilibrium conditions occurring in a galvanic chain used to measure the Gibbs free energy of formation of compound semiconductors.

1.5 Equilibrium Conditions of Multicomponent Systems Which Do Not React Chemically

For multicomponent systems, the dependence of every thermodynamic function of state or of every extensive property (as an example the volume and the density) on the composition is normally expressed in terms of the atomic/molar fraction $x_1 = \dfrac{n_i}{\sum\limits_i n_i}$, where n_i is the number of atoms/moles of the component i.

For heterogeneous mixtures of two phases, each consisting of an elemental component A or B, and each assumed fully insoluble in the other,[5] the Gibbs free energy, like any other extensive property of the system, is given by the sum

$$G = x_A G_A^o + x_B G_B^o = x G_A^o + (1-x) G_B^o \tag{1.46}$$

where G_A^o and G_B^o are the molar free energies of the components A and B, at the temperature of the experiment.

The dotted curve in Figure 1.11 fits the linear dependence of the Gibbs free energy of the heterogeneous mixture of A and B on composition, according to Eq. (1.46).

In the same figure the hypothetical[6] dependence of G on the composition x is reported for the case of a system exhibiting complete mutual solubility of the components A and B. Depending on the temperature, the solution will be either liquid or solid. We define as the free energy of mixing $\Delta G_{mix} (T, P, x)$ the free energy excess associated with the formation of a solution from the pure elements, whose value is supposed to depend on the composition, temperature and pressure.

The dependence of $\Delta G_{mix} (T, P, x)$ on the composition at constant temperature and pressure might be formally deduced, starting from the definition of the chemical potential given by the Eq. (1.27), written for a system of n components.

$$\mu_i = \left(\frac{dG}{dn_i} \right)_{T,P,n_1 \dots n_j} \tag{1.27}$$

Therefore, the molar free energy G_{soln} of a solution is given by the sum

$$\sum_1^i x_i \mu_i = G_{soln} \tag{1.47}$$

where the sum is extended to all the components of the solution.

For a binary system we can write for the molar free energy of the solution

$$G_{soln} = x_A \mu_A + x_B \mu_B \tag{1.48}$$

---

[5] Full insolubility is not an equilibrium condition, as a limited solubility of A in B and B in A, respectively, increases the thermodynamic stability of the phase due to an entropic mixing contribution to the free energy. Instead of pure A and B, we should therefore use as components A saturated in B and B saturated in A, respectively.

[6] The actual shape of the free energy curve depends on the properties of the solution.

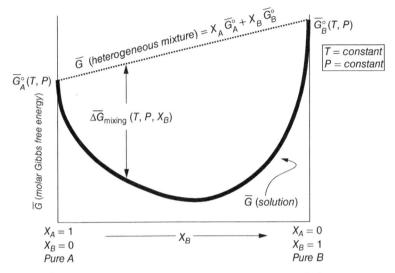

Figure 1.11   *Composition dependence of the Gibbs free energy G of a heterogeneous mixture of the components A and B (dotted curve) and of a solution of A and B (solid curve). Courtesy of W. Craig Carter, MIT*

and derive a relationship between G_{soln} and the chemical potentials, for any value of x between $x_A = 1$ and $x_B = 1$, by writing first

$$dG_{soln} = \mu_A dx_A + \mu_B dx_B = (\mu_B - \mu_A)dx_B \qquad (1.49)$$

where

$$x_A + x_B = 1; x_B = 1 - x_A; dx_A = -dx_B \qquad (1.50)$$

It is then possible to write

$$\frac{dG_{soln}}{dx_B} = \mu_B - \mu_A \qquad (1.51)$$

and

$$\mu_B = \mu_A + \frac{dG_{soln}}{dx_B} \qquad (1.52)$$

Therefore, from Eq. (1.48)

$$G_{soln} = (1 - x_B)\mu_A + x_B \left[\mu_A + \left(\frac{dG}{dx_B} \right) \right] = \mu_A + x_B \left(\frac{dG}{dx_B} \right) \qquad (1.53)$$

and

$$\mu_A = G_{soln} - x_B \left(\frac{dG_{soln}}{dx_B} \right) \qquad (1.54)$$

$$\mu_B = \mu_A + \frac{dG_{soln}}{dx_B} = G_{soln} - x_B \left(\frac{dG_{soln}}{dx_B} \right) + \frac{dG_{soln}}{dx_B} = G_{soln} + (1 - x_B) \left(\frac{dG_{soln}}{dx_B} \right) \qquad (1.55)$$

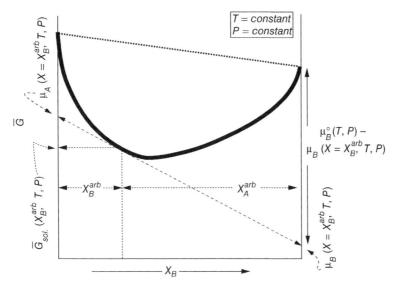

Figure 1.12 *Geometrical determination of the values of the chemical potentials of the components of a binary solution using the common tangent procedure (dotted line in the figure). The composition dependence of the Gibbs energy, as shown by the solid line, is arbitrary. Courtesy of W. Craig Carter, MIT*

where $\frac{dG_{soln}}{dx_B} = \frac{G_{soln} - \mu_A}{x_B}$ is the equation of the tangent to the curve $G_{soln}(x)$ for any arbitrary value of the composition x_B between $x_B = 0$ and $x_B = 1$ and the terms $(1 - x_B)\left(\frac{dG_{soln}}{dx_B}\right)$ and $x_B\left(\frac{dG_{soln}}{dx_B}\right)$ are the values of the intercepts of the tangent to the curve $G_{soln}(x)$ for a defined value of composition x_B, as is shown in Figure 1.12.

The equilibrium composition of two coexisting phases, both solids or one liquid and the other solid, may be easily found by imposing the condition (1.30) for each component of the system.

In the case of a binary A_xB_{1-x} system, the equilibrium condition among a liquid and solid phase should be written

$$\mu_A{}^l(x_A{}^l) = \mu_A{}^s(x_A{}^s) \tag{1.56}$$

$$\mu_B{}^l(x_A{}^l) = \mu_B{}^s(x_A{}^s) \tag{1.57}$$

To determine the equilibrium composition of the solid and liquid phases at a given temperature (and pressure) it is necessary to solve Eqs. (1.56) and (1.57) simultaneously on the basis of the available information on the dependence of the chemical potentials on the composition, as deduced from measurements or by solution models, as will be shown in the next section. Alternatively, it is possible to use the geometrical procedure illustrated in Figure 1.12 called the common tangent procedure, when the composition dependence of the Gibbs free energy G^l and G^s of the liquid and solid phases, respectively, is experimentally available or can be suitably modelled.

One can see that the equilibrium condition of the system of Figure 1.13 implies the coexistence of a liquid phase of composition $x^l{}_B$ and a solid phase of composition $x^s{}_B$. A liquid phase of composition x^l would be thermodynamically metastable because there will be a

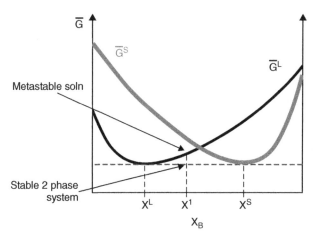

Figure 1.13   *Graphical representation of the coexistence equilibrium conditions of a solid and liquid phase: the black curve pertains to the liquid phase and the blue one to the solid phase. Adapted from http//ocw.mit.edu licensed under the Creative Commons Attribution 3.0 Unported license*

decrease in free energy ΔG_{x1} if this metastable solution spontaneously decomposes into two phases.

To proceed further it is necessary to find some quantitative relationship between the chemical potentials and the composition of a phase or other measurable variables proportional to the concentration of the components.

This is immediate for the case of the equilibrium between a condensed and a vapour phase, under the assumption that the vapour phase behaves as an ideal gas. This assumption could be applied to a dilute vapour phase provided each *i* component of the condensed phase sublimes without chemical changes arising from incongruent sublimation, pyrolysis, chemical reactions or polymerization, a condition violated in many practical circumstances, as for example during crystal growth processes of semiconductors from metal-organic precursors [57], but ignored in the following analysis.

Under these conditions it is possible to find a direct relationship between the Gibbs free energy, the chemical potential and the pressure p of the vapour phase using the equation

$$\left(\frac{\mathrm{d}G}{\mathrm{d}p}\right) = V = \frac{RT}{p} \tag{1.58}$$

which at constant temperature can be written as

$$\mathrm{d}G = RT\frac{\mathrm{d}p}{p} \tag{1.59}$$

After integration

$$\Delta G = \int_{p^o}^{p} \mathrm{d}G = RT \int_{p^o}^{p} \frac{\mathrm{d}p}{p} = RT\ln\frac{p}{p^o} \tag{1.60}$$

$$\Delta G = \mu - \mu^o = RT\ln\frac{p}{p^o}; \mu = \mu^o + RT\ln\frac{p}{p^o} \tag{1.61}$$

where p^o is an arbitrary reference pressure.

The equilibrium conditions between a condensed (liquid or solid) phase s and its vapour phase v may be then deduced applying to each component i of the system the equilibrium condition (1.30)

$$\mu_i^{\alpha} = \mu_i^{\beta} \tag{1.30}$$

and writing

$$\mu_i^s = \mu_i^v = \mu_i^{0,v} + RT\ln\frac{p_i}{p_i^0} \tag{1.62}$$

where p_i is the partial pressure of the component i for an arbitrary value of the composition of the condensed phase, p_i^0 is the partial pressure of the pure component i and $\mu_i^{0,v}$ is the chemical potential of the pure component i in the vapour phase.

It is then possible to define a new function, λ, the absolute activity

$$\lambda = e^{\mu/RT} \tag{1.63}$$

which allows one to redefine the equilibrium conditions with the following equation

$$\lambda_i^s = \lambda_i^v \tag{1.64}$$

After substitution of $\mu_i^v = \mu_i^{ov} + RT \ln \frac{p}{p^0}$ in λ, we obtain

$$\lambda_i^v = \lambda_i^{ov} \left(\frac{p}{p^0}\right) \tag{1.65}$$

and

$$\frac{\lambda_i^v}{\lambda_i^{ov}} = \frac{p}{p^0} \tag{1.66}$$

If we define the *chemical activity* a_i of the component i in solution as the ratio $\frac{\lambda_i^s}{\lambda_i^{os}}$

$$a_i = \frac{\lambda_i^s}{\lambda_i^{os}} = \frac{p}{p^0} \tag{1.67}$$

It is possible, according to Eqs. (1.61) and (1.62), to express the chemical potential of the component i in the condensed phase(s) as a function of its activity, which is a measurable, composition dependent, property of the system

$$\mu_i^s = \mu_i^v = \mu_i^{0,s} + RT\ln a_i = \mu_i^{0,v} + RT\ln\frac{p_i}{p_i^0} \tag{1.68}$$

where $\mu_i^{0,s}$ is the chemical potential of a condensed phase consisting of the pure component i for which the equilibrium partial pressure is p_i^0.

It would, therefore, be sufficient to know from experimental measurements or from theoretical analysis the composition dependence of the vapour pressures of the components of a solution to get the values of their chemical potentials and activities in the condensed phase and to draw the corresponding Gibbs free energy curves as a function of the composition.

Because the partial pressure p_i of a solute i in a solution is proportional to its concentration x_i (with the linear Raoult's or Henry's law holding strictly for ideal or ideally diluted liquid solutions), we might also write

$$p_i = \gamma(x)x_i p_i^0 \tag{1.69}$$

thus finding a direct relationship between the chemical potential and the concentration x_i of the i-solute

$$\mu_i = \mu_i^\circ + RT \ln \gamma(x) x_i \tag{1.70}$$

where $\gamma(x)$ is a composition dependent activity coefficient. Ideal solutions are those for which $\gamma(x) = 1$ for each component of the solution.

We can apply the same procedure for the equilibrium conditions relative to the ith component between two generic condensed phases α and β, which may be written

$$\mu_i^{\circ,\alpha} + RT \ln \gamma_i^\alpha(x) x_i^\alpha = \mu_i^{\circ,\beta} + RT \ln \gamma_i^\beta(x) x_i^\beta \tag{1.71}$$

1.6 Thermodynamic Modelling of Binary Phase Diagrams

1.6.1 Introductory Remarks

The availability of phase diagrams is not only the prerequisite for carrying out materials research in the field of crystal growth, solid state reactions, phase transformations and dry corrosion processes, but phase diagrams are also by themselves a thermodynamics-based powerful tool for material design and optimization processes.

While a very large number of binary and some ternary systems, these last often limited to narrow composition regions, have been experimentally investigated, the experimental approach is almost impossible for most ternary and multicomponent systems.

The real challenge has been to develop phenomenological models, using as an example the Calphad route, to calculate phase diagrams of interest for technological developments [58]. Concerning binary and pseudo-binary systems, these have not only been the subject of experimental studies but also, and for about a century, the subject of thermodynamic modelling studies based on various solution theories, with the aim of finding formal, but predictive, correlations between the energetics of the bonding interactions in solution and the phase diagram topology.

The first, successful, approach was carried out in the framework of the ideal or regular solutions approximations [59, 60], which would require only the preliminary availability of the thermodynamic functions of the pure components, their structural and electronic properties and the energetics of their mutual interactions in solution.

This approach was demonstrated to be useful in a significant number of cases, especially (but not always, as will be shown in the next section) for liquid solutions, but rarely in the case of solid semiconductor systems, for which the assumption of a rigid lattice and the absence of elastic constraints often leads to conclusions which are contradicted by experience [61].

When elastic constraints were first accounted for, severe approximations were used for the calculation of the elastic energy contributions and the interplay between elastic and chemical ordering energies was often underestimated, making it impossible to reconcile, for example, order–disorder transitions with thermodynamic data [61]. To overcome these difficulties, a *first principle statistical mechanics* approach has been applied to study the phase diagrams of a number of tetrahedral, FCC pseudo-binary semiconductors [34], such as the arsenides, phosphides and tellurides.

The QCA has been instead used [33] to compute the partition functions of several binary alloys and then the chemical potentials of the species involved. Once the chemical potentials are known, all the thermodynamic functions are easily obtained. With the QCA, the system under study is considered equivalent to a grand ensemble of statistically independent, few-atom clusters, in equilibrium with the solution, which behaves as a reservoir. In FCC lattices the typical configuration of the clusters is the tetrahedral one.

The QCA is based on the quasi-chemical model, originally applied to metal solutions and to ionic salts, where only pair-like interactions are assumed to be relevant for the properties of a solution and pairs are distributed randomly [60].

It is not the aim of this section to go deeply into the formal aspects of these theories, considering the vast literature available on the subject and the vast number of applications [32, 59, 60, 62–64], but rather to discuss the physico-chemical background of these thermodynamical models and to report the results of their application to semiconductor alloys of technological interest. The intelligent manipulation of conditions of total or incomplete miscibility, order–disorder transitions, the stress effect on phase stabilities, plays a particularly important role in the application and modelling of crystal growth, doping and refining processes, as will be seen in Chapter 4 and also in the next section.

1.6.2 Thermodynamic Modelling of Complete and Incomplete Miscibility

The equilibrium distribution of components among coexisting phases as a function of the temperature and composition and the topological aspects of the phase diagrams, commonly reported in temperature–composition (T, x) coordinates, are the expected output of these modelling studies.

It should be noted here that in a vast number of semiconductor systems decomposition may occur if the pressure of the environment is not taken into proper account. In these cases the phase diagrams are reported in T, x, P diagrams or in their T, x projections.

The equilibrium distribution of a component i among two coexisting phases, of which one is liquid and the other is solid, could be, however, given also in terms of its *distribution* or *segregation coefficient* $k_i(x, T)$, defined as the ratio $k = \frac{x_i^s}{x_i^l}$ where x_i^s is the equilibrium atomic fraction of i in the solid phase and x_i^l is the equilibrium atomic fraction of i in the liquid phase.

Segregation coefficients may be experimentally determined using appropriate analytical techniques, but the need for knowledge about equilibrium segregation of impurities in semiconductors is essential, as will be seen in Chapter 4, because it allows a proper handling of semiconductor doping and semiconductor refining processes.

It is important, therefore, to show the relationships occurring between the segregation coefficient and the thermodynamic properties of the biphasic system where the repartition of a solution component occurs, in order to understand the physico-chemical grounds.

Using Eq. (1.71) it is possible to write the following general equation for the segregation coefficient

$$\ln k = \ln \frac{x_i^s}{x_i^l} = -\frac{\mu_i^{o,s} - \mu_i^{o,l}}{RT} - \ln \frac{\gamma_i^s}{\gamma_i^l} = -\frac{\Delta G_i^{o,f}}{RT} - \ln \frac{\gamma_i^s}{\gamma_i^l} = \frac{\Delta S_i^{o,f}}{R} - \frac{\Delta H^{o,f}}{RT} - \ln \frac{\gamma_i^s}{\gamma_i^l} \quad (1.72)$$

which formulates the case of the equilibrium distribution of a generic solute i among two coexisting solid and liquid phases as a function of the chemical potentials of the pure solute i and the activity coefficients of i in the solid and liquid state, over the whole range of temperatures where the two phases coexist. The ratio $\frac{\gamma_i^s}{\gamma_i^l}$ is a solution mixing term, whose value depends on the specific interaction of i with the solvent in the liquid and solid phases, as will be seen later in this section, and, thus, on the temperature and the composition of the solution.

The difference $\mu_i^{o,s} - \mu_i^{o,l}$, in turn, is the Gibbs free energy of transformation (i.e. the Gibbs free energy of melting $\Delta G_i^{o,f} = \Delta H_i^{o,f} - T\Delta S_i^{o,f}$ of the pure component i at the melting temperature T_i^f, which takes positive or negative values depending on whether $T < T_i^f, T > T_i^f$.

The segregation coefficient is 1, and $x_i^s = x_i^l$ when the following relationship holds for the activity coefficients and the Gibbs free energy of melting, in the case of a solid/liquid equilibrium

$$\ln\frac{\gamma_i^s}{\gamma_i^l} = -\frac{\Delta G_i^{o,f}}{RT} = \frac{\Delta S_i^{o,f}}{R} - \frac{\Delta H_i^{o,f}}{RT} \tag{1.73}$$

We will see in Section 1.7.8 a particular case of this conditions which is presented by the nanocrystalline Si-Ge alloys, for which $k \sim 1$ from pure Ge to pure Si. The physical sense of this condition is that for $\Delta G_i^{o,f} \to 0$ and $\frac{\gamma_i^s}{\gamma_i^l} = 1$, a distinction between the solid and liquid phase is thermodynamically impossible.

The segregation coefficient is also unitary for the segregation of i at the melting point T_i^f of i, because in this condition the free energy of fusion $\Delta G_i^{o,f} = 0$ and $\mu_A^{o,s} = \mu_A^{o,l}$

For the equilibrium of i among a couple of ideal solutions, for which, by definition, $\gamma_i(s) = 1$ and $\gamma_i(l) = 1$, the segregation coefficient is instead given by the equation

$$\ln k_i = \ln\frac{x_i^s}{x_i^l} = -\frac{\Delta G_i^f}{RT} \tag{1.74}$$

where $\Delta G_i^{o,f}$ is again the Gibbs free energy of fusion of the pure component i.

Except for these particular cases, the equilibrium distribution of a generic component i among two condensed phases or a condensed phase and a gaseous or vapour phase, given by Eq. (1.72), depends on the specific thermodynamic properties of the coexisting solutions, which should be experimentally available or theoretically determined.

A systematically practised solution to arrive at the theoretical evaluation of a phase diagram is modelling the temperature, composition and pressure dependence of the Gibbs free energy $G(T, P, x)$ functions of the coexisting liquid and solid solutions and of the potentially stable intermediate compounds. In the case of binary or pseudo-binary solutions, this corresponds to the evaluation of the features of the hypersurfaces $G(T, p, x)$, or, in isobaric conditions, those of the surfaces $G(T, x)$ or of their isothermal sections $G(x)$, associated with the solid and liquid phases of the system under consideration, assuming, at a first approximation, that both the liquid and the solid phases are in internal mechanical equilibrium.

Let us first consider the simplest case of two elemental components A and B with very close physical, chemical and structural properties, that is very close values of the lattice constants, atomic radii and electronic properties in both the solid and liquid state. Such features hypothetically grant conditions of mutual solubility in all proportions in the liquid and solid state and, furthermore, of *ideal behaviour* of both the liquid and solid solutions. In this case, the thermodynamic activities of the components coincide with their atomic fractions, that is $a_i = x_i$.

The molar Gibbs free energy of both the liquid and solid solution is, therefore, given by the equation

$$G_{soln} = x_A \mu_A + x_B \mu_B = x_A \mu_A^o + x_B \mu_B^o + RT(x_A \ln x_A + x_B \ln x_B) \qquad (1.75)$$

and the excess Gibbs free energy of mixing $\Delta G_{mix} = \Delta H_{mix} - T \Delta S_{mix}$ is given by

$$\Delta G_{mix} = G_{soln} - (x_A \mu_A^o + x_B \mu_B^o) = RT \, [x_A \ln x_A + x_B \ln x_B] \qquad (1.76)$$

Therefore, the Gibbs free energy of mixing has only a purely configurational entropy character and $\Delta H_{mix} = 0$

The assumption about a purely configurational entropy character of the Gibbs free energy of mixing can be easily demonstrated, starting from the use of the Boltzmann equation

$$S = k \ln W \qquad (1.77)$$

where k is the Boltzmann constant and W is the total number of possible random configurations of a system consisting of N_A atoms of A and N_B atoms of B

$$W = \frac{(N_A + N_B)!}{N_A! N_B!} = \frac{N!}{N_A! N_B!} \qquad (1.78)$$

and $N_A + N_B = N$. Then one can see, using the Stirling approximation, that a pure configurational entropy term is given by the following equation

$$\Delta S = Nk(x_A \ln x_A + x_B \ln x_B) = R(x_A \ln x_A + x_B \ln x_B) \qquad (1.79)$$

For solutions displaying significant deviations from ideality, the activities $a = \gamma x$ should be used instead of the concentrations in Eq. (1.76), and the excess Gibbs free energy of mixing is then given by

$$\Delta G_{mix} = -RT(x_A \ln \gamma_A x_A + x_B \ln \gamma_B x_B) \qquad (1.80)$$

In this case is possible to preliminarily apply the so-called regular model, originally proposed by van Laar [65, 66], Hildebrand [67], Fowler and Guggenheim [68] and Guggenheim [69], where the excess Gibbs free energy of a regular solution is ruled by pair-like interactions between neighbour atoms in solution and is given by the following general equation

$$\Delta G_{mix}(x, T) = x_A x_B \Omega - T S_{conf} - x_A x_B \eta T \qquad (1.81)$$

where $\Omega = ZN \left[H_{AB} - \left(\frac{H_{AA} + H_{BB}}{2} \right) \right]$ is an interaction coefficient, taken as the difference between the pair-binding enthalpy of unlike species in solution and the average of the binding enthalpies of like pairs and ZN is the product of the coordination number Z and the total number N of atoms. Here the term $x_A x_B \, \Omega$ is the enthalpy of mixing ΔH_{mix}, now different from zero and $x_A x_B \, \eta$ is a non-configurational entropy term.

The model indirectly assumes the absence of internal elastic energy contribution to the enthalpy of mixing, arising from different atomic volumes or lattice mismatch of the components, as well as a random distribution of components atoms in a rigid lattice.

The excess Gibbs free energy of mixing, with respect to that of an ideal solution, is given by the equation

$$\Delta G^*_{mix}(x, T) = x_A x_B \Omega - x_A x_B \eta = x_A x_B (\Omega - \eta T) \tag{1.82}$$

The composition dependence of the solution enthalpy term $\Delta H_{mix} = x_A x_B \, \Omega$ of Eqs. (1.81) and (1.82) may be understood by noting first that the enthalpy of solution H_{soln} is given by the following sum

$$H_{soln} = P_{AA} H_{AA} + P_{BB} H_{BB} + P_{AB} H_{AB} \tag{1.83}$$

where P_{XX} is the number of XX pairs in solution and H_{XX} is their binding energy. The evaluation of the P_{XX} terms is straightforward for systems where each X atom has the same number Z of neighbours, as would be the case for Si and Ge, where $Z = 4$.

In the case of a solution of A and B species, with a total number N_A of atoms A and N_B of atoms B

$$N_A = \frac{P_{AB}}{Z} + 2\frac{P_{AA}}{Z} \tag{1.84}$$

$$N_B = \frac{P_{AB}}{Z} + 2\frac{P_{BB}}{Z} \tag{1.85}$$

$$P_{AA} = \left(N_A - \frac{P_{AB}}{Z}\right)\frac{Z}{2} \tag{1.86}$$

$$P_{BB} = \left(N_B - \frac{P_{AB}}{Z}\right)\frac{Z}{2} \tag{1.87}$$

$$
\begin{aligned}
H_{soln} &= H_{AA}\left(N_A - \frac{P_{AB}}{Z}\right)\frac{Z}{2} + H_{BB}\left(N_B - \frac{P_{AB}}{Z}\right)\frac{Z}{2} + H_{AB} P_{AB} \\
&= \frac{Z N_A H_{AA}}{2} - H_{AA}\frac{P_{AB}}{2} + \frac{Z H_{BB} H_{BB}}{2} - \frac{H_{BB} P_{AB}}{2} + H_{AB} P_{AB} \\
&= \frac{Z}{2} N_A H_{AA} + \frac{Z}{2} N_B H_{BB} + P_{AB}\left(H_{AB} - \frac{1}{2}(H_{AA} + H_{BB})\right)
\end{aligned} \tag{1.88}
$$

$$\Delta H_{mix} = H - \frac{Z}{2} N_A H_{AA}\frac{Z}{2} N_B H_{BB} = H - H^\circ_A - H^\circ_B = P_{AB}\left(H_{AB} - \frac{1}{2}(H_{AA} + H_{BB})\right) \tag{1.89}$$

As the total number of pairs is $\frac{1}{2} ZN$, the number of AB pairs $P_{AB} = \frac{N_A}{N}\frac{N_B}{N} ZN = x_A x_B ZN$ and

$$\Delta H_{mix} = x_A x_B ZN\left(H_{AB} - \frac{H_{AA} + H_{BB}}{2}\right) = x_A x_B \Omega \tag{1.90}$$

To account for the non-configurational entropy term, the vibrational and electronic contributions should be known or calculated with a quantum mechanical approach [70]. It is common practice, however, when accounting for the properties of a regular solution, to neglect the non-configurational entropy term and consider for the Gibbs free energy of the solution and the excess Gibbs free energy of mixing the sole enthalpic term

$$G_{soln}(x, T) = x_A x_B \Omega - T S_{conf} \tag{1.91}$$

$$\Delta G_{mix} = \Delta H_{mix} = x_A x_B \Omega \tag{1.92}$$

Table 1.5 *Domains for invariant transformations in binary systems*

Liquid	$\Omega > 0$	$\Omega < 0$	$\Omega > 0$	$\Omega < 0$	$\Omega = 0$	$\Omega = 0$	$\Omega > 0$	$\Omega < 0$
Solid	$\Omega > 0$	$\Omega < 0$	$\Omega < 0$	$\Omega > 0$	$\Omega > 0$	$\Omega < 0$	$\Omega = 0$	$\Omega = 0$

Sarma *et al.*, 2003, [32]. Reproduced with permission from Springer Science and Business Media.

It should be noted that the interaction coefficient Ω is assumed here to be temperature independent, but it will be shown in the next section that the modelling of compound semiconductor alloys often requires the use of temperature-dependent interaction coefficients.

It will also be shown that a mixing enthalpy term with a quadratic composition dependence might be foreseen in the case of II–VI compound semiconductor alloys (see Section 1.7.10) where misfit lattice strain dominates the interaction, although within the limit of validity of Hooke's law.

As only the solid solution or only the liquid solution or both might be regular, and for both positive and negative Ω values, the modelling of systems with Ω values different from zero requires a solution for all the eight domains shown in Table 1.5 [32].

The shape of the free energy curves depends strongly on the absolute value of the $x_A x_B \Omega$ term. Positive values of this term will balance the configurational entropy contribution (see Eq. (1.91)) at low temperatures, with the onset of miscibility gaps in the temperature range of solid solutions. For very large positive values of Ω^α the formation of an ordered α' phase might be foreseen. Negative values of the $x_A x_B \Omega$ term, corresponding to large bonding energy of unlike atoms, add to the entropy term, thus increasing the stability of the solid solution.

If the thermodynamics of a regular solution are dominated only by the enthalpic term $\Delta H_{\mathrm{m}} = x_A x_B \Omega$ (i.e. the non-configurational entropy terms are negligible), it is easy to get the values of the activity coefficients of the components in solution and thus to evaluate the chemical potentials from the experimental or calculated values of ΔH_{mix}.

In fact, from Eq. (1.80)

$$\frac{\delta \Delta G_{\mathrm{mix}}}{\delta x_B} = RT \ln \gamma_B \tag{1.93}$$

and Eq. (1.92) using the Gibbs-Duhem relationship [7] we obtain

$$\frac{\partial \Delta G_{\mathrm{mix}}}{\partial x_B} = \frac{\partial \Delta H_{\mathrm{mix}}}{\partial x_B} = (1 - x_B)^2 \Omega \tag{1.94}$$

from which the equations for the activity coefficients can be obtained

$$\ln \gamma_B = \frac{(1 - x_B)^2 \, \Omega}{RT} \tag{1.95}$$

and

$$\ln \gamma_A = \frac{x_B^2 \, \Omega}{RT} \tag{1.96}$$

and the chemical potentials

$$\mu_A(x, T) = \mu_A^\circ + x_B^2 \Omega + RT \ln x_A \tag{1.97}$$

$$\mu_B(x, T) = \mu_B^\circ + (1 - x_B)^2 \Omega + RT \ln x_B \tag{1.98}$$

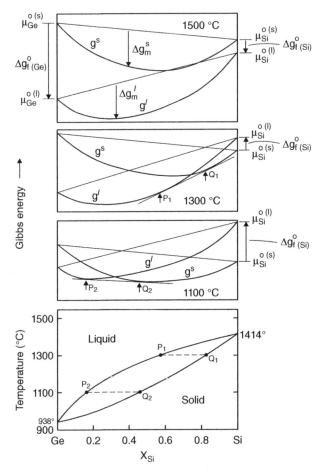

Figure 1.14   *Modelling of the isomorphous phase diagram of the Ge-Si system. Pelton, 2001, [71]. Reproduced with permisssion from John Wiley & Sons*

Conversely, these equations may be used to calculate the interaction coefficient Ω from the experimental values of the activities or the chemical potentials, determined by suitable thermochemical or electromotive force (EMF) measurements on electrochemical cells (see Appendix).

A classical example of the application of thermodynamic modelling to a semiconductor system is that reported in Figure 1.14 for the Ge-Si alloys, which was proposed by Pelton [71] under the assumption that both the solid and liquid solutions are ideal ($\Omega^s = \Omega^l = 0$).

The $G(x)$ isotherms, calculated according to Eq. (1.48), are displayed in a ΔG versus x graph, where the difference between the Gibbs free energy of the pure component Ge and Si in the liquid and solid state, at any temperature between the melting temperatures of Ge and Si, is given by their Gibbs free energy of fusion $^e\Delta G_f^{oGe} = \mu^o\mathrm{Ge(l)} - \mu^o\mathrm{Ge(s)}$ and $\Delta G_f^{oSi}\Delta G_f^{oSi} = \mu^o\mathrm{Si(l)} - \mu^o\mathrm{Si(s)}$, respectively.

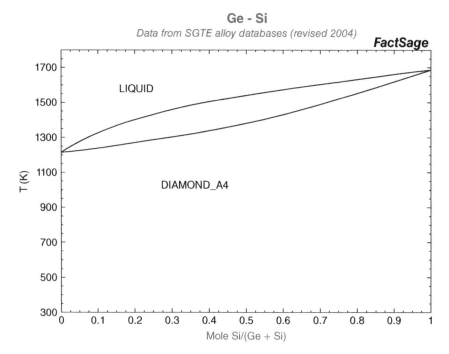

Figure 1.15 *Experimental phase diagram of the binary Ge-Si system. Scientific Group Thermodata Europe (SGTE). Reproduced with permission from SGTE*

As the free energy of fusion of the pure components Ge or Si[7] can be written

$$\Delta G_f^o = \Delta H_f^o - T \Delta S_f^o \tag{1.99}$$

so far the enthalpy and the entropy of fusion remain independent of the temperature (which is the actual case, considering the relatively narrow difference between the melting temperature of Ge and Si), the Gibbs energy of fusion is linearly dependent on the temperature and takes negative values for $T > T_f$ and positive values for $T > T_f$, as indicated by the vertical arrows in Figure 1.14.

Then, from the $G(x)$ isotherms at 1100 and 1300 °C, the equilibrium composition of the coexisting phases at these temperatures can be obtained, given by the intercepts of the common tangents at the points P_1 and Q_1 and P_2 and Q_2. Using this procedure for a sufficient number of temperatures, the isomorphous, lens-shaped, phase diagram at the bottom of Figure 1.14 may be obtained, which compares well with the experimental one displayed in Figure 1.15.

Using the regular solution model, van Laar [65, 66] demonstrated that many of the observed types of simple phase diagrams could be obtained by a systematic variation of the Ω values (Ω^β and Ω^α for liquid and solid phases, respectively, in the positive domain ($\Omega > 0$) with $\Omega^\beta \leq \Omega^\alpha$.

---

[7] See Ref. [72] for the actual values of free energy of fusion.

By varying Ω^α from zero to very large positive values (keeping the liquid solution as ideal with $\Omega^\beta = 0$), it was possible to obtain lens-shaped isomorphous, simple eutectic, peritectic and isomorphous with congruent maximum or minimum diagrams.

A few examples of the procedure which should be used are displayed in Figures 1.16–1.18 [73].

Figure 1.16a refers to the case of two pure components A and B crystallizing with different structures, α and β. For this system it is supposed that the liquid solution is ideal and that two different regular solid solutions, one with the structure of the α phase and the other with the structure of the β phase, are thermodynamically stable, both with Ω values <0, over the entire range of composition.

The relative Gibbs energy curves are calculated for a number of temperatures, ranging from values above the melting temperatures of both components down to low temperature.

Using for the evaluation of the phase equilibria the common tangent procedure (dotted curves in the three panels), the shape of a eutectic-type of phase diagram may be obtained, with a wide miscibility gap.

The case of Figure 1.16b is that of a system of two components A and B crystallizing with the same α structure. For this system it is supposed that the liquid solution is again ideal and that the solid solution of A and B is regular, with Ω values $\gg 0$. Consequently, the Gibbs energy curves present two nodes and a positive maximum due to the interplay of the configurational entropy contribution $T\Delta S = RT(x_A \ln x_A + x_B \ln x_B)$, which decreases with decrease in temperature, and the mixing enthalpy $\Delta H_{mix} = x_A x_B \Omega$ term. Also in this case a eutectic type of diagram could be obtained using the same geometrical procedure adopted in Figure 1.16a. It is supposed that the A-rich α_1 phase and the B-rich α_2 phase have close structural relationships with the phase α.

The origin of a solubility gap, with the consequent two phase decomposition processes of the homogeneous solid solution is instead illustrated in Figure 1.17 for the case of an isomorphous system with a congruent minimum. Here, slightly positive $x_A x_B \Omega^s$ terms dominate the shape of the Gibbs energy curves at very low temperatures, when the configuration entropy term contribution $T\Delta S = RT(x_A \ln x_A + x_B \ln x_B)$ begins to be small or negligible.

Some key features of the phase decomposition processes will be discussed in Section 1.6.4 for their crucial role in the epitaxial growth from the vapour phase of several compound semiconductors at temperatures or compositions close to or within the miscibility gap.

The formation of a low-temperature, ordered phase is eventually shown in Figure 1.18, which occurs in the case of a system where the thermodynamics of the solid solution are ruled by negative values of Ω^s, and a phase with a composition centred at $x_A = 0.5$ becomes stable at low temperature.

To complete the analysis, Figure 1.19 shows the variety of phase diagrams that can be obtained by progressive changes of the interaction coefficients of the solid Ω^S and of the liquid solution Ω^l from negative to positive values.

These results confirm the ability of the regular solution model in forecasting the structure of the most common phase diagrams as a sole function of the interaction parameters Ω for the liquid and solid solutions. It is also shown that minute changes, around $10 \, kJ \, mol^{-1}$ of the interaction parameter Ω, have a significant impact on the mutual solubility relationships.

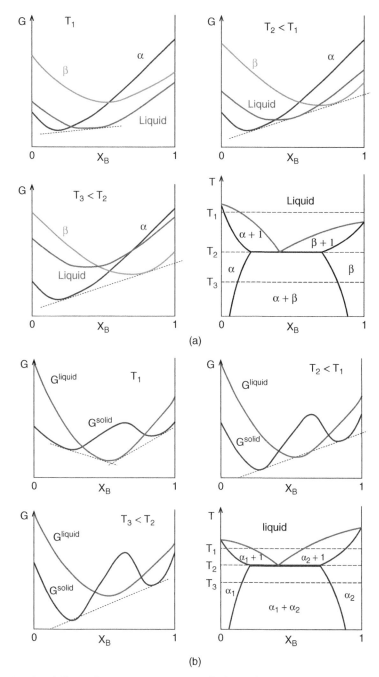

Figure 1.16 *Modelling of (a) a eutectic type of phase diagram for the case of a solution of pure components having different structures and regular solution interaction coefficients $\Omega^s < 0, \Omega^l = 0$; (b) a eutectic type of phase diagram for the case of a solution of pure components having equal structures and regular solution interaction coefficients $\Omega^s \gg 0, \Omega^l = 0$. Courtesy of L. Zhigilei*

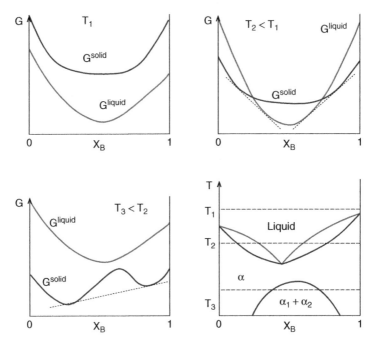

Figure 1.17   *Modelling of an isomorphous diagram with congruent minimum $\Omega^s > 0, \Omega^l = 0$. Courtesy of L. Zhigilei*

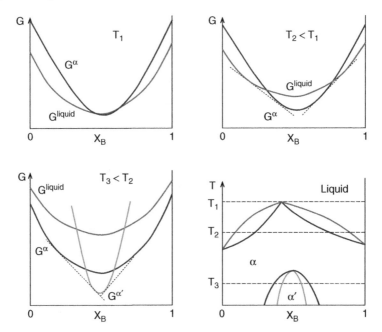

Figure 1.18   *Modelling of the phase diagram of an isomorphous system with congruent maximum in the presence of a low temperature ordered, α' phase region. $\Omega^s < 0, \Omega^l = 0$. Courtesy of L. Zhigilei*

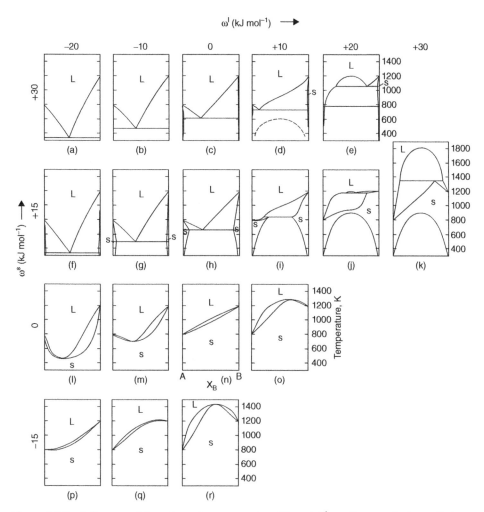

Figure 1.19 *Influence of the interaction parameters Ω^s and Ω^l on the morphology of phase diagrams. Pelton, 2001, [71]. Reproduced with permission from John Wiley & Sons*

In the last 30 years, among others, Pelton [71], Pelton and Thompson [74], Oonk [75], DeHoff [76] and Chang and Chen [77] demonstrated the occurrence of several other types of simple binary phase diagrams for different combinations of regular solution parameters, while Sarma *et al.* [32] extended the analysis to most of the possible alternatives.

These last authors confirmed, as an example, that isomorphous-type of phase diagrams ruled by negative Ω^s values display an ordered region at low temperatures, as experimentally demonstrated in the case of the copper-gold system which will be dealt with in the next section. They showed also that the isomorphous diagrams with a congruent minimum, ruled by positive values of Ω^s, display, instead, a miscibility gap [32].

The predictive ability of the regular solution approximation fails, however, in a large number of cases of binary and pseudo-binary semiconductor systems, where structural misfits and consequent misfit strains play a crucial role in the solution thermodynamics.

The regular solution scheme leaves open a number of conceptual challenges, especially in the case of disordered phases of compound semiconductors of general composition A_nB_m, as it assumes the presence of a fixed coherent AB lattice and the absence of lattice mismatch and ignores the existence of thermodynamically stable atomic arrangements different from neighbour pairs. The assumption of a rigid lattice precludes consideration of the effects of elastic energy contribution, with the consequence of excluding the contemporary condition of phase separation (the opening of a solubility gap) and ordering, as the first condition calls for Ω values >0 and the second for Ω values <0 [61].

We will show in Section 1.7 how the QCA, an advanced version of the regular solution model, and other models, allowing an interplay between chemical and elastic factors, permits computation of the thermodynamic properties of several binary alloys with satisfactory precision.

1.6.3 Thermodynamic Modelling of Intermediate Compound Formation

An important family of binary systems shows the presence of an intermediate compound, as is the case in the diagram of Figure 1.20. It is apparent that this diagram may be empirically discussed as the superposition of two eutectic-type phase diagrams, one having as the components pure Mg and the intermediate compound Mg_2Si, and the other one the compound Mg_2Si and pure Si.

1.6.4 Retrograde Solubility, Retrograde Melting and Spinodal Decomposition

At temperatures above the eutectic temperature, see Figure 1.16a,b, the solubility of the solute is expected to decrease with increasing temperature after having reached its

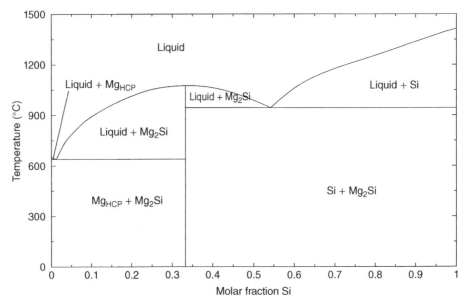

Figure 1.20   *Phase diagram of the Mg-Si system, showing the presence of the intermediate compound Mg₂Si. Scientific Group Thermodata Europe (SGTE). Reproduced with permission from SGTE*

maximum value at the eutectic point. In several binary systems, however, the solubility continues to increase at temperatures well above the eutectic temperature until a solubility maximum is reached, as shown schematically in Figure 1.21, for the solubility of an impurity M in Si on the Si-rich corner.

This phenomenon is called retrograde solubility and known to be common to several binary (Transition Metal) TM-Si or TM-Ge systems [78] but also to compound semiconductor systems [79, 80]. An example is given by the Ni-Si system, where the eutectic temperature is 964 °C while the maximum solubility is achieved at 1300 °C [81].

In spite of the vast number of cases where retrograde solubility occurs, this phenomenon has been considered, erroneously, thermodynamically anomalous [82, 83]. Actually [71], the common assumption of a maximum solubility at the eutectic temperature is contradicted by the regular solution theory, see Figure 1.19d, where one can observe that the solubility maximum for the species B lies well above the eutectic temperature for the equilibrium between regular liquid and solid solutions, characterized by very large, positive interaction coefficients ($\Omega^l = +10; \Omega^s = +30$).

Therefore, retrograde solubility is not an anomalous phenomenon, but one which should be considered in the case of potential occurrence of relevant interaction processes between solution components and solution components and point defects [80], as will be discussed in Chapter 2.

Retrograde melting is a possible consequence of retrograde solubility and is expected to occur at temperatures above the eutectic, as was recently discussed by Buonassisi *et al.* [30] and successfully confirmed by Hudelson *et al.* [27] with a delicate experimental approach, using μ-XAS.

In fact, if the phase diagram of a Si-M binary system has the shape of Figure 1.21, and a homogenous alloy of composition x_1 at point A is quenched from the temperature T_1 to T_2, the system is brought to thermodynamically unstable conditions in one phase configuration and spontaneously decomposes in two phases, one of which is a liquid of composition x_2. Independently of the thermodynamic aspects of this phenomenon, the formation of liquid droplets of known composition upon cooling a homogeneous semiconductor alloy might be of interest for the engineering of quantum devices, if the droplets size was nanometric.

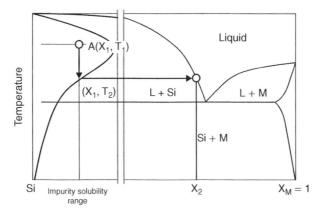

Figure 1.21 *Schematic diagram of a Si-M binary system exhibiting retrograde solubility. Adapted from Hudelson et al., [27], with permission from John Wiley & Sons*

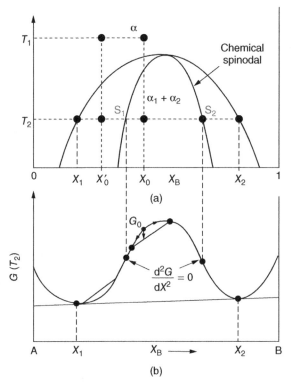

Figure 1.22   *Binodal and spinodal decomposition processes. Porter* et al., *2009 [265]. Reproduced with permission from Taylor & Francis*

Spinodal decomposition is another, frequently observed, phenomenon in semiconductor growth and processing, which occurs when a binary system presents a miscibility gap (see Figure 1.22a), its Gibbs free energy curves have the typical shape of Figure 1.22b for any temperature inside the range of existence of the miscibility gap and the process must be operated in the temperature/composition range of the miscibility gap for compelling technological reasons.

If an alloy of composition within X_1 and X_2 is quenched from temperature T_1 to the T_2, the system is brought from its homogeneous α-phase configuration to thermodynamically unstable conditions and the condition, therefore, occurs for its decomposition into two phases α_1 and α_2 of limiting composition X_1 and X_2, given by the common tangent procedure.

The curve drawn using a set of points X_1 and X_2 kept at the nodes of the Gibbs energy curves at different temperatures is the *coexistence* curve and represents the set of the isothermal boundaries of the miscibility gap in a T versus x diagram. Instead, a set of points S_1 and S_2, corresponding to the coordinates of the zeroing of the second derivative $\frac{\delta^2 G}{\delta x^2} = 0$ of the Gibbs free energy defines the shape of the *spinodal* curve, represented in Figure 1.22a.

Where the two branches of the coexistence and spinodal curve meet, one finds the critical point above which the system is thermodynamically stable in the homogeneous α-phase configuration.

Inside the miscibility gap the phase separation process $\alpha \rightarrow \alpha_1 + \alpha_2$ should occur spontaneously as it implies a Gibbs free energy variation $\Delta G < 0$, but it requires a suitable (thermal) activation in order to occur at a reasonable rate.[8] Under the assumption that the α_1 phase has the same, or a very close, structure as the original homogeneous α phase, the hypothesis should be made that the decomposition process may be either heterogeneous or homogeneous. Heterogeneous unmixing occurs by a nucleation process of the α_2 phase at suitable heterogeneous nucleation sites, which could be GBs, dislocations or point defect complexes.

Homogeneous phase decomposition is supposed, instead, to be initiated, in the absence of heterogeneous nuclei, by two different mechanisms, corresponding to two types of statistical concentration fluctuations, the heterophase- or the homophase-local fluctuation. The concentration width of these fluctuations is between the spinodal and the coexistence curves for the heterophase mechanism, which occurs when the concentration of the precursor phase is X_o, that is for a system only slightly supersaturated and requires overcoming a nucleation barrier typically larger than $5\,kT$ ($>0.5\,eV$ at 1000 K) [84]. This process may be treated as a homogeneous nucleation and growth mechanism, initiated with the formation of a distribution of ordered domains (nuclei) of the α_2 phase, which will then grow to macroscopic size.

The concentration width of these fluctuations is, instead, between the two branches of the spinodal curves for the homophase process, which occurs when the concentration is X_o, deep into the solubility gap. This process is called *spinodal decomposition* and, differently from that initiated by nucleation, involves composition fluctuations in the whole bulk of the material [84, 85]. Both processes are controlled by the same mechanism, the isothermal diffusion of solvent and solute atoms driven by a gradient of the chemical potential, but the consequences, in terms of material quality and properties, are quite different. Spinodal decomposition of metallic alloys has the practical advantage of producing a very disperse microstructure which could improve the mechanical properties of the alloy. In the case of semiconductors, however, it has the drawback of producing local composition inhomogeneities, with consequent electrical and optical inhomogeneities which could be dramatic from the view point of devices. The role of spinodal decomposition in growth processes of compound semiconductors carried out in the region of the solubility gap will be discussed in the next section.

1.7 Solution Thermodynamics and Structural and Physical Properties of Selected Semiconductor Systems

1.7.1 Introductory Remarks

This section deals with the analysis of the thermodynamic properties of several binary and pseudo-binary semiconductor systems of technological interest, with the aim to discuss the relevant correlations between structural, physical, topological and solution thermodynamics aspects.

In this regard, while there is a well-settled empirical knowledge that the solid solubility of an element or of a compound in another one is generally favoured by equal crystal

---

[8] This is the reason why strained layers may survive in a metastable homogeneous single phase configuration.

structures, similar atomic sizes, close lattice parameters and equivalent stable charge states or electronic properties, the macroscopic correlations between thermodynamic behaviour and solution structure present important criticalities.

As an example, contrary to the common belief, complete mixing and ideality in binary or pseudo-binary systems, even belonging to the same crystal system, is not necessarily associated with the fulfillment of Vegard's law [35]

$$a^o_{A_xB_{1-x}} = xa^o_A + (1-x)a^o_B \qquad (1.11)$$

which was discussed in Section 1.3.

A number of binary metallic and non-metallic alloy systems present complete solid state miscibility but important deviations from this law. The possible presence of bimodal distributions of lattice constants, with values intermediate between those of the pure compound and the average Vegard values [70] and of very different distributions of bond distances as a function of the composition, has also been seen, as is the case for the Ge-Si systems which will be discussed later in this section.

The main reason for this behaviour is that the accomplishment of this law is based on the fulfilment of a continuum elasticity model, while the effective physics is fundamentally a problem of quantum mechanics and physical chemistry [86], on *which there already now begins to exist a sound theoretical background.*

Deeper information on structural relationships could be obtained using the EXAFS technique, which is particularly powerful for the determination of the distance between selected atomic pairs. It was, in fact, demonstrated to be capable of giving supplementary information on the evolution of the nearest neighbour distances for both ionic solutions (such as the pseudo-binaries $K_{1-x}Rb_xBr$, $RbBr_{1-x}\,I_x$, $KCl_{1-x}Br_x$ compounds) and semiconducting alloys (such as $In_{1-x}Ga_xAs$, $GaAs_{1-x}P_x$ and $Si_{1-x}Ge_x$), as will be seen later.

As an example, the configurationally averaged lattice parameters versus composition for mixed nitrides ($In_xGa_{1-x}N$ and $Al_xGa_{1-x}N$) have been calculated using total energy minimization procedures, showing that they fulfil Vegard's law. Instead, for the experimental bond lengths measured using EXAFS measurements as a function of the composition, Vegard's law fails [64, 87, 88].

Furthermore [86], the size effects have often been studied from an idealized point of view and the limited success of the continuum elasticity model depends on the fact that the electronic interactions between the outermost quantum shells of the solute and solvent atoms are too complex to be only accounted for by a simple size effect.

By using an elasticity inclusion model Lubarda [86] calculated the apparent size of the solute atoms for a number of binary systems and carried out a comparison with the experimental data of selected systems, with a modest general success, which emphasizes the need to apply non-linear elasticity theories and *ab initio* methodologies.

More recently, Jacob *et al.* [89] showed with pure solution thermodynamics arguments that the conformity with Vegard's law is not an indication of ideal solution behaviour. For non-ideal solutions it is shown that positive Ω values increase the positive deviations from Vegard's law and that for very negative values of Ω the positive deviations of the law caused by lattice mismatch may be compensated by the large effect associated with the solute–solvent interaction. The conclusion is that Vegard's law, due to the absence of a sound theoretical background, might be downgraded to a simple approximation, at least until deeper experimental evidence becomes available.

1.7.2 Au-Ag and Au-Cu Alloys

To analyse the descriptive and predictive ability of solution thermodynamics, associated and implemented by a comprehensive knowledge of structural and physico-chemical properties, it is convenient to start with the analysis of two binary metal alloys (Ag-Au and Au-Cu), deeply discussed in the literature, for which there is an almost complete set of physical and structural information.

The experimental phase diagrams of these two binary systems are displayed in Figure 1.23. Both present a continuous series of solid solutions [90–92], while the Au-Cu system shows also the presence of low-temperature ordered phases [91].

First should be noted the very narrow liquidus–solidus gaps presented by both systems, which may be understood considering that the shape of an isomorphous type of phase diagram depends on the Gibbs free energy of fusion of the components and on the mixing terms Ω^s and Ω^l. In the hypothetical case of ideal behaviour and of close values of entropy of fusion, the liquid–solidus gap decreases with decrease in the entropy of fusion, as shown in Figure 1.24 [71] for the hypothetical case of an AB alloy.

The properties of the silver-gold system have been critically reviewed by Okamoto *et al.* [91], who argued that the very narrow liquid–solidus gap, of the order of 2 °C and less, leads to a segregation coefficient close to 1, almost independent of the temperature. Their study on the thermodynamic properties of the liquid solutions, carried out with calorimetric and EMF measurements, showed that the liquid solution is not ideal and found an interaction coefficient Ω^l value of -11 to 16 kJ mol^{-1} over the entire composition change, leading to a negative value of the enthalpy of mixing.

On the basis of the known values of the melting enthalpy for silver and gold ($\Delta H^f_{Ag} = 11.3$ kJ mol^{-1} and $\Delta H^f_{Au} = 12.5$ kJ mol^{-1}), the entropy of fusion at 1200 K, a temperature intermediate between the melting point of Ag and Au, may be calculated as $\Delta S^f = \Delta H^f/T$, with $\Delta G_f = 0$ at the melting temperature. The resulting values of fusion entropy are 9.53 J K^{-1} for silver and 10.4 J K^{-1} for gold, so close to each other as to justify a very small liquid–solidus gap.

Moreover, the solid solutions are not ideal, as is suggested from accurate bond energy values for the Ag-Ag, Au-Au and Ag-Au pairs calculated by Jian-Jun *et al.* [93], for whom the Ag2 pair has a binding energy of 1.55 eV, the Au2 pair an energy of 1.88 eV and the Ag-Au pair an energy of 1.92 eV, with the Ag-Ag bond energy well comparable with the experimental value of 1.65 ± 0.03 eV [94] and with the theoretical values of 1.64–1.98 eV obtained by Kobayashi *et al.* [95]. On that basis we could estimate for the solid solution a value of the interaction coefficient $\Omega^S = -10$ kJ mol^{-1}.

As could be inferred from an inspection of Figure 1.19q, these Ω values should lead to a positive curvature of the phase diagram, in extremely good agreement with the experimental shape of the Ag-Au phase diagram.

The thermodynamic properties of silver-gold alloys, therefore, are satisfactorily described by the regular solution model, which also foresees the set-up of conditions of local order in correspondence to the $Ag_{0.5} Au_{0.5}$ composition due to the strong Ag–Au interactions, in good agreement also with the results of Flanagan and Averbach [96], who interpreted the non-linear dependence of the Hall effect on the composition as a deviation from a free electron behaviour, due to the formation of new Brillouin zones in the solid, induced by local ordering.

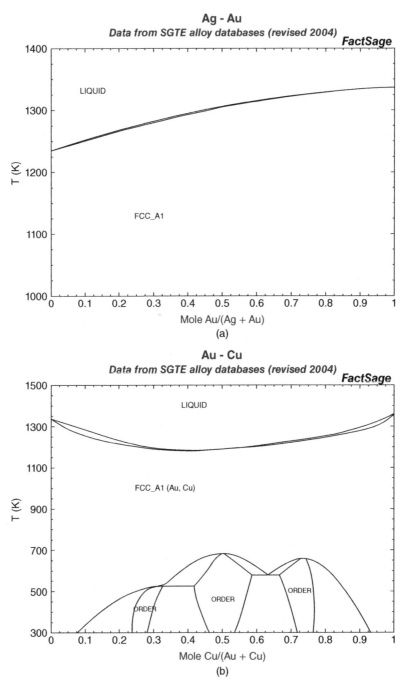

Figure 1.23 *Phase diagrams of (a) the silver-gold system and (b) the copper-gold system. Scientific Group Thermodata Europe (SGTE). Reproduced with permission from SGTE*

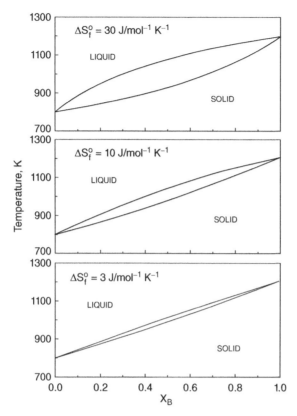

Figure 1.24 *Influence of the entropy of fusion on the width of the liquid-solidus gap. Pelton, 2001, [71]. Reproduced with permisssion from John Wiley & Sons*

The properties of the Au-Cu system, for which we expect, from an inspection of Figure 1.19p, regular behaviour for the solid solutions, with an Ω^s value around -15 kJ mol^{-1} and regular behaviour also for the liquid solutions, with an Ω^l value around -20 kJ mol^{-1}, have been critically reviewed by Okamoto *et al.* [90]. These authors give for the interaction parameter of the liquid solution a value of $\Omega^l(x) = -21.740 - 16.614\ x$(kJ mol^{-1}) and a value of $\Omega^s(x) = -11.053-22.878x$ (kJ mol^{-1}) for the solid solutions, where x is the atomic fraction of copper. As the values of Ω correspond to the ratio $\frac{\Delta H_{mix}}{x(1-x)}$, a value of about 5.6 kJ mol^{-1} is obtained for ΔH_{mix} of the equimolar mixture.

The enthalpy of mixing values calculated using the interaction parameters given above fit well with the experimental values of heat of mixing reported by Weinberger *et al.* [97], who investigated the dependence of the heat of mixing on the composition over the whole range of Cu-Au solid solutions, see Figure 1.25.

The authors note that the shoulder at $x_{Cu} = 0.66$ may suggest the formation of the ordered phase with composition AuCu$_3$, actually present in the phase diagram of Figure 1.23b [91].

The calculated phase diagram using the interaction parameters given by Okamoto *et al.* [90] is reported in Figure 1.26, which shows a very good fit between the calculated solid–liquidus curve and the experimental results.

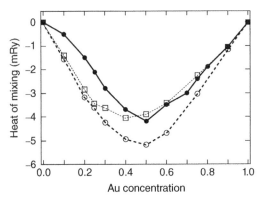

Figure 1.25 *Composition dependence of the heat of mixing of the Au-Cu system (1 mRy = 1.31 kJ mol⁻¹). Weinberger et al., 1994, [97]. Reproduced with permisssion from American Physical Society and P. Weinberger*

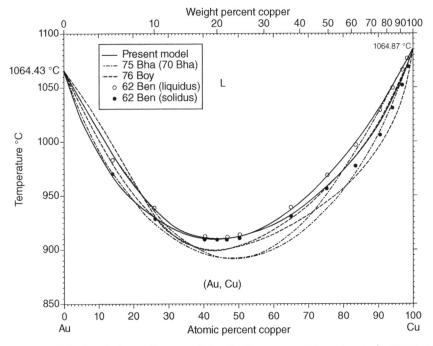

Figure 1.26 *Calculated phase diagram of the Au-Cu system. Okamoto et al., 1987, [92]. Reproduced with permisssion from Springer Science and Business Media*

Okamoto's experimental values of the interaction coefficients Ω^s are composition dependent and a factor of 10 higher (-22.97 kJ mol⁻¹ for the equimolar mixture) than those calculated using the bond energies determined by Szoldos [98], giving a value of 0.595 eV for the Au-Au bond, 0.590 eV for the Cu-Cu bond and 0.623 eV for the Au-Cu bond, and thus an Ω^s value of -2.91 kJ mol⁻¹.

Apparently, the composition dependence of the Gibbs free energy of mixing in these alloys is not only ruled by chemical mixing contributions, but also by elastic energy contributions. It will be shown in Section 1.7.10 that an $\Omega^s x_A x_B$ relationship for the mixing enthalpy, according to Eq. (1.91), has a purely phenomenological character, as it accounts also for elastic energy contribution terms.

The analysis of the structural properties of these systems, which were studied, among others, by Lubarda [86], Bessiere *et al.* [91] and Weinberger *et al.* [97], add little, but significant, supplementary information about their overall behaviour.

The composition dependence of the experimental lattice parameters of Ag-Au alloys reported in Figure 1.27a shows a strong deviation from Vegard's law and a bimodal distribution of the lattice parameters, centred at $x_{Au} = 0.5$, which could be taken as a confirmation of the start of ordering in this composition range, suggested by the Hall effect measurements [96] discussed above.

The presence of slight positive deviations from Vegards law, with a maximum centred at $x_{Au} = 0.5$ (see Figure 1.27b) for the Cu-Au alloys could instead be correlated with the onset of ordered domains at low temperature (see Figure 1.23b), in agreement with the results of Hall effect measurements which show here an important decrease in the Hall effect with a minimum corresponding to $Au_{0.5} Cu_{0.5}$. This effect was interpreted as due to the formation of new Brillouin zones, corresponding to the establishment of structural ordering and superlattice stabilization [95].

The analysis carried out on the properties of these metallic alloys shows the tremendous value of the complementary use of structural, physical and theoretical methods to implement the information coming from thermodynamics, a research methodology that will be demonstrated to be even more important in the case of semiconductor systems. The analysis shows also that in the case of these metallic alloys the regular solution model apparently works well, but that the model should be taken as a phenomenological rather than a physical model. In addition, as will be shown more clearly in the next sections, the deviations from Vegard's law have little to do with the thermodynamics of the solid solutions.

1.7.3   Silicon and Germanium

Silicon is known as the material of choice for microelectronic and photovoltaic applications and the study of its structural, physical and physico-chemical properties has been the subject of systematic experimental and theoretical investigations over the last 60 years. It has also been also as a model system for advanced simulations using *ab initio* MD calculations, as will be shown later in this section. Only recently, germanium gained a strategic role in microelectronic technology, after the development of strained Si-Ge structures presenting enhanced carrier mobility. The main physical properties of silicon and germanium, together with those of other elemental semiconductors which will be dealt with in this and the following chapters, are reported in Table 1.6.

In spite of being one of the most extensively studied semiconductors [99], the complete phase diagram of silicon is only known approximately because of the challenge of measurements in extreme temperature and pressure conditions.

It has, however, been demonstrated that certain tight binding models [100] and *ab initio* MD calculations provide an accurate description of the thermal properties of this system and the correct simulation of phase transformations, respectively. In the case of *ab initio*

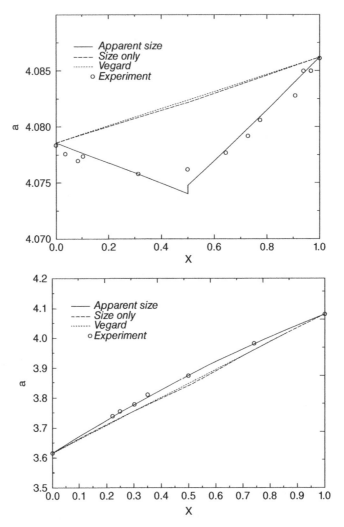

Figure 1.27   *Composition dependence of the lattice constants of the Au-Ag (a) and the Cu-Au system (b). Lubarda, 2003, [86]. Reproduced with permisssion from Elsevier*

MD calculations, the advantage is that the combination of the Car–Parrinello with the Car–Rahman method [101] allows a correct quantum-mechanical description of the inter-atomic forces and internal stresses of the material.

Concerning the properties experimentally available, both silicon and germanium present a negative Clapeyron slope dt/dP along the melting line of their diamond-cubic, low pressure polymorphs (see Figure 1.28a,b) due to the higher density of the liquid phases and both present a number of different high-pressure polymorphs.

The most studied high-pressure polymorphs of silicon and germanium are the metallic β-Sn phases (Si II and Ge II in Figure 1.28), stable above ∼12 GPa, but other Si and Ge phases are stable or metastable at increasing pressures. As an example, Table 1.7 reports

Table 1.6 *Physical properties of Group IV and VI semiconductors*

	Silicon	Germanium	Carbon	Selenium	Tellurium
Melting temperature (T_m (K))	1685	1210	3823	494	722.66
Structure	Diamond cubic	Diamond cubic	Diamond cubic	Monoclinic	Trigonal
Lattice constants (nm)	0. 5431	0.5658	0.35668	$a : 0.9054$ $b : 0.9083$ $c : 1.1601$ $\alpha = \gamma = 90°$ $\beta = 90.810°$	$a : 0.4457$ $b : 0.4457$ $c : 0.5299$ $\alpha = \beta = 90°$ $\gamma = 120°$
Density (g cm^{-3}) at 298 K	2.3296	5.323	3.52	4.809	6.240
Density of the solid at T_m	2.29	5.22	3.51	—	—
Density of the liquid at T_m	2.54	5.51	1.2	3.99	6.2
Electron mobility (cm^2 V^{-1} s^{-1})	1400	3900	2200	—	—
Hole mobility (cm^2 V^{-1} s^{-1})	450	1900	1600	—	—
Band gap (eV)	1.12	0.661	5.45	2.25	—

Data from Glazov and Shchelikov, 2000, [99].

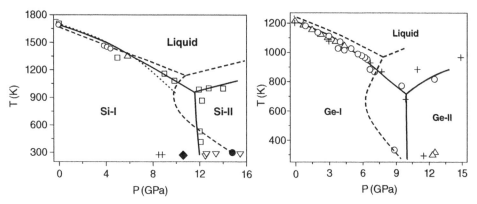

Figure 1.28 *(a) Phase diagram of silicon. Solid, dashed and dotted lines: theoretical pre-dictions, symbols: experimental results. Yang et al., 2004, [102]. Reproduced with permisssion from Elsevier. (b) Phase diagram of germanium. Solid, dashed and dotted lines: theoretical pre-dictions, symbols: experimental results. Yang et al., (2004) [103]. Reproduced with permission from Elsevier*

the crystal data for 6 high-pressure phases of silicon [105], but the actual situation is more complex as at least 11 silicon high-pressure phases are known to be stable.

It has been observed that the high-pressure Si phases never convert directly back to the diamond phase by pressure release, but undergo transformation into several metastable phases. As an example, the β-Sn phase of Si undergoes by pressure release first a transition to an intermediate rhombohedral R8 phase which then transforms to the metastable BC8 phase. On heating at ambient pressure the BC8 phase transforms to another phase which has the hexagonal diamond (HD) phase structure.

Table 1.7   *Crystal data structures of main silicon polymorphs*

Designation	Structure	Pressure region (GPa)
I	Cubic (diamond)	$0 \rightarrow 12.5$
II	Body-centred tetragonal (β-Sn)	8.8–16
III	Body-centred cubic (BC8)	$\sim 10 \rightarrow 0$
V	Primitive hexagonal	$\sim 14 \rightarrow 40$
VII	Hexagonal close packed	40–78.3
VIII	FCC cubic	≥ 78.3

EMIS Data Review Series, 1988, [104]. Reproduced with permission from the Institution of Engineering and Technology.

The corresponding β-Sn phase of Ge undergoes instead a transformation to a short-lived metastable BC8 phase by rapid depressurization or to the tetragonal ST12 phase by slow depressurization [105, 106]. It should be noted, however, that the experimental transition pressures do not necessarily correspond to the equilibrium ones because kinetic factors are involved.

The phase diagrams of silicon and germanium were originally studied by Bundy [107] early in 1964 and recently revisited by Yang *et al.* [102, 103] and by Kaczmarski *et al.* [100], see Figure 1.28. It can be seen that in both cases the agreement between the theoretical predictions and the experiments is particularly good in the relatively narrow pressure/temperature range of the equilibria involving the cubic diamond and β-Sn type solids with the liquid phase.

The Car–Parrinello and Car–Rahman methods have been used by Bernasconi *et al.* [101] and Car and Parrinello [108] for the simulation of the transition of silicon from the cubic diamond- to β-Sn structure, which is complicated by the high energy barrier inherent to this transition and yielded large overpressurizations (42 GPa vs 12 GPa). To overcome the problem, Behler *et al.* [109] adopted a modified Parrinello–Rahman method, based on the introduction of a bias-potential to overcome the large energy potential of the transformation. Using this method several other high-pressure transitions were successfully studied [109].

The theoretical study of solid–liquid transitions in silicon and germanium is very challenging, as liquid silicon and germanium present some unusual properties. Upon melting, different from metals, a density increase of $\sim$10% occurs in silicon and $\sim$5% in germanium [50] (see Table 1.6) and the coordination rises from 4 to 6–7 at the phase transition. This is a lower coordination than would be expected if the liquid were entirely metallic, and is interpreted as an indication of persistence of covalent bonding [110], in agreement with XR-diffraction measurements carried out close to the melting temperature [111] and density measurements [112].

A recent re-examination of the properties of liquid silicon carried out by the Car and Parrinello group, aimed at getting direct information of the temperature dependence of the liquid Si density [113] not only supported the hypothesis of persistence of covalent bonding in liquid silicon [110, 113], but also indicated that liquid silicon is a mixture of two polymorphs, one covalently bonded and the other behaving as metallic silicon [114]. It has been found using inelastic X-ray scattering, that a population of 17% of covalent-bonded silicon atoms coexists at 1787 K with metallic-bonded silicon atoms in liquid silicon [115].

Similar behaviour has been observed in liquid germanium, where the persistence of a strong covalent bonding in the liquid state has been demonstrated [116–118], as well as a decrease in the fraction of the covalently bonded Ge with increase in pressure.

It will be shown also that carbon with the structure of diamond presents a similar, even more complicated behaviour, which is typical of Group IV semiconductors.

1.7.4 Silicon-Germanium Alloys

Few binary elemental semiconductors systems of Group IV present conditions of complete solid solubility, among which the case of silicon-germanium alloys is particularly interesting for the role played by these alloys in microelectronic and optoelectronic technology, together with their elemental precursors (see Table 1.6), as has been preliminarily mentioneded in the last section.

Silicon-germanium alloys, thanks to the elevated electron mobility of germanium, cover a technology sector where silicon is in decline, as they enable faster, more efficient devices to be manufactured using smaller, less noisy circuits than conventional silicon, through greater integration of components onto the chips.

The key advantage of silicon-germanium alloys over their rival technologies is their compatibility with mainstream complementary metal-oxide semiconductor (CMOS) processing. In addition, silicon-germanium alloys provide ultra high frequency capability (well over 100 GHz) on the identical silicon platform where baseband, memory and digital signal processing functions can also be integrated. This is the rationale and the reason why silicon-germanium alloys appear prominently on the technology roadmaps around the world and why they attract increasing attention.

The phase diagram of this system has been studied by Olesinski and Abbashian [119] and reported in a revised form in Figure 1.15: it looks qualitatively very close to the calculated one of Figure 1.14, based on the assumption of the ideality of both the liquid and solid solutions.

On the basis of the bond energies of solid alloys reported in the literature ($E_{Si-Si} = -2.32$ eV, $E_{Ge-Ge} = -1.94$ eV and $E_{Si-Ge} = -2.12$ eV [120] one obtains an interaction coefficient for the solid solutions close to zero ($\Omega^s(eV) = +0.01$ eV). This value is consistent with the assumption of ideality of the solid solutions and with the experimental values of the heat of mixing, which are small ($\Delta H_{max} \sim 1.3$ kJ mol^{-1}) and positive (see Figure 1.29) [121].

These results also agree well with those obtained by Jivani *et al.* [121], see again Figure 1.29, who carried out the theoretical evaluation of the Gibbs free energy and of the heat of mixing of $Ge_x Si_{1-x}$ solutions using a pseudo-potential theory of covalent crystals [122].

They confirmed that the heat of mixing of the solid solution is small and positive, predicting small deviations from the ideality and un-mixing features at low temperature, in agreement with the results of an earlier study of Dunweg and Landau [123]. The modelling of the liquid solution is more challenging, in view of the presence of two phases, one metallic and the other covalent, as shown in the last section.

We could suppose, therefore, that the liquid $Si_x Ge_{1-x}$ alloys are heterogeneous mixtures of covalent- and metallic-bonded Si and Ge, making the evaluation of the liquid–solid equilibria questionable when carried out in terms of a simple regular solution model.

The structural aspects of these alloys are also challenging. There is experimental evidence from XR diffraction measurements of slightly negative deviations of the lattice

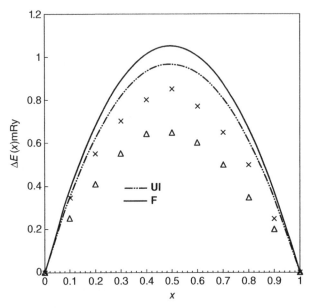

Figure 1.29 *Dependence of the heat of mixing on the composition of the* Ge_xSi_{1-x} *alloys.*
Jivani et al., 2005 [121]

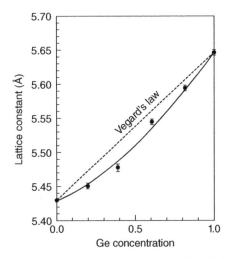

Figure 1.30 *Dependence of the lattice parameters on the* Si_xGe_{1-x} *alloy composition.*
Kajiyama et al., 1992, [124]. Reproduced with permission from American Physical Society

distances from Vegard's law (see Figure 1.30) and of almost complete relaxation of Ge-Ge
and Ge-Si bonds, from EXAFS measurements (see Figure 1.31). These latter show that the
bond lengths are independent of the solution composition and closely follow the Pauling
limit [124]. The authors did not investigate directly the Si-Si bond distances, but it was
supposed that these bonds are also completely relaxed close to a bond length of 0.235 nm,
which is the sum of two Si atomic radii.

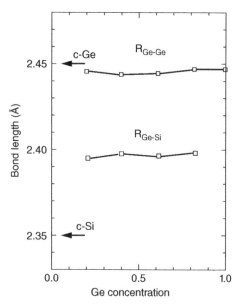

Figure 1.31 *Ge-Ge and Ge-Si bond lengths (R in the figure) as a function of the Ge concentration in* $Si_{(1-x)}Ge_x$ *alloys. Kajiyama et al., 1992, [124]. Reproduced with permission from American Physical Society*

The explanation given is that bond angle distortion accounts for the monotonic variation of the lattice distances and that their non-linear variation might be considered a consequence of the bond lengths invariance.

Shen *et al.* [125] showed that the experimental values of the relaxed bonds could be significantly well predicted using a simple radial force model, while models based on the valence force field fail when the bond-bending forces are included.

Eventually, it was also found from the study of the coordination ratio of Ge in Ge-Si solutions that the distribution of Si and Ge atoms in the lattice is random [124].

The final conclusion is that close lattice parameters, random mixing and close bond energies values are sufficient to make Si-Ge alloys near to ideal solid solutions.

1.7.5 Silicon- and Germanium-Binary Alloys with Group III and Group IV Elements

The solubility of Group III and IV elements in germanium and silicon is generally low or negligible and, consequently, eutectic type phase diagrams are representative of these binary systems.

The solubility of boron in solid germanium is $5 \cdot 10^{18}$ at cm^{-3} at 850 °C [126], but it is known that its solubility might be enhanced, as occurs for B in silicon [127], by point defects engineering, a technique which will be discussed in Chapters 2 and 5.

The phase diagram of the B-Si system is reported in Figure 1.32 [128], which shows that the solid solubility of B at the eutectic temperature is around 2% in mass units or $3 \cdot 10^{21}$ at cm^{-3}, as also shown by later studies [129].

Carbon also exhibits a limited solubility in solid Si and Ge, see Figure 1.33 [130], as a consequence of the very large difference in lattice parameters (see Table 1.6) and of the

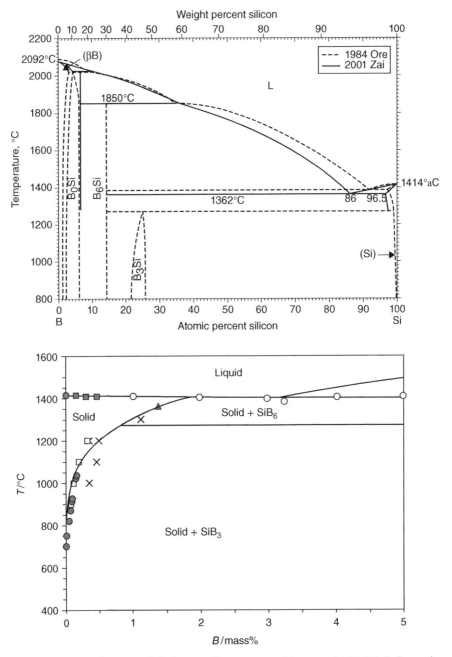

Figure 1.32   *Phase diagrams of the boron-silicon system. Okamoto, 2005, [128]. Reproduced with permission from Springer Science and Business Media*

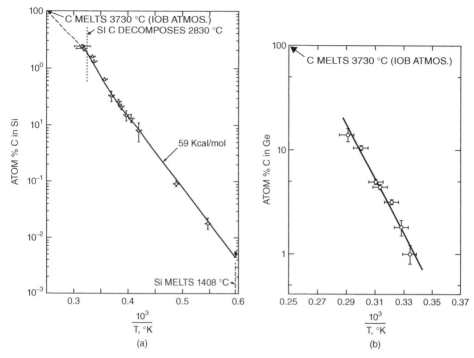

Figure 1.33 *Solubility of carbon in silicon (a) and germanium (b). Scace, 1959, [130]. Reproduced with permission from AIP Publishing*

large Si-C and Ge-C bond energies (318 and 238 kJ mol^{-1}, respectively), which leads to the stability of intermediate phases. While the C-Si system has been the subject of detailed studies in the last few decades, with thousands of published papers, in view of the role of C as an unwanted and detrimental impurity in semiconductor silicon and as the precursor of SiC precipitates, as will be dealt with in Chapters 2 and 3, very limited information is available on the C-Ge system [131].

For both binaries the phase diagrams are available, in full detail for the C-Si system, see Figure 1.34a, as originally investigated by Olesinki and Abbaschian [132], and in less detail for of C-Ge, which was originally investigated by Scace and Slack [130] (see Figure 1.34b). It is worth noting that all the Ge-C phase diagrams, including that recently calculated by Hu *et al.* [131] ignore the presence of the intermediate GeC phase.

Actually, different from the case of silicon carbide, only recently germanium carbide (GeC) has attracted experimental and theoretical interest [133, 134], mostly because it is now being considered as a promising alternative material for photovoltaic and electro-optical applications. The carbon addition in germanium results in an increase in its band gap and a reduction in the lattice parameters, thus allowing its ordered growth on Si substrates [135].

In spite of the relatively close atomic radii ($r_{Si} = 0.1319$ nm, $r_{Ge} = 0.1369$ nm, $r_{Al} = 0.1432$ nm), which should favour larger solid solubilities, the phase diagrams of the Al-Ge

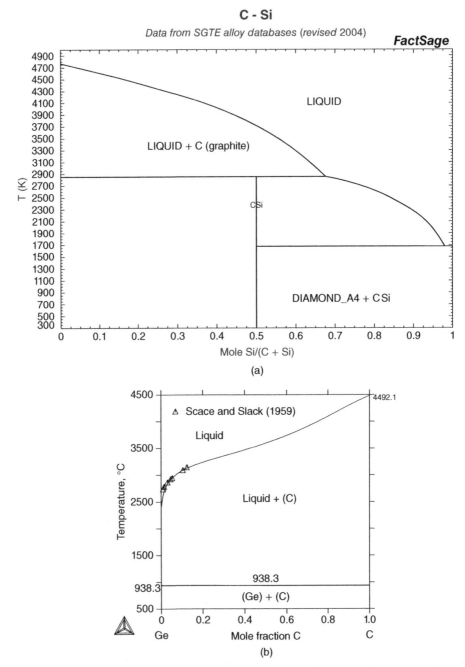

Figure 1.34 *(a) Phase diagram of the C-Si system. Scientific Group Thermodata Europe (SGTE). Reproduced with permission from SGTE (b) Calculated phase diagram of the C-Ge system. Hu et al., [131]. Reproduced with permission from Y. Du*

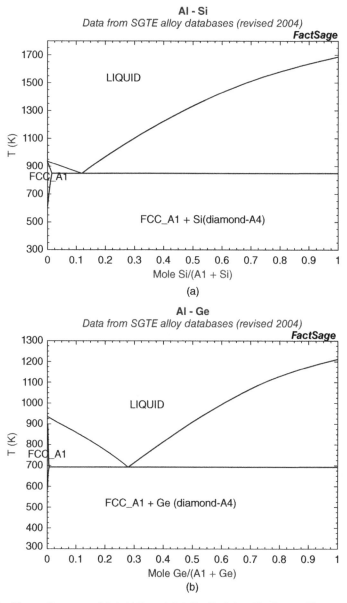

Figure 1.35 *Phase diagrams of (a) Al-Si and (b) Ge-Si. Scientific Group Thermodata Europe (SGTE). Reproduced with permission from SGTE*

and Al-Si systems are typical of metal-semiconductor systems presenting limited recipro-cal solubility (see Figure 1.35). It is interesting to note, however, that the solubility of Al in Ge finds its maximum (1% in weight) at the eutectic temperature [136], while that of Al in silicon shows a retrograde behaviour (see Figure 1.36), with a solubility maximum (450 ppmw) at 1450 K [129, 137].

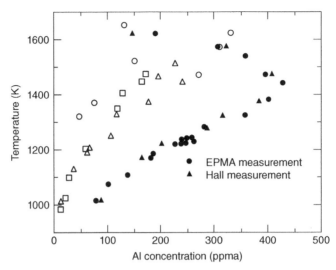

Figure 1.36   *Solid solubility of Al in silicon (EMPA = electron probe microanalysis). Yoshikawa and Morita, 2003 [137]. Reproduced with permission from Electrochemical Society*

The limited solid solubility of Si in Al is a technological obstacle to the growth of homogeneous, Si-rich Al-Si alloys, but the possibility of inducing a larger solid solubility of Si (and Ge) in Al was demonstrated experimentally using a rapid quenching process of Si-Al or Ge-Al melts under pressure (10 GPa) [138].

This last effect might be qualitatively accounted for by the increase in the melting temperature of aluminium with increase in pressure (see Figure 1.7) [53] and by the contemporaneous decrease in the melting temperature of silicon (see Figure 1.8) [42] (see Section 1.4.5 for details).

It was also explained theoretically [54, 139] using an extension to solid solutions of Lindeman's melting law [51] already discussed in Section 1.4.5. The agreement between the experimental and calculated values is reasonably good, as shown in Figure 1.37.

As the solubility of silicon in liquid Al (see again Figure 1.37) is appreciable at temperatures well below the melting point of silicon, Al has been proposed for use industrially as a refining medium for metallurgical (MG) silicon [140], in spite of potentially severe problems of minority carriers lifetime degradation caused by the presence of traces of Al in silicon crystallized from an Al-rich silicon melt [140].

It is known that Group III elements (B, Al, Ga) behave as electron acceptors, but that Al also behaves as recombination centre in silicon, thanks to the presence of two deep levels at 0.315 and 0.378 eV [141, 142], of which the second is a hole trap and the first is a recombination centre, both possibly associated to the Al–O centres in single crystal silicon, as will be discussed in Chapter 2.

A significant advantage of Al refining over other refining routes comes, however, from the very low segregation coefficients of metallic impurities in Al-Si solutions, which are even lower than the corresponding segregation coefficients in liquid silicon [143], thus granting an effective purification from B and TM elements, which are the main and the most deleterious contaminants of MG silicon.

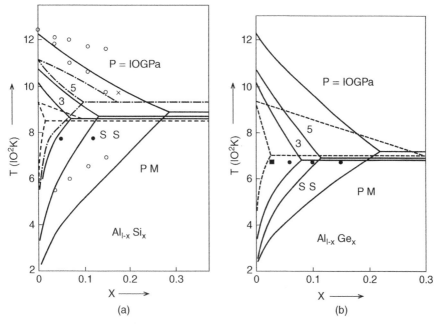

Figure 1.37 *Effect of the pressure on the solubility of (a)silicon and (b)germanium (b) in aluminium. • experimental points. Kagaya et al., 1998 [139]. Reproduced with permission from Elsevier*

By a suitable quenching of saturated liquid solutions of Si in Al one can segregate virtually pure silicon lamellae which can be filtered from the liquid solution [143, 144] and further processed by ingot growth processes, which could bring the Al-content in silicon to acceptable levels, as will be discussed in Chapter 4.

1.7.6 Silicon-Tin and Germanium-Tin Alloys

The other Group IV elements (Sn and Pb), present a limited solubility in Ge and Si, leading to eutectic type diagrams. It should be mentioned here that Ge-Sn alloys are receiving growing attention because it is predicted that, at 10% Sn content, unstrained $Ge_{1-x}Sn_x$ should exhibit a direct band gap leading to important optoelectronic applications.

In strained, cubic diamond $Ge_{1-x}Sn_x$ theoretical studies suggest that the crossover concentration towards a direct band gap should be even lower [145–148], stressing the importance of Sn doping for advanced microelectronic applications of Ge.

Chapter 3 will deal with details concerning the preparation of strained materials, but here it is sufficient to note that although thermodynamic constraints would forbid the preparation of Ge-Sn alloys at equilibrium compositions above the saturation (the maximum solubility of Sn in Ge is 0.5 at%), the use of the epitaxial deposition of a thin epitaxial layer of a Ge-Sn alloy on a substrate presenting a proper lattice mismatch would provide the right solution. It has be shown that strained metastable epitaxial layers of $Sn_{0.07}Ge_{99.3}$, 40–100 nm thick, could be prepared on a silicon substrate with an intermediate, 1 μm thick, Ge buffer layer [24], opening the way to the use of these alloys in advanced optoelectronic applications.

1.7.7 Carbon and Its Polymorphs

Diamond is not only unique among semiconductors for its extreme values of hardness[9] and energy gap, which is of interest for the development of semiconductor devices working in harsh environments (i.e. very high temperature and high γ-rays level), but has been also selected as one of the materials of possible use for inertial confinement fusion (ICF) applications [149]. Graphite, on the other hand, is used as the precursor of diamond and, more recently, of graphene and is a material of common use in high-temperature applications as a mechanical support, as a crucible or as a heating element.

Due to the extreme difficulty of carrying out experiments even in the pressure/temperature range of the coexistence equilibria involving the two common solid polymorphs of carbon (graphite and diamond) and liquid carbon, which occur at temperatures between 4000 and 6000 K and pressures between 100 and 1000 GPa, the thermodynamic properties of this system are not yet well known and most, though not all, of the known properties are the result of theoretical studies.

So far, only the graphite/diamond and the graphite/liquid phase boundaries, that occur at relatively low temperatures and pressures, have been located experimentally with reasonable accuracy [150–155].

The phase diagrams of carbon, displayed in Figure 1.38a,b, are the result of the pioneering experimental studies of Bundy [150–153], who succeeded in defining the thermodynamic properties of the system in the temperature/pressure range of the graphite/diamond/liquid carbon coexistence.

It should be noted that in both diagrams the Clapeyron dT/dP slope of the graphite melting line is positive, in good agreement with the experimental values of the density of the liquid at low applied pressures (1.2 g cm^{-3}) compared to graphite (2.26 g cm^{-3}). The negative slope above the melting point of diamond shown in the original diagram of Figure 1.38a was, however, based on a hypothetical analogy of diamond with silicon and germanium, for which the density of the liquid is higher than that of the solid. This conjecture was abandoned after the 1980s when the results of experimental [153] and theoretical work [156] demonstrated that the slope is actually positive. Boundy *et al.* [153] also showed that the liquid in equilibrium with solid diamond is metallic but that there is also evidence of a transformation between a non-conducting and a conducting form of liquid carbon.

This conclusion is in good agreement with the results of Togaya [154], who suggests the onset of a first order transition in liquid carbon along the melting line of graphite at 5 GPa. The same author also showed that liquid carbon is 10 times more resistive than liquid silicon (80 μΩ cm) and germanium (75 μΩ cm) (see Figure 1.39) and proposed, as a preliminary qualitative explanation, that it could arise from the basically covalent structure of liquid carbon, with three of the four valence electrons involved in covalent bonding and the conducting electron being strongly localized. It is in any case apparent that liquid carbon, as silicon and germanium (see Section 1.7.3), is stable also in a non-metallic form, but that the larger C-C bond energy (347 kJ mol^{-1} vs 222 kJ mol^{-1} for Si-Si and 188 kJ mol^{-1} for Ge-Ge bonds) favours the persistence of a larger amount of covalently bonded C in liquid carbon.

---

[9] A recent work [149] shows that C60 molecules solvated with m-xylene give rise to an ordered structure phase that becomes harder than diamond at pressures above 32 GPa.

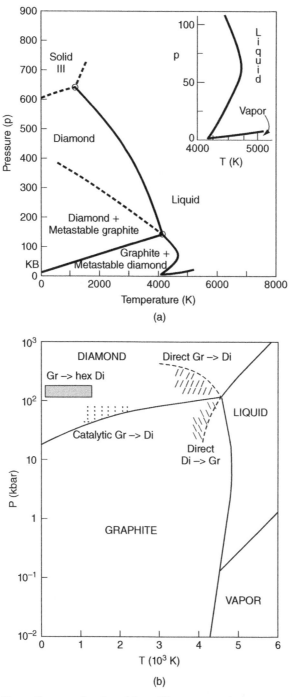

Figure 1.38 *(a) Phase diagram of carbon (dotted lines: extrapolations or predictions). Bundy, 1963 [151]. Reproduced with permission from AIP Publishing (b) Updated diagram. Bundy, 1989, [152]. Reproduced with permission from Elsevier*

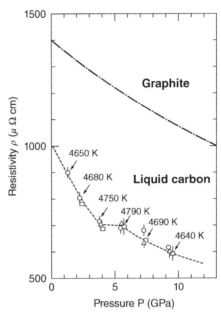

Figure 1.39   *Experimental dependence of the resistivity of graphite and liquid carbon along the melting line of graphite. Togaya, 1997, [154]. Reproduced with permission from American Physical Society and M. Togaya*

Additional information on the properties of the carbon system might be inferred from the phase diagram of Figure 1.38b, which shows that the triple point where diamond, graphite and liquid phases coexist occurs in a temperature range around 4500–5000 K for pressures close to 12 GPa. It also shows the presence of a region where a hexagonal polymorph of diamond can be found as a metastable phase and indicates that the equilibrium graphite → diamond transformation occurs under catalytic enhancement. In the same diagram it is also shown (dotted curves on the top region of the diagram) that the direct transformation of diamond (D) to graphite (G) and graphite to diamond (dotted line at the top of the figure) does not occur spontaneously at the coexistence pressures, but requires much higher pressures in the absence of catalysts.

This apparently happens notwithstanding graphite and diamond having very close Gibbs free energies of formation and, consequently, that the transformation enthalpy and entropy are very small ($\Delta H_{G \rightarrow D} = +1.897$ kJ mol^{-1}, $\Delta S_{G \rightarrow D} = -3.36$ J mol^{-1} K^{-1}). It can be demonstrated that excess temperatures and/or pressures are required to overcome the high activation energy for the transformation (728 kJ mol^{-1}), a process which implies the rupture of graphitic bonds and their rebonding via sp3 hybridization.

At lower synthesis temperatures, the formation of HD, a metastable polymorph of diamond, from graphite is favoured instead of the stable cubic diamond phase. Static compression of graphite at 1200–1700 K results in the formation of HD and further compression results in the formation of diamond, but never at pressures lower than 12 GPa, much higher than the coexistence pressure in this temperature range.

Moreover, the transformation of the metastable HD to graphite does not occur sponta-neously, but requires either the assistance of catalysts or its enhancement by size effect factors [157], which were examined in detail by Khaliullin *et al.* [158], as will be seen later in this section.

These uncommon features have been the subject of a number of theoretical studies, aimed at the development of models capable of explaining the physical grounds of this behaviour.

Among the considerable amount of literature available on the subject, one has to deal, at first, with the work of Wang *et al.* [159] who calculated the diamond melting line using an *ab initio* Car–Parrinello MD scheme (see Figure 1.40a) and proposed the phase diagram shown in Figure 1.40b.

It can be observed that the Clapeyron slope is positive up to 600 GPa, turning then to slightly negative values. Using the density-functional theory based on MD *ab initio* simu-lations, Ghiringhelli *et al.* [161] obtained the phase diagram of Figure 1.41 which confirms the positive Clapeyron slope for the diamond melting line in the low-pressure range.

The same authors calculated the solid and liquid densities as a function of the pressure (see Table 1.8 [161]) and showed that the density of both solid and liquid increase with pressure, with a tendency for the liquid to become denser than the solid, in good agreement with the results of Wang *et al.* [159] and of Grumbach and Martin [162]. These last authors observed a change from fourfold to sixfold coordination in the liquid when the pressure var-ied from 400 to 1000 GPa, just in the range where the liquid becomes denser than the solid.

An important aspect of the physical chemistry of carbon is the thermodynamics of the phase transformations of carbon polymorphs, which have been studied by among others, Dmitriev *et al.* [163] and Khaliullin *et al.* [158]. According to Dmitriev *et al.* [163] the graphite → diamond transition may be interpreted as a transition between two low-symmetry, ordered phases, which are formed from a latent, parent disordered phase

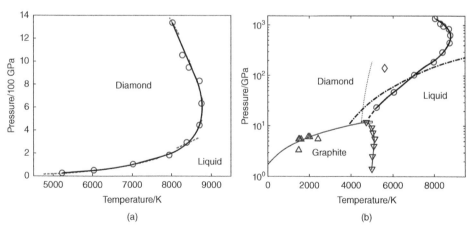

Figure 1.40 *(a) Calculated melting line of diamond (open circles are calculated values, dot-ted line are the Clapeyron slopes). (b) Phase diagram of carbon (solid line from F.P. Bundy et al. [160]). Wang et al., 2005, [159]. Reproduced with permission from American Physical Society and S. Scandolo*

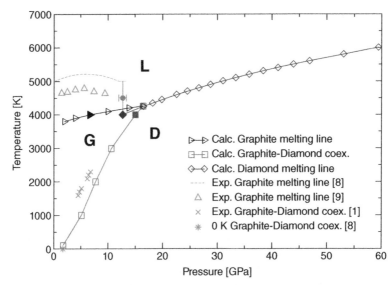

Figure 1.41   *Phase diagram of carbon. Ghiringhelli et al., 2005, [161]. Reproduced with permission from American Physical Society and L. Ghiringhelli*

Table 1.8   *Effect of the pressure on the density of solid and liquid carbon*

P (GPa)		T (K)	ρ_s (g cm^{-3})	ρ_l (g cm^{-3})
2.0	Graphite	3 800	2.134	1.759
6.7		4 000	2.354	2.098
16.4		4 250	2.623	2.414
16.4	Diamond	4 250	3.427	2.414
25.5		4 750	3.470	2.607
43.9		5 500	3.558	2.870
59.4		6 000	3.629	3.043
99.4		7 000	3.783	3.264
148.1		8 000	3.960	3.485
263.2		10 000	4.286	3.868
408.1		12 000	4.593	4.236

Data from Ghiringhelli *et al.*, 2005, [161].

by a combined mechanism of ordering plus shift for the diamond structure and ordering plus decompression for the graphite structure.

Khaliullin *et al.* [158] studied the same process using first principle arguments and concluded instead that the slow stage of the phase transformation of graphite to diamond is a nucleation process of an embryo of the diamond phase, which forms by sp3 hybridization in defective regions of the graphite lattice, whose activation enthalpy strongly depends on the pressure (see Figure 1.42).

This result agrees well with the experimental features of the same transformation, which occurs under catalytic support. They could also demonstrate that the formation of the hexagonal polymorph of diamond during the transition of graphite to diamond is associated

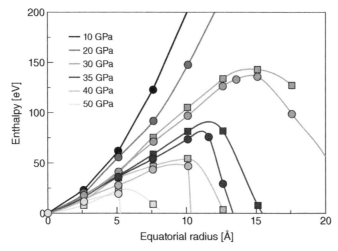

Figure 1.42 *Pressure dependence of the nucleation barriers for the graphite-diamond transition RG → CD (circles) and HG → HD (squares). Khaliullin et al., 2011, [158]. Reproduced with permission from Macmillan Publishers*

with similar activation energies for the transition of rhombohedral graphite (RG) to CD and hexagonal graphite (HG) to HD.

The hexagonal and the cubic phases of diamond are not its sole polymorphs. A body centred cubic phase, labelled BC8, is predicted to be stable by, among others, Correas *et al.* [164] and Grumbach and Martin [162] (see Figure 1.43). The latter, as mentioned before, also predict the stability of several liquid carbon phases at high pressures, presenting different local coordination, of which the fourfold one is stable at pressures around 10^2 GPa and lower, while the sixfold one is that typical in the $1–4 \cdot 10^3$ GPa range. They also mention, as a theoretical prediction, that one of the liquid phases of carbon is insulating and the other is metallic.

The stability of at least two different liquid carbon phases is also predicted by several other authors [165–167], who speculate about the presence of a metallic liquid carbon phase, whose density depends on the pressure, being equal to 1.2 g cm^{-3} at low pressures and increasing to 1.8 g cm^{-3} at 5.4 GPa at the melting line [168], definitely lower than those of diamond and graphite at low pressures (3.51 and 2.26 g cm^{-3}, respectively). Knudson *et al.* [52]show, using a shock wave technique, a drastic increase in the density of the liquid, up to 7.2 g cm^{-3} at pressures in the 700–1000 GPa range, where liquid carbon is in equilibrium with the diamond and BC8 phases. It should be noted that these densities agree well with those obtained by linear extrapolation of the data reported in Table 1.9.

1.7.8 Silicon Carbide

Silicon carbide (SiC), thanks to its hardness (surface microhardness 2900–3100 kg mm^{-2}: Mohs values 9.2–9.3), second only to diamond (surface microhardness $8–10 \cdot 10^3$ kg mm^{-2}:10 on the Mohs scale), is used as an abrasive in cutting tools and is very popular in the silicon semiconductor industry where it is employed (now together with diamond) in the wire sawing technology of single crystal and multicrystalline silicon.

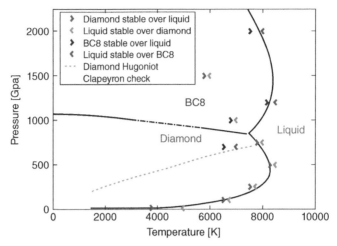

Figure 1.43 *Calculated phase diagram of the carbon system (the graphite phase has been omitted for simplicity). Correa et al., 2006 [164]. Reproduced with permission from National Academy of Sciences, U.S.A*

Table 1.9 *Physical properties of the most common polytypes of silicon carbide*

SiC polytype	Structure	Lattice constant (nm) 300 K	Thermal conductivity (W cm^{-1} C^{-1})	Surface energy (J m^{-2})	Density (g cm^{-3}) 300 K	Heat of formation (kJ mol^{-1})	Energy gap (eV)	Electronic mobility (cm^2 V^{-1} s^{-1})
3C	Cubic:ZB	0.43590	3.6	1.724	3.166	−64	2.36	900
6H	Hexagonal:W	a : 0.3073	4.9	1.767	3.211	−63.5	3.23	a:370
		b : 1.0053						b:50
4H	Hexagonal:W	a : 0.3073	3.7	1.8	3.21	−66.6	3.0	a:720
		b : 1.0053						b:650

ZB = zinc blende and W = wurtzite.

Since the end of the 1980s, when single crystals of SiC became available, SiC has also been employed as the substrate of high voltage-high power transistors and thyristors, as it offers (see Table 1.10), in comparison with silicon, a lower intrinsic carrier concentration, a higher electric breakdown field and a higher thermal conductivity. It is also used as a substrate for the hepitaxial growth of III–V semiconductors [170–173].

As can be seen in the phase diagram of the C-Si system of Figure 1.34a, SiC is the stable, intermediate phase, which, however, cannot be formed by liquid crystallization techniques as it segregates from the liquid at the peritectic decomposition temperature (3103 ± 40 K and 3.5 MPa), together with carbon.

It could, obviously, be prepared in the polycrystalline form by sintering equimolar quantities of silicon and carbon in a reducing atmosphere at temperatures below the decomposition temperature, although the industrial route, known as the Acheson process, is to react silica (SiO$_2$) powder with carbon in an electric arc furnace at temperatures between 1600 and 2500 °C.

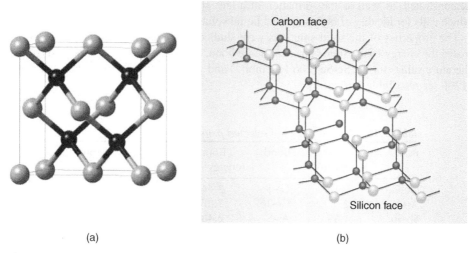

(a)   (b)

Figure 1.44   *(a) Structure of the cubic SiC Materials Design, 2014.(b) Structure of the 4H-SiC polytype. Madar, 2004 [175]. Reproduced with permission from Macmillan Group*

The growth of single crystal samples, which will be described in Chapter 4, is carried out by seeded sublimation, a process often referred to as physical vapour transport (PVT) growth [174]. Seeded crystal growth is needed, as SiC is stable in more than 200 known polytypes, of which only a few are, however, of technological interest.

The polytypism of SiC is a kind of polymorphism which arises from the different order of stacking layers of Si-C tetrahedra in the SiC lattice. Three types of layers exist, labelled A, B, C, which allow a comfortable packing, and the difference between the different polytypes depends on the stacking sequence adopted. As an example, the common 6H-SiC polytype is formed by an infinite repetition of the sequence ABCACB and the cubic 3C-SiC results by an infinite repetition of the sequence ABC, see Figure 1.44.

Some structural and thermodynamic properties of the most common SiC polytypes [176] are reported in Table 1.9. The Gibbs energy of formation of 3C- and 6H-silicon carbide was determined electrochemically between 1623 and 1898 K [177] by measuring the silicon solubility in liquid gold in equilibrium with a mixture of silicon carbide and graphite. The behaviour of the Au-Si system provides a wide compositional range for the small difference in the activity of silicon exhibited by the two polytypes, thus offering greater experimental precision over previous similar methods. For 6H-SiC, a Gibbs free energy value ΔG (kJ mol^{-1}) = $-116\,900(\pm 7.2) + 38.2(\pm 4.1)T$ was obtained.

Major problems in the operation of SiC-based electronic devices arise from perturbations of the ideal layer stacking, with the formation of stacking faults, which are known as the origin of deterioration of SiC devices after long operation periods. This issue will be discussed in Chapter 3.

1.7.9   Selenium-Tellurium Alloys

In spite of more than half a century of investigations, the properties of the Se-Te system remain not entirely settled. The full miscibility of their hexagonal structures has been demonstrated, as well as the formation of a lens-shaped isomorph-type phase diagram, which calls for ideality of both solid and liquid solutions (see Figure 1.45).

The properties of the solid solutions were studied by Pattanaik *et al.* [178], who estimated the energy of the single covalent Se-Te bond, as 201.1 kJ mol^{-1}, on the basis of the literature values for the Se-Se (206.1 kJ mol^{-1}) and Te-Te (158.8 kJ mol^{-1}) bond energies [179], see also Table 1.10 [169].

Table 1.10   *Bond energies of selected pairs*

Bond	Bond energy (eV)	Bond	Bond energy (eV)	Bond	Bond energy (ev)
Ge-Ge	2.13	Ge-Se	2.44	Bi-Te	1.69
As-As	2.07	Ge-Te	1.87	Pb-S	2.55
Sb-Sb	1.55	As-S	2.48	Pb-Se	2.40
Bi-Bi	1.32	As-Se	2.26	Pb-Te	1.90
Pb-Pb	0.89	As-Te	1.99	Sn-Se	2.36
Sn-Sn	1.57	As-Sb	2.06	Sn-Te	1.86
S-S	2.69	Sb-Se	2.25	Se-Te	2.00
Se-Se	2.14	Sb-Te	1.73		
Te-Te	1.65	Bi-Se	2.17		

Rao and Mohan, 1981, [69]. Reproduced with permission from Elsevier.

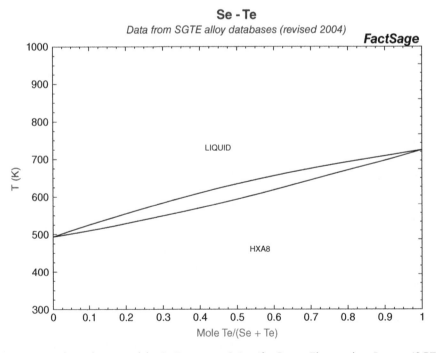

Figure 1.45   *Phase diagram of the Se-Te system. Scientific Group Thermodata Europe (SGTE). Reproduced with permission from SGTE*

The problem here is that the Se-Te bond has a partial heteropolar character [180] and, therefore, a simple chemical model could not be directly applied for evaluation of the interaction coefficient. Within this uncertainty, the interaction coefficient takes a value $\Omega = -18.55$ kJ mol^{-1}.

The thermodynamics of the liquid solutions are even more difficult to be settled, because the Se-Te solutions present a semiconductor character at high Se concentrations and undergo a transition to metallic character above 80% Te [181]. Qualitative information concerning the bond energies in the liquid phase might be inferred from the properties of calchogenide Bi-Se-Te glasses, considered as metastable liquids, from which a value of 1.43 eV for the Te-Te pair, 1.89 eV for the Se-Se pair and 1.87 eV for the Se-Te pair might be deduced [182].

From these values the interaction coefficient in the liquid solutions Ω^l might be estimated to range around 0.225 eV or 21.701 kJ mol^{-1}, within the errors of the bond energies [182]. Both the solid and the liquid solutions, therefore, have non-ideal behaviour.

The good fit with Vegard's law observed for the lattice distances along the *c*-axis in contrast with deviations observed along the *a*-axis, see Figure 1.46 [183] are issues which offer additional information on the overall physico-chemical behaviour of this system. The deviations along the *a* axis have been tentatively explained [183] considering that the chain structure of these materials would make the description of the microscopic structure of the Se-Te alloys different from that of conventional systems, where a simple substitutional or interstitial solution would form. The hypothesis was suggested that the formation of chains containing Se-Te covalent bonds, aligned along the *a*-axis, is favoured against that of homogeneous chains formed only by Se and Te atoms, in view of the larger energies of the Se-Te bonds.

1.7.10 Binary and Pseudo-binary Selenides and Tellurides

Sulfides, selenides and tellurides of Group II elements (Cd, Zn, Hg) are families of binary and pseudo-binary compound semiconductors of relevant scientific and technological interest, of which the selenides and tellurides are the most important.

As an example, thin film CdTe is a material of excellent photovoltaic properties [184, 185] in terms of efficiency, cost and pay-back time, in spite of the environmental problems

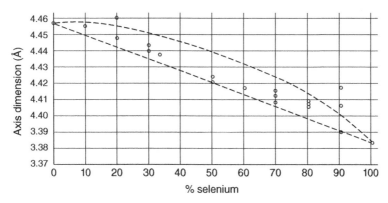

Figure 1.46 *Plot of the lattice constant, a, of Te-Se alloys as a function of the alloy concentration. Grison, 1951, [183]. Reproduced with permission from AIP Publishing*

associated with the toxicity of Cd and Se and Cd vapours or CdO spills in the case of fire. On the other hand, CdZnTe (CZT) is unrivalled for IR, X-rays and γ-detectors [186–189] and is also used as a lattice-matched substrate for the epitaxial growth of HgCdTe, another widely employed semiconductor used for the fabrication of detectors in the long-wavelength infrared spectral region.

Thin films of CuInSe (CIS) and of quaternary alloys belonging to the same family are strong competitors of CdTe as photovoltaic materials, in virtue of their high photovoltaic efficiency[10] and for being Cd-free and thus avoiding the main drawback of CdTe.

It will be shown in Chapter 3 that a problem concerning these materials is their ingot growth, which requires a proper control of the growth atmosphere, considering that the melt consists of highly volatile components. In the case of the growth of $Cd_{1-x}Zn_xTe$ crystals, as an example, one has a steady loss of the constituents from the vapour phase above the melt or through the porous walls of the crucible during the growth. The loss of the constituents might be suppressed by the application of an external, inert gas pressure, typically $\approx 15\,GPa$ of argon.

Since Cd has the highest vapour pressure among the Cd ZnTe melt constituents, the vapour phase predominantly consists of Cd atoms. Although the external Ar pressure greatly reduces sublimation, it does not completely eliminate Cd loss from the crucible [190], which induces the formation of non-stoichiometric deviations and the presence of Cd-vacancies and Te-antisites in the crystal lattice, as will be discussed in Section 2.4.3.

Knowledge of the pressure–temperature–composition $(P - T - x)$ phase equilibria gives, therefore, the necessary thermodynamic basis for the crystal growth of materials with controlled composition.

As an example, detailed studies of the $P - T - x$ phase diagram for the Cd-Te system and of the non-stoichiometry in CdTe have been reported [191, 192] and information on the vapour pressure measurement and estimates of solubility of the components in ZnTe are also available [193, 194].

A key property of these semiconductors is the presence of a second order transition between their stable cubic zinc blende phase and the hexagonal wurtzite phase (Figure 1.47) [195] and a further transition to a NaCl structure at pressures of the order of several GPa [196].

Polytype formation is proposed to be associated with the effect of a progressive distortion of the lattice due to the addition of ordered staggered layers and to internal-rotation forces, while configurational and vibrational entropy effects are not found to contribute to the stabilization of polytype lattices [195]. Two typical examples of this kind of phase transition will be given at the end of this section.

The challenges with the thermodynamic study of II–VI and IV–VI compounds are the marked deviations from the ideality of the liquid solutions, which cannot be accounted for by the QCA nor by the pseudo-regular solution model, based on the assumption of a linearly temperature-dependent interaction coefficient $\Omega(T)$. In these systems the large difference in electronegativity produces more ionic and stronger interactions among unlike atoms, evidenced by the onset of liquid immiscibility and by a step temperature peak in correspondence to the melting temperature of binary systems, as shown in Figure 1.48 [197].

---

[10] According to a January 2013 press release, EMPA, the Swiss Federal Laboratory for Materials Science and Technology, achieved a record efficiency of 20.4% with CIGS on polymer foils.

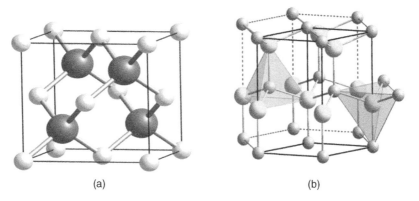

(a)   (b)

Figure 1.47   (a) Zinc blende structure B. Mills and (b) wurtzite structure. Solid State

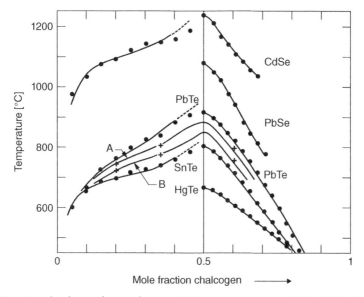

Figure 1.48   *Liquidus lines of some binary II–VI systems. Laugier, 1973, [197]. Reproduced with permission from EDP Sciences*

Better results were obtained using the so-called surrounded atoms model, in which the basic entity is an atom surrounded by all its neighbours, not a simple atom pair or the associated solution model, postulating the existence of stable complexes, and the polyassociative model, postulating the stability of a multiplicity of complexes in solution [198, 199], each of them representing an attempt at a better thermodynamic description of the studied system.

Later in this section will be seen the advantages of these techniques, after having gained insight on some details of the phase diagrams of II–VI compounds.[11]

---

[11] A recent comprehensive analysis has been given by Adachi [200].

A number of II–VI semiconductors (CdTe, CdS, CdSe, PbTe, SnTe, to mention only the most important from the viewpoint of their technological applications) but also GaAs and InP, present significant deviations from the stoichiometry, whose onset might be described with the following defect reactions, taking CdTe as an example

$$Cd_{Cd} \rightleftharpoons Cd^g + V_{Cd} \tag{1.100}$$

$$Te_{Te} + Cd_{Cd} \rightleftharpoons Te_{Cd} + Cd^g \tag{1.101}$$

$$Te_{Te} \rightleftharpoons Te_i + V_{Te} \tag{1.102}$$

and whose extent depends on the Gibbs free energy of formation of the different point defects [201], as will be seen in Chapter 2.

Although the deviations from the stoichiometry are generally modest, and range between $\pm 10^{-4}$ molar fraction for GaAs and $\pm 10^{-3}$ for PbTe, their growth from the melt could be troublesome, as will be seen in Chapter 4.

The phase diagram of CdTe is reported in Figure 1.49 as a typical example. One can see in Figure 1.49a that the solid melts congruently [199, 202] but exhibits marked non-stoichiometry with decreasing temperature (see Figure 1.49b), below the melting point, with a maximum Te-excess around 1000 °C [201, 203].

It is also interesting to mention that the substitution of Se with S leads to an improvement in the stoichiometry [203]. This effect plays a beneficial role in photovoltaic materials based on Cd compounds and depends on the different Gibbs energy of formation of Cd vacancies in CdS and CdSe, as will be discussed in Chapter 2. Both defects are deep level and responsible for minority carriers lifetime degradation effects.

The phase diagrams of CdSe, ZnSe and ZnTe show close topologies with congruent melting and liquid immiscibility features [204–206], which has stimulated the interest of many authors [207–209] in the updating of the thermodynamic properties of these alloys.

As anticipated above, the problem found in the theoretical analysis of these binary systems is the description of the liquid phases, which behave neither as regular nor pseudo-regular solutions.

Good results were, however, obtained in the case of ZnTe using the polyassociative model, as shown in Figure 1.50 [206], where the calculated points are superimposed on the experimental diagram [205], in spite of some crude assumptions on the equilibrium partial pressures of Zn and Te.

The basic approach of this model is to consider the presence in the liquid solution of undissociated Zn and Te together with the complexes ZnTe, Zn_2Te, Zn_2Te_3, $ZnTe_2$. Each complex has a dissociation constant $K_{xy}(T)$, which in the case of the complex Zn_pTe_q could be written [206]

$$K_{pq}(T) = \frac{x_{Zn}^q x_{Te}^p}{x_{Zn_p Te_q}} \tag{1.103}$$

(where $x^q{}_{Zn}$ and $x^p{}_{Te}$ are the molar fractions of the undissociated Zn and Te in solution) which can be calculated using numerical procedures on the basis of the experimental values of the vapour pressures of Zn and Te.

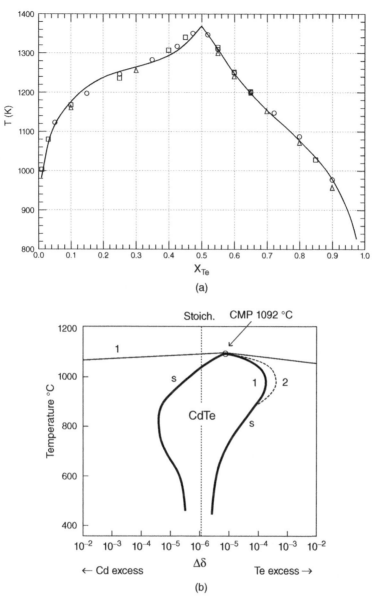

Figure 1.49 (a) Phase diagram of Cd-Te system. Brebrick, 2010, [199]. Reproduced with permission from Elsevier. (b) Non-stoichiometry of the CdTe phase. Rudolph, 2003, [201]. Reproduced with permission from John Wiley & Sons

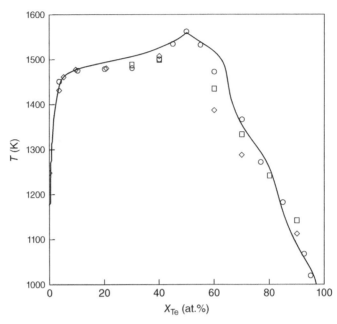

Figure 1.50 *Phase diagram of the Zn-Te system. Moskvin, et al., 2011, [206]. Reproduced with permission from Centrum Informatyczne TASK*

The concentration of the complexes in solution is related to the overall concentration of the components in solution via mass balance equations and the T, x phase equilibrium is defined by the equation [206]

$$\Delta S^f_{ZnTe} \left(\frac{T^f_{ZnTe} - T}{RT} \right) + \ln \frac{x_{Zn} x_{Te}}{x^{soln}_{Zn} x^{soln}_{Te}} = 0 \tag{1.104}$$

where ΔS^f_{ZnTe} is the entropy of fusion of ZnTe and T^f_{ZnTe} is the melting temperature of $Zn_{(1 \pm x)}Te_x$ for a couple of values of x^{soln}_{Zn} and x^{soln}_{Te} in the liquid solution. Eventually, the experimental data relative to the temperature dependence of the vapour pressures of Zn and Te in equilibrium with the solid solution, under the assumption that only the free (dissociated from the complex) Zn and Te atoms determine the vapour pressure, allows one to obtain the dissociation constants for the ZnTe, Zn_2Te, Zn_2Te_3 and $ZnTe_2$ complexes and thus the phase equilibria.

Using the associated solution model, a simpler version of the polyassociative model, the dissociation coefficients β of a number of binary systems exhibiting complete miscibility in the liquid and solid state were also calculated [197]. The results are shown in Table 1.11, from which it is possible to conclude that 100% association in the liquid state occurs in most cases, except binaries involving tellurium (PbTe, SnTe and CdTe).

The values of the interaction coefficients Ω in Table 1.11 were instead evaluated by fitting the experimental solidus curves with the equations describing the solid–liquid equilibria, showing that CdTe, CdSe, SnTe and PbTe present the largest interaction coefficients and, therefore, the largest deviations from ideality.

Table 1.11 Interaction parameter Ω and dissociation coefficient β of the stoichiometric liquids according to the associated solution model

System	ZnTe (Te-rich)	ZnTe (Zn-rich)	CdTe (Te-rich)	CdTe (Cd-rich)	HgTe (Te-rich)	PbTe (Te-rich)	PbTe (Pb-rich)	SnTe (Te-rich)	SnTe (Sn-rich)	PbSe (Se-rich)	CdSe (Se-rich)	CdSe (Cd-rich)
Ω (kJ mol^{-1})	7.95	71.1	7.11	45.6	0	−25.94	23.01	−13	35.56	−39.33	32.64	50.2
β	0.00	0.065	0.00	0.055	0.04	0.04	0.14	0.04	0.10	0.00	0.02	0.12

For a completely dissociated liquid $\beta = 1$ and in this case Ω corresponds to the interaction coefficient of a regular solution [197].

Laugier, 1973, [197]. Reproduced with permission from EDP Sciences.

Table 1.12 Regular solution interaction coefficients for some pseudo-binary solid solutions

System	Melting temperature range (K)	Ω (kJ mol^{-1})
ZnTe-CdTe	1560 to 1365	5.60
ZnTe-HgTe	1560 to 943	13
ZnSe-ZnTe	1803 to 1560	6.48
CdTe-HgTe	1365 to 943	5.86
CdSe-CdTe	1528 to 1365	6.28
HgSe-HgTe	1073 to 943	2.93
PbSe-PbTe	1353 to 1200	6.28
PbTe-SnTe	1200 to 1080	0.73
MnTe-GeTe	1430 to 1013	3.38

Laugier, 1973, [197]. Reproduced with permission from EDP Sciences.

It should be noted that the actual interaction coefficient value and sign may depend on the stoichiometry of the melt, while here only metal-rich alloys were considered.

Using the regular solution model, the temperature-independent interaction coefficients Ω were also calculated for several pseudo-binary systems [197]. The results are reported in Table 1.12, which shows that all these systems present positive deviations from the ideality.

The phase diagram of the $Cd_{1-x}Zn_xTe$ system is reported in Figure 1.51 [210, 211].The high-temperature section of the phase diagram (Figure 1.51a) has been fitted, taking for the interaction coefficient of the solid solutions a value of $\Omega^s = -0.33$ kJ mol^{-1}, deduced from electrochemical measurements of the Gibbs energy of mixing and a value of $\Omega^l = 0.21$ kJ mol^{-1} for the liquid from a best fitting procedure of the liquid line.

This conclusion contradicts the occurrence of a solubility gap at low temperatures (Figure 1.51b) which would require a positive value of Ω^s, as that reported in Table 1.12. This conclusion is also in contradiction with the results of the concentration dependence of the Gibbs energy of mixing, reported in Figure 1.52, which show a small negative excess of Gibbs energy of mixing over the ideal one, calling for negative values of Ω^s [212].

In conclusion, the accuracy of the thermodynamic data available for this system and the use of regular solution approximations does not allow a reasonable fit of the phase diagram of this system, which, however, presents features very close to ideal. This conclusion is well supported by the rather close lattice parameters of CdTe and ZnTe (see Table 1.13), which

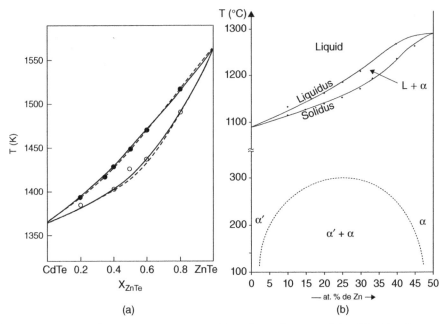

Figure 1.51 (a) High-temperature section of the phase diagram of the Cd$_{1-x}$Zn$_x$ Te system. Zabdyr, 1984, [210]. Reproduced with permission from Electrochemical Society. (b) Phase diagram of the Cd$_{1-x}$Zn$_x$ Te system including the low-temperature biphasic region. Haloui et al., 1997, [211]. Reproduced with permission from Elsevier

Table 1.13 *Physical properties of sulfides, selenides and tellurides of Cd, Zn and Hg*

	CdS	CdSe	CdTe	ZnS	ZnSe	ZnTe	HgSe	HgTe
T_m (K)	2023	1623	1314	2123	1373	1513	1063	943
	(100 atm)			(150 atm)				
Structure	ZB	ZB	ZB	ZB^a	ZB	ZB	ZB	ZB
Lattice constant (nm)	0.582	0.608	0.648	0.541	0.567	0.610	0.6085	0.6453
Structure	W^b	W^b	—	W^a	—	—	—	—
a (nm)	0.4135	0.430	—	0.3811	0.398	0.427	—	—
c (nm)	0.6749	0.702	—	0.6234	0.653	0.799	—	—
Energy gap (eV)	2.5 (direct)	1.714 (direct)	1.474 (direct)	3.91 (direct)	2.82 (direct)	2.39 (direct)	8^c	0.01– 0.02

^{a}Both zinc blende (ZB) and W phases are stable.
^{b}Wurtzite (W) is the stable phase.
^{c}HgSe is a semimetal with a valence and conduction overlap of 0.07 eV.

lead us to assume a negligible influence of elastic energy contribution to the free energy of mixing, which should be dominated by the contribution of complexes in solution.

The very symmetric phase diagram of the ternary Hg$_{1-x}$Cd$_x$Te alloys is reported in Figure 1.53 which exhibits the typical features of ideality of both the liquid and solid solutions [213], in good agreement with the structural data of the components (see Table 1.13).

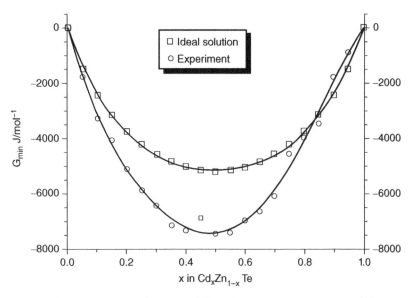

Figure 1.52   *Gibbs free energy of mixing of the CdZnTe system at 900 K. Alikhanian et al., [212]. Reproduced with permission from Elsevier*

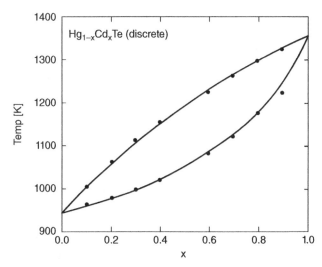

Figure 1.53   *Phase diagram of the $Hg_{1-x}Cd_xTe$ system. Patrick et al., 1988, [213]. Reproduced with permission from AIP Publishing*

Also in this case the thermodynamic data of Table 1.12 indicate only close to negligible deviations from the ideality of the solid solutions, with a value of the interaction coefficient $\Omega = 5.6 \text{ kJ mol}^{-1}$.

A complementary approach to the analysis of these systems is to consider the role of the polytype equilibria relative to the binaries CdS, CdSe and ZnS on the structure of the corresponding pseudo-binary solutions.

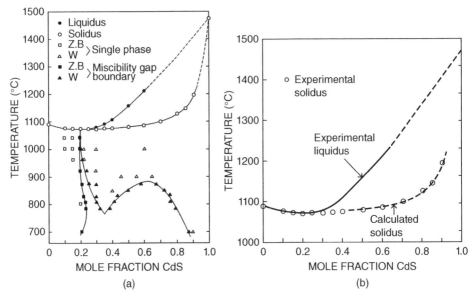

Figure 1.54 *Phase diagram of the CdS-CdTe system (a) experimental phase diagram (b) calculated under the approximation of Eq. (1.95) for the excess strain energy. Ohata, et al., 1973, [215]. Reproduced with permission from The Japan Society of Physics*

As an example, room-temperature lattice parameters of the systems CdZnTe and CdTeSe, measured over the full range of their solid solutions, show that the CdZnTe system remains cubic over its whole range of compositions, while the CdTeSe system shows a transition from the cubic phase to the hexagonal wurtzite structure at high Se contents [214].

A more significant example is given by the system CdS-CdTe [215, 216]. As shown in Figure 1.54, the structure of the CdTe-rich solid solutions is cubic (zincblende) but is hexagonal (wurtzite) at high CdS concentrations. In correspondence with the minimum at 1071 °C the zinc blende–wurtzite boundary coincides. At this temperature there is also the onset of a miscibility gap, whose size increases with decreasing temperature. Ohata *et al.* [215] succeeded in accounting for the thermodynamics of this system by considering that the S-Te bond is not the nearest bond and that the S-Te repulsive energy should not be too high, whereas the strain energy should be expected to be very large in view of the large difference in the covalent radii of S (0.105 nm) and Te (0.136 nm).

As in this system Vegard's law was demonstrated to hold both for the zinc blende (ZB)-type and W-type of solutions, the strain energy contribution W to the mixing enthalpy was evaluated by assuming the validity of Hooke's law over the whole range of compositions. Within these crude approximations we have

$$L_x = L_A x_A + L_B x_B \tag{1.105}$$

$$W = \tfrac{1}{2}\, k\, x_A (L_x - L_A)^2 + \tfrac{1}{2}\, k\, x_B (L_x - L_B)^2 = \tfrac{1}{2}\, k\, (L_A - L_B)^2 x_A x_B = \tfrac{1}{2}\, k\, x_A (1 - x_A) \tag{1.106}$$

where L_x is the average lattice parameter of the solid solution and L_A and L_B are the corresponding lattice distances in CdS and CdTe, leading eventually to the equation

$$\Delta H^s = C\,x(1-x) \tag{1.107}$$

for the strain contribution ΔH^s to the mixing enthalpy, where C is a fitting constant taken equal to 17.11 kJ mol^{-1}. A good fitting of the solidus curve over the whole range of compositions was obtained, as shown in Figure 1.54b.

It should be remarked that the composition dependence of the strain-induced enthalpy of mixing of Eq. (1.107) is formally equivalent to that of the enthalpy of mixing of a regular solution Eq. (1.89).

This formal equivalence shows that the interaction coefficient Ω of a system which follows the regular solution model may contain an elastic contribution which adds to the chemical energy terms or even dominates the interaction, in the approximation of validity of Hooke's law. Therefore, it would be misleading to deduce from the experimental evidence of a quadratic dependence of the enthalpy of mixing on composition a true regularity, as this quadratic dependence is a pure phenomenological law.

1.7.11 Arsenides, Phosphides and Nitrides

The success of modern optoelectronics and the potential of SSL for general illumination is closely related to the use of III–V compounds, to the development of specific technologies for their growth and processing and to a half century of dedicated research.

The availability of lasers and LEDs emitting in a bright range of wavelengths, from the mid-IR to the near-UV and of HEMTs (high electron mobility transistors) working in the gigahertz to terahertz range, is eminently due to the peculiar electronic properties of these compounds. They present direct gap features and high electronic mobilities (see Table 1.14), with a peak at $7.7 \cdot 10^4$ (cm^2 V^{-1} s^{-1}) in the case of InSb, and, therefore, different from silicon and germanium, which are indirect gap semiconductors, presenting efficient light emission capabilities over a wide range of wavelengths.

Arsenides and phosphides were the first direct gap semiconductors used for micro- and opto-electronic purposes, because their melting properties are not so extreme as those of the nitrides, but they still present problems for their bulk single crystal growth due to their large decomposition pressures [46, 217, 218]. The benefits of GaAs and other III–V compounds over silicon for optoelectronic applications are similar to those given by the use of silicon instead of germanium as a transistor material [219].

However, though the GaAs transistor performance is better than that of silicon, silicon is still the material of choice for applications where physics is not against it, as is the case of SSL, and will probably remain unrivalled for the years to come until Moore's law will reach its limit at the silicon atoms size.

Among III–V compounds, Group III nitrides (InN, GaN, AlN and their alloys) are unique, because they share the benefits of a wide, direct gap range (6.2, 3.4 and 0.7 eV for AlN, GaN and InN, respectively) and strong chemical bonds. They share also the drawback of high melting temperatures (2500 °C for GaN) and high decomposition pressures (45.000 bar for GaN) that make their direct crystallization from liquids practically impossible. These drawbacks hindered for years their growth in single crystalline form,

Table 1.14 Physical properties of III–V compounds of technological interest

	GaAs	InAs	InP	InSb	GaP	InN	GaN	AlN	Si
T_m (K)	1523	1215	1333	800	1738	2173[a]	2773[a]	3473[a]	1688
P_{eq} (atm)	15	<1	<1	—	30	> 60.000 [b]	45.000[b]	200[b]	—
Band gap (eV)	1.424	0.354	1.344	0.17	2.24	0.7[c]	3.44	6.2	1.12
μ_e (cm^2 V^{-1} s^{-1})	8500	$4 \cdot 10^4$	5400	$7.7 \cdot 10^4$	<250	3200	<1000	300 (calc)	1400
μ_h (cm^2 V^{-1} s^{-1})	<400	$5 \cdot 10^2$	<200	≤850	150	—	<350	14	<450
Structure	ZB	ZB	ZB	ZB	ZB	W	W	W	Diamond
Lattice constant (nm)	0.5653	0.6058	0.5869	0.6479	0.5450	$a = 0.35446$	$a = 0.3187$	$a = 0.311$	0.5430
						$c = 0.57034$	$c = 0.5186$	$c = 0.4982$	

For comparison the properties of silicon are also reported.
[a]Calculated.
[b]Extrapolated.
[c]The energy gap value of InN is still under discussion.
ZB = zinck blende and W = wurtzite.

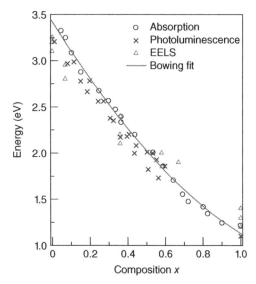

Figure 1.55 *Band gap of $In_{1-x}Ga_x$ N nanowires as a function of the gallium concentration. Kuykendall, et al., 2007, [221]. Reproduced with permission from Macmillan Publishers*

needed for their use as substrates for microelectronic and optoelectronic devices since the discovery of Nakamura *et al.* in 1986 [220], who succeeded in the heteroepitaxial growth of crystalline multilayer structures of InGaN using MOCVD on a GaN buffer layer deposited on a sapphire substrate.

The GaN buffer layer has the scope to reduce the lattice misfit, which would be intolerable in the case of the direct growth of ternary nitrides on sapphire or SiC, being the origin of exceedingly high dislocation densities.[12]

The value of a GaP-buffer layer is particularly evident in the case of the $Al_{0.83}In_{0.17}N$ alloy, as the GaN-buffer is lattice-matched and allows deposition of a low dislocation density material.

Today small single crystals of GaN might be grown in high pressure furnaces or autoclaves,[13] but all the nitride devices present in the market are manufactured from GaN substrates grown heteroepitaxially on sapphire, 6H SiC and, more recently, on silicon.

Blue, nitride-based LEDs working as a pump for suitable phosphors show promise for the solid state lamps which will be used in the next decade for general illumination. An additional advantage of nitrides is that the $In_xGa_{1-x}N$ solutions allow the complete tunability of the light emission from the near-UV to the near-IR region by changing the composition, as seen in Figure 1.55 for the case of InGaN nanowires [221], but this feature is typical also of InGaN thin films.

The discussion on the understanding of the physico-chemical properties of these systems and the modelling of their phase diagrams is carried out in the next section, while their growth processes are discussed in Chapter 3.

---

[12] The effect of dislocations on the electronic properties of semiconductors will be discussed in Chapter 3.

[13] Details about growth processes of bulk GaN will be discussed in Chapter 3 [217].

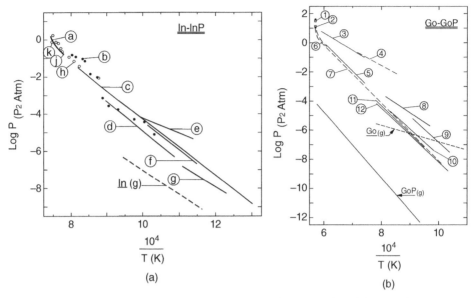

Figure 1.56   *Equilibrium vapour pressures of P$_2$ (solid line), In (dotted line) and Ga over (a) the In-InP and (b) the Ga-GaP systems. For the references indicated in the figures see original paper. Tmar et al., 1984, [217]. Reproduced with permission from Elsevier*

1.7.11.1   *Phosphides*

Concerning the basic thermodynamic properties of phosphides, they exhibit, in general, high saturation pressure values, with their consequent decomposition if the system is not properly overpressurized. In what follows, only the properties of the binary In-InP and Ga-GaP and ternary InGaP systems will be discussed, since their properties are typical for all phosphide phases.

The equilibrium partial pressures of P$_2$ over InP and GaP are reported in Figure 1.56, while their binary phase diagrams are reported in Figure 1.57, which shows that only the intermediate stoichiometric phases segregate from the In-rich melts [217].

The phase diagram of the pseudo-binary Ga$_x$In$_{1-x}$P system [222] is reported in Figure 1.58. The properties of this system are dominated by the large elastic energy required to overcome the lattice mismatch (see Table 1.14) of the pure components [223], which results in the opening of a miscibility gap below a critical temperature of 923.7 K at $x_c = 0.6$ according to [222] (or 830 K at $x_c = 0.4$), in good agreement with the results of a theoretical evaluation [223].

Another key feature of this system is the calculated and experimental bimodal distribution of bond lengths (see Figure 1.59) at 1000 K, a temperature at which, according to the phase diagram of Figure 1.58, the alloy is homogeneous. The solid lines were calculated using a Monte Carlo simulation [223] which also indicates a random distribution of metallic atoms in the alloy. Here, as in the case of the Si-Ge alloys, the individual bond lengths maintain values close to those of the pure compounds and very different from the Vegard's law values of the lattice distances (dotted line).

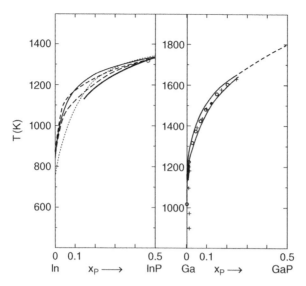

Figure 1.57 *Phase diagrams of the In-InP and Ga-GaP systems under their own equilibrium vapour pressures. Tmar et al., 1984, [217]. Reproduced with permission from Elsevier*

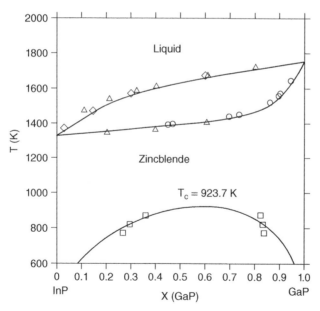

Figure 1.58 *Phase diagram of the pseudo-binary $Ga_xIn_{1-x}P$ alloys. Ch-Li et al., 2000, [222]. Reproduced with permission from ASM International*

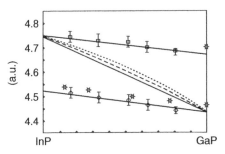

Figure 1.59 *Calculated and experimental bond lengths as a function of the composition of the In_xGa_{1-x} P alloy at 1000 K. Empty dots with error bars are calculated values, stars are experimental results from literature. The average lattice parameters, scaled to the bond lengths, are given by the dotted and dashed lines. Marzari, et al. 1994, [223]. Reproduced with permission from American Chemical Society*

The thermodynamic modelling of these alloys has been carried out by a number of authors [222, 224–226], with the aim of obtaining the temperature-dependent interaction coefficients.

This approach has been used, as an illustrative example, by Li *et al.* [222] to calculate the Gibbs free energies of the Ga_xIn_{1-x} P system, where the largest contribution to the mixing enthalpy for the solid solutions is the elastic energy term.

Here the molar Gibbs free energy of the ternary liquid phase G^l is given by the following equation

$$G^l = \sum_{i=1}^{3} y_i G_i^{o,l} + RT \sum_{i=1}^{3} x_i \ln x_i + G^l_{Ga,P} + G^l_{In,P} + G^l_{Ga,In} + G^l_{Ga,In,P} \tag{1.108}$$

where $G_i^{o,l}$ is the Gibbs free energy of the pure element i (Ga, In, P) and x_i is the molar fraction of i in the liquid phase. The $G_{i,j}^l$ terms are the binary excess Gibbs free energies expressed by the Redlich-Kister equation

$$G_{i,j}^l = x_i x_j \sum_{n}^{k} L_{i,j}^k (x_i - x_j)_{i,j}^k \tag{1.109}$$

which could be calculated using the model parameters $L_{i,j}^k$ given by Ansara *et al.* [226] for 15 III–V semiconductor binary systems.

The Gibbs energy of the solid phase is given by the following equation

$$G^s = G_{GaP}^{o,s} x_{GaP} + G_{InP}^{o,s} x_{InP} + RT(x_{GaP} \ln x_{GaP} + x_{InP} \ln x_{InP})$$

$$+ x_{GaP} x_{InP} \sum_{k=0}^{n} L_{GaP-InP} (x_{GaP} - x_{InP})^k \tag{1.110}$$

where $G^{o,s}_{GaP}$ and $G^{o,s}_{InP}$ are the Gibbs free energies of formation of the solid GaP and InP and the $L_{GaP-InP}$ terms are the interaction coefficients for the binary systems, which were optimized by Li *et al.* [222], including elastic energy correction terms.

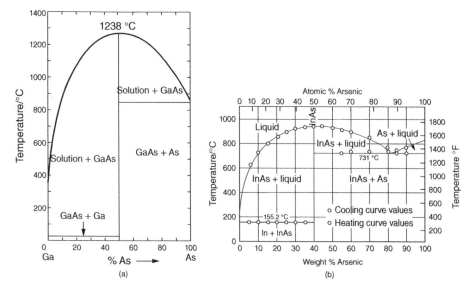

Figure 1.60   *(a) Phase diagrams of GaAs [227] (b) Phase diagram of InAs*

The results of this simulation are the solid lines in the phase diagram of Figure 1.58, which fit well the experimental points of the liquid and solid lines, as well as those of the solubility gap.

1.7.11.2   Arsenides

InAs and GaAs-based multicomponent semiconductor alloys are widely used for the manufacture of lasers, LEDs[14] and photodiodes operating in the 2–5 μm range at room temperature. The quaternary alloys (GaInAsSb, InAsSbP, GaAlAsSb) have the advantage, overpseudo-binary ones, of allowing a better tuning of the emission wavelength by varying the alloy composition. A drawback of these systems is, however, the presence of a miscibility gap, which limits the application of liquid phase epitaxy to a narrow temperature range and induces regions of modulated composition due to spinodal decomposition.

In this section we will limit attention to bulk GaAs, InAs and Ga_xIn_{1-x} As pseudo-binary alloys and also to the features of epitaxial layers of the quaternary alloys deposited on InP or GaAs substrates by liquid phase epitaxy,[15] where large effects are expected due to misfit strain.

The phase diagrams of GaAs and InAs are displayed in Figure 1.60, which shows that the stoichiometric compounds GaAs and InAs segregate by crystallization of the understoichiometric and overstoichiometric melts, these last only at temperatures above the eutectic temperature.

The general approach to fit the experimental equilibrium data of bulk crystals is to use the regular or quasi-regular solution model with adjustable parameters. More recently,

---

[14] Typically employed for traffic lights and automotive applications.

[15] Liquid phase epitaxy (see details in Chapter 4) is typically used for these types of compounds, which can be deposited, almost stoichiometrically, from the liquid phase at temperatures much lower than the melting temperature of the compound.

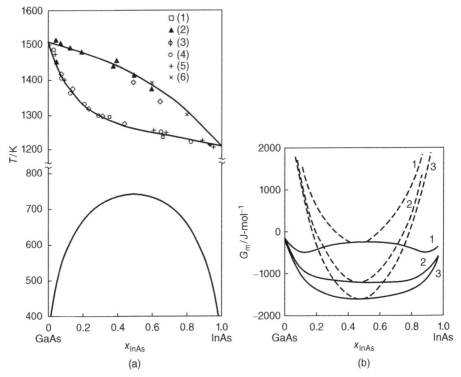

Figure 1.61   *(a) Phase diagram of the free standing system* $Ga_xIn_{1-x}As$*: solid curves are the calculated solid and liquid lines, symbols are experimental results according to references reported in the original paper. (b) Free energy curves for the same system deposited on an InP substrate: solid curves, without elastic energy contribution; dotted curves, with elastic energy contributions 1.* $T = 600$ *K, 2.* $T = 741$ *K and 3.* $T = 800$ *K. Quiao et al., 1994, [229]*

however, the excess thermodynamic functions have been calculated applying the model of linear combinations of chemical potentials (EFLCP) [228].

The phase diagram of the pseudo-binary Ga_xIn_{1-x} As alloy is reported in Figure 1.61a [229], where it is easy to see the good fit between the experimental results and those arising from a simulation (solid lines). The simulation was carried out making use of a model similar to that adopted by Li *et al.* [222] for the GaInP alloys and discussed in the previous section, with optimized, temperature-dependent, interaction parameters. It should be noted that the fit is good in spite of having neglected the possible influence of internal elastic energy contributions, due to the different lattice parameters of the components (see Table 1.14). One can also see that the simulation foresees the opening of a miscibility gap, with a critical temperature of 742 K and a critical composition $x_c \sim 0.5$, whose isothermal composition widths were evaluated from the excess Gibbs free energy curves reported as solid lines in Figure 1.61b.

The effect of elastic energy contributions on the phase equilibria was instead evaluated for the same quaternary system, epitaxially deposited on the (100) surface of an InP substrate, at temperatures slightly above or within the miscibility gap.

The excess molar Gibbs free energy due to the lattice elastic energy was evaluated, in this case, using the following equation

$$\Delta G^{mix} = -(Ax^2_{GaAs} + Bx^2_{InAs})$$ (1.111)

where A and B are the elastic energy contribution terms associated with the misfit strain. The calculated Gibbs energy curves are reported as dotted lines in Figure 1.61b.

It is apparent that in the presence of elastic energy contributions the system remains homogeneous over the whole range of temperatures considered, including that corresponding to the miscibility gap.

These results are supported by a more recent work [230], addressed at the study of unstrained and strained thin films of III–V semiconductor alloys. The thermodynamic properties of the unstrained alloys are evaluated by assuming the regularity of the liquid solution and calculating its Gibbs energy with the following equation

$$G^l = G^{o,l}_A x_A + G^{o,l}_B x_B + G^{o,l}_C x_C + RT(x_A \ln x_A + x_B \ln x_B + x_C \ln x_C) + \Omega^l_{AB} x_A x_B + \Omega^l_{AC} x_A x_C$$
$$+ \Omega^l_{BC} x_B x_C + \Omega^l_{ABC} x_A x_B x_C$$ (1.112)

which holds for a generic III–V ($A_x B_{1-x} C$) alloy, where the G^o terms are the standard Gibbs energies of the pure components and the Ω terms are the composition-dependent interaction coefficients for the binary components and for the ternary alloy.

The Gibbs energy of unstrained $In_x Ga_{1-x}$ As thin films (0.1–1 μm) deposited on a GaAs substrate was calculated by means of the following equation

$$G^s = G^{o,s}_{AC} y_{AC} + G^{o,s}_{BC} y_{BC} + RT(y_{AC} \ln y_{AC} + y_{BC} \ln y_{BC}) + L^s_{AC-BC} y_{AC} y_{BC}$$ (1.113)

where the G^o terms are the Gibbs energy of the binary compounds and L^s is the interaction coefficient for the ternary compound.[16]

The results of this simulation (solid lines in Figure 1.62) agree well with the experimental and theoretical results reported in Figure 1.61, although the calculated onset of the miscibility gap occurs at a lower critical temperature of 344 °C.

It is also apparent that the effect of misfit strain is very limited, and almost independent of the film thickness, when the calculated values of bulk samples (dotted curves in Figure 1.62) are compared to those of the films, although with a limited increase in the critical temperature.

For the $In_x Ga_{1-x} As$ alloys deposited in thicker (1–10 μm) layers on (100)- or (111)-oriented InP substrates the effect of misfit strain is severe [230], see Figure 1.63. In this case, a phase with the zinc blende structure segregates in correspondence of equimolar solid solutions and the two phase separation below the solidus temperature should prevent homogeneous growth.

This problem is of critical importance in the epitaxial growth of III–V semiconductors, where it is known that elastically constrained alloys are stabilized against phase separation [232], although in practice phase separation and ordering often also occurs. We will not discuss this further, but it is worth mentioning here that phase separation and ordering in

---

[16] The thermodynamic parameters used for the calculations were obtained from tabulated SGTE (Scientific Group Thermodata Europe) data [231].

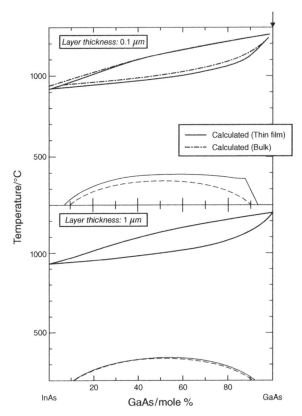

Figure 1.62 *Calculated phase diagram of the InGaAs system as deposited on a GaA substrate. Ohtani et al., 2001, [230]. Reproduced with permission from ASM International*

bulk alloys depend on the sign of the heat of mixing, and that their simultaneous appearance in thin film samples can only be explained by surface energy effects [232]. Phase separation and ordering are features of direct concern for all III–V compounds, but of relevant interest for Group III nitrides, as will be shown in the next section.

1.7.11.3   Nitrides

As already mentioned, due to the extreme thermal properties of these materials, single crystal samples are only available for GaN, and the properties of 'bulk' materials can only be experimentally investigated on unstrained, free standing films separated from the substrate by suitable chemical or mechanical techniques. Due to these experimental challenges, very little is known about the thermodynamic properties of bulk nitride alloys.

The discussion here will be limited to two nitride systems, the InGaN alloys and the InAlN alloys, of which the seconds are very attractive because they could be used as the active layers of LEDs emitting from the red to the near-UV, depending on the alloy composition. In addition, the $Al_{0.83}In_{0.17}N$ alloy is lattice matched with the GaN buffer layer, thus leading to a substantial reduction in defects arising from lattice mismatch, when a single crystal of sapphire, SiC and, more recently, silicon are used as substrates.

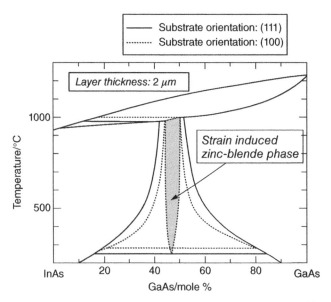

Figure 1.63 *Effect of misfit strain on the phase equilibria of InGaAs thin films deposited on InP substrates. Ohtani et al., 2001, [230]. Reproduced with permission from ASM International*

The growth of the pseudo-binary nitride films is carried out from the vapour phase, at relatively low temperatures, normally using the MO-CVD technique. Therefore, the thermodynamic properties of interest are those concerning the stability and the homogeneity of the phases deposited in the temperature range of the solid phase domain, where the possible opening of a miscibility gap and phase separation processes are typical issues. As spinodal decomposition may involve clustering, phase separation and ordering, knowledge of the features of the phase decomposition processes is important for their direct influence on the optoelectronic properties of the devices which could be manufactured thereof.

Phase separation and ordering have been experimentally observed for $In_xGa_{1-x}N$ samples with $x \geq 0.25$, prepared at 800 °C with atomic layer-MOCVD on sapphire substrates [233]. Phase separation is supposed to be associated with spinodal decomposition and ordering is interpreted to occur via stacking of In and Ga atoms along the c-axis of their wurtzite structure. Ruterana *et al.* [234] also succeeded in showing ordering features in these alloys. Yamaguchi *et al.* [235] studied the $Al_xIn_{1-x}N$ and $Al_{1-y-z}Ga_yIn_z$ N ternary and quaternary alloys grown on a buffer layer of GaN on a sapphire substrate and showed that the Al_xIn_{1-x} N layers could be grown in the $0.01 < x < 0.58$ range without macroscopic phase decomposition, although the tendency to a microscopical phase decomposition was observed. It was also shown that the lattice matched $Al_{0.83}In_{0.17}N$ alloy presented the best crystallinity.

In view of the experimental difficulties associated with the study of phase diagrams of nitride alloys and in the practical absence of experimental results, theoretical studies are well suited and of great support for the design of device growth processes.

As an example, Teles *et al.* [236–238] calculated the phase diagram of the In_xGa_{1-x} N system (see Figure 1.64) which indicates the presence of a very wide region of immiscibility at temperatures lower than 1250 K, with phase separation or spinodal decomposition, in

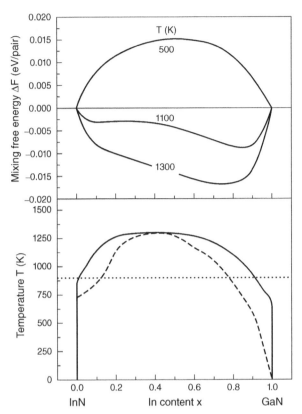

Figure 1.64 *Gibbs energy curves and phase diagram of the InGaN alloys (dotted curves define the region of spinodal decomposition) Teles et al., 2004, [236]*

good agreement with the experimental results [237] which show that phase decomposition effects are active in this system, leading to In-rich crystalline inclusions in the layer.

A better thermodynamic description of the In_xAl_{1-x} N alloys was obtained by an *ab initio* calculation [238], using the Generalized QCA. This model [64] describes the temperature and composition dependence of the Helmholtz free energy $\Delta F(x, T)$ within a cluster approximation, that considers the crystal divided into an ensemble of clusters, statistically and energetically independent of the surrounding atoms. Each class of clusters has the same energy E_j. Each cluster fraction is unknown, but it can be taken as a variational parameter. At the mesoscopic scale the system is assumed to be spatially homogeneous.

Using this approach it was possible to show the presence of an extended miscibility gap in this system with a very high critical temperature, which is even higher (1485 K) in the case of $In_xGa_{1-x}N$ alloy. A substantial composition invariance of the In-N and Al-N bond lengths was also demonstrated in the In_xAl_{1-x} N alloy, which causes Vegard's rule to fail also for the $In_xGa_{1-x}N$ alloys.

Finally, the most complete theoretical investigation concerning the phase diagram of $In_xGa_{1-x}N$ alloys was carried out by Gan *et al.* [240], considering both the zinc blende and wurtzite structure of these alloys. The effect of lattice vibration was also included, which required additional computing time, but which was supposed to have a non-negligible contribution to the free energy.

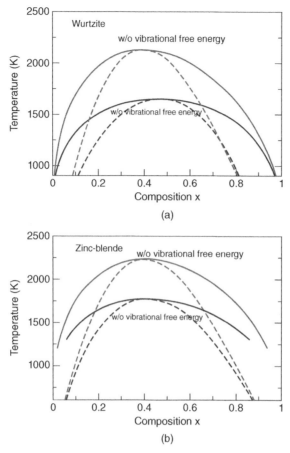

Figure 1.65 *Phase diagram of the InGaN system: the solid lines correspond to the equilibrium bimodal curves, the dotted ones the spinodal ones. Gan, et al., [240]. Reproduced with permission from American Chemical Society and C. K. Gan*

The results of this study (see Figure 1.65) show that the effect of including the vibrational contribution is to decrease the critical temperature and the width of the two phase region, that is to increase the computed solubility range at low temperatures, thus making a notable difference in terms of predictivity compared to less sophisticated models. On that basis, the doping with suitable impurities capable of influencing the lattice vibration spectrum could be an effective strategy capable of extending the useful composition range of nitride alloys.

1.8 Size-Dependent Properties, Quantum Size Effects and Thermodynamics of Nanomaterials

The physical properties of semiconductor nanomaterials (quantum dots, quantum wells, nanowires) depend, as is well known, on their sizes and on quantum effects (increase in the energy gap, quantum localization effects) that do appear when the size is comparable with the Bohr radius of the excitons in the particular semiconductor under study [241] (see Table 1.15 for a number of selected materials). Quantum confinement of strong carriers,

Table 1.15 *Bohr radii of excitons in selected semiconductors*

Material	Exciton Bohr radius (nm)	Electron mass $(m_n/m°)$	Hole mass $(m_h/m°)$
Si	4.2	0.26	0.49
CdSe	6	0.8	0.45
GaP	6	0.82	0.5
CdTe	8	0.14	0.35
GaAs	15	0.068	0.50
Ge	36	0.12	0.30
PbS	20	—	0.15
InAs	34–74	0.023	0.023
PbSe	46	—	—
InSb	86–138	0.014	0.4
PbTe	1700	0.14	0.20

however, only occurs when the radius of the nanoparticle is much lower than the Bohr radius of the exciton ($R \ll a$) and it comes out that strong confinement would be very difficult to obtain with most of the semiconductors reported in Table 1.15, perhaps with the exception of In- and Pb-compounds.

Also the density of states (DOS) depends on the dimensionality of the nanoparticle. For a spherical particle it has the form

$$\frac{dN}{dE} \approx \frac{d}{dE} E^{3/2} \approx E^{1/2} \tag{1.114}$$

for a 2D system, such as a quantum well, the DOS is a step function

$$\frac{dN}{dE} \approx \frac{d}{dE} \sum_{\varepsilon_i < E} (E - \varepsilon_i) \approx \sum_{\varepsilon_i < E} (1) \tag{1.115}$$

for a 1D system (a quantum wire) the DOS is given by

$$\frac{dN}{dE} \approx \frac{d}{dE} \sum_{\varepsilon_i < E} (E - \varepsilon_i)^{1/2} \approx \sum_{\varepsilon_i < E} (E - \varepsilon_i)^{-1/2} \tag{1.116}$$

while for a 0D system (a quantum dot) the DOS is given by

$$\frac{dN}{dE} \approx \frac{d}{dE} \sum_{\varepsilon_i < E} \theta(E - \varepsilon_i) \approx \sum_{\varepsilon_i < E} \delta(E - \varepsilon_i) \tag{1.117}$$

There is, however, a problem that remains debated, concerning the role of the surface atoms, whose amount increases with the reduction in size, as seen in Table 1.16 [242] and that of surface reconstruction, passivation and oxidation, which makes the evaluation of 'intrinsic' quantum effects difficult, and strongly dependent on the nature of the material.

As an example, while in the case of CdSe a dot of 1.8 nm, with a relative size of 0.3 Bohr excitons, has 90% of the atoms as surface atoms, in the case of PbSe a dot of 13.8 nm has the same relative size of 0.3 Bohr excitons but only 15% of atoms as surface atoms. It is therefore questionable whether the luminescence of a quantum dot of CdSe comes from quantum confinement effects or from surface atoms. In other wordss, one could argue that when the size dimensions of the Bohr exciton are approached by a nanoparticle, the

Table 1.16 Density, D (%), of surface atoms as a function of the relative size (R/a_B) and of the absolute size(R) of CdSe and PbSe nanoparticles

R/a_B	R (nm)	D (%) (CdSe)	R (nm)	D (%) (PbSe)
1	6	30	46	5
0.3	1.8	90	13.8	15
0.1	—	—	4.6	45

Wise, 2000, [242]. Reproduced with permission from American Chemical Society.

distinction between bulk atoms and surface atoms can no longer be ignored with respect to their influence on the optical and electronic properties of the whole system.

It is also worth considering that when the particles are so small, their electronic structure changes and a continuum of energy states is substituted by discrete bonding and antibonding orbitals, in such a way that the properties of these entities in covalent semiconductors are closer to a cluster of molecules than to an extended solid. Therefore, strong quantum confinement effects are better explained using a hybrid molecular language than a semiconductor language [243, 244].

It should finally be noted that the confinement of carriers (electrons and holes) in a nanoparticle calls for a decrease in the thermodynamic stability of the system [243], as it is associated with a decrease in the entropic contribution to its Gibbs free energy.

Besides optoelectronic properties, thermodynamic properties are therefore expected to change when approaching the nanosize. This is the case for the decrease in the eutectic temperatures of nanocrystalline Si-alloys [245] and of the different melting behaviour of metallic, semiconductor and organic nanocrystals, which presents significant differences from that of the corresponding microscopic or macroscopic phases [246], as was already seen in Figure 1.9 for the case of silicon [49].

The decrease in the eutectic temperature[17] of binary metal-Si alloys might be formally evaluated taking into account the contribution of the surface energy to the total energy of the system

$$\Delta G_{tot} = \Delta G_{bulk} + \Delta G_{surf} \qquad (1.118)$$

where the first term of Eq. (1.118) may be calculated with the procedures illustrated in Section 1.6.2 and the surface excess term may be evaluated using the following equation

$$\Delta G_{surf} = \frac{2}{R} \left[x_A \left(\sigma_l^A V_l^A - \sigma_s^A V s_s^A \right) + x_{Si} \left(\sigma_l^{Si} V_l^{Si} - \sigma_s^{Si} V_s^{Si} \right) \right] \qquad (1.119)$$

where the σ and V terms are the surface tension and molar volume values for silicon and the metallic solute in the liquid and solid state.

The results of an evaluation for three Si alloys are reported in Figure 1.66 [245], which shows that the decrease in the eutectic temperature is substantial and should be taken into account for the design of growth processes of silicon nanowires seeded with gold or silver.

Concerning the size effect of the melting temperature, the information now available relates only to the properties of single components, not that of alloys. It is, however, known that the melting temperatures T_m and melting enthalpies decrease with the reduction in size when the nanocrystals are deposited on inert matrices or are free standing.

---

[17] The evaluation of the size dependence of the eutectic temperature of binary Si alloys is important because metallic dots are used to initiate the growth of silicon nanowires, as will be shown in Chapter 4.

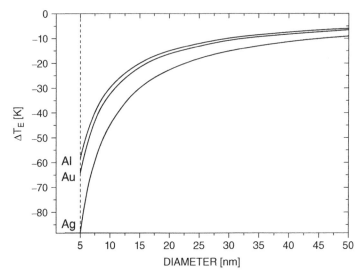

Figure 1.66 *Calculated decrease in the eutectic temperature for Al-Si, Au-Si and Ag-Si alloys with the decrease of the diameter of the nanoparticle. Schmidt, et al., 2010, [245]. Reproduced with permission from American Chemical Society*

As nanophases are systems in metastable equilibrium, the experimental study of size-dependent phase diagrams for binary systems is very difficult, if not impossible. An empirical model without adjustable parameters [247] based on the Lindeman's criterion predicts, however, an exponential relationship between the melting temperature $T_m(D)$ and $1/D$ for metallic nanocrystals

$$T_m(D) = T_m^o \exp -\frac{2S_m^o(D^o - 1)}{3RD} \tag{1.120}$$

where D is the nanocrystal size, T_m^o is the bulk melting temperature, S_m^o is the bulk melting entropy and D^o is a critical size at which almost all the atoms are located at the surface, which corresponds approximately to a value of $D^o = 6\,h$, where h is the atomic/molecular diameter.

A model concerning the size dependence of the interaction coefficient $\Omega(D)$ of regular solutions is also available [248, 249]. As all thermodynamic properties are roughly dependent on 1/D, that is on the surface to volume ratio, it is supposed that Ω should also follow the same relationship. It is assumed, therefore, that the following equation should hold for the interaction coefficient

$$\frac{\Omega(D)}{\Omega^o} = 1 - \frac{2D^o}{D} \tag{1.121}$$

where Ω^o is the bulk value.

This assumption is based on the concept that spontaneous surface bond contraction occurs with reduction of the coordination number, a concept that has been experimentally proven with Raman measurements on several inorganic materials and, also, on AlGaN [250].

On the basis of the known values of the bulk interaction coefficients Ω^o for Ge-Si solutions and on the calculated values of $\Omega(D)$ reported in Table 1.17 [249], the calculated

Table 1.17 *Thermodynamic functions of the Ge and Si systems as a function of the crystal size*

D (nm)		Ge	Si
∞	T_m (K)	1210.4	1685
	H_m (kJ mol^{-1})	36.940	50.550
	Ω^s (kJ mol^{-1})	7.666	
	Ω^l (kJ mol^{-1})	7.715	
10	T_m (K)	1096.7	1472.2
	H_m (kJ mol^{-1})	24.542	33.135
	Ω^s (kJ mol^{-1})	4.437	
	Ω^l (kJ mol^{-1})	4.465	
5	T_m (K)	1096.7	1472.2
	H_m (kJ mol^{-1})	24.542	33.135
	Ω^s (kJ mol^{-1})	1.208	
	Ω^l (kJ mol^{-1})	1.215	

Liang *et al.*, 2003, [249].

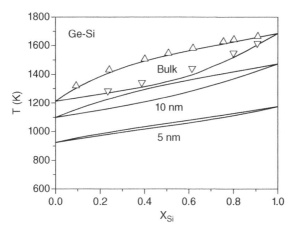

Figure 1.67 *Evolution of the Ge-Si phase diagrams as a function of the nanocrystallite size. Liang* et al., *2003, [249]. Reproduced with permission from Institute of Physics and Q. Jiang*

evolution of the phase diagram of the Ge-Si system as a function of the size is reported in Figure 1.67. As the size of the crystals decreases to a critical value, which for the case of Si-Ge alloys is close to the excitonic radius, the liquid and the solid phase are almost indistinguishable.

Another size-dependent transformation occurs inside the CdTe-CdSe system. It was shown [251] that the structure of spherical CdTe/CdSe quantum dots having a size ~8 nm, synthesized in a two-step procedure with CdTe nanoparticles (4 nm) as nuclei, have the structure of a CdTe-rich core enclosed by a CdSe shell. However, spherical CdSe/CdTe quantum dots with a size of about 9 nm, synthesized via a similar two-step procedure but with smaller CdSe nanoparticles (3 nm) as nuclei, appear single phase with a relatively uniform distribution of CdSe and CdTe.

Another, notable example of size-dependent thermodynamic properties is given by the phase transformation of bulk diamond to graphite which, in the absence of catalysts occurs at pressures larger than 10 GPa around 4000 K (see Figure 1.38), well above the equilibrium solidus line. When this phase transition is studied on diamond nanocrystallites having a grain size of 18 nm, it occurs at a pressure of about 5 GPa, as demonstrated by X-ray diffraction and Raman experiments, at 1623 K, close to the equilibrium conditions [252].

The effect of pressure on the transformation of silicon nanocrystals from the semiconducting- to the metallic β-Sn phase has also been considered, as well as that of the pressure on the transformation of CdSe nanocrystals from the wurtzite to the rocksalt structure [253].

In this last case it was demonstrated that the transformation involves single domains and only one nucleation event per crystallite and that the transformation pressure increases with decrease in the size of the nanocrystals [253], different from the case of phase transitions in bulk semiconductors, which are highly hysteretic, are nucleated by defects and typically involve multiple nucleation centres and domain fracture.

The result of this work leads to conclusions of serious impact for the study of phase transformations. As nanocrystallites are smaller than the domain fragments generated in a solid–solid transformation, nanocrystallites are ideally suited for the study of phase transformations, because they can be grown almost defectless, though with a large surface to bulk ratio, where the surface provides the nucleation sites.

Moreover, the increase in the transformation pressure with the decrease in nanocrystal size demonstrates that this effect is associated with a decrease in available nucleation centres at the surface and that the process is controlled by kinetic, not thermodynamic factors. The observed changes in nanocrystal shape during the transformation, induced by interior atoms motion towards the nucleation sites, provides direct evidence that homogeneous deformations play a role in phase transitions in semiconductors.

APPENDIX

Use of Electrochemical Measurements for the Determination of the Thermodynamic Functions of Semiconductors

EMF measurements with galvanic cells using solid electrolytes have been so widely used for the determination of the thermodynamic functions of inorganic solids at high temperatures in the last half century [254–256] that this method may be considered a standard thermochemical technique [257].

As is well known, the EMF of a galvanic cell with the following configuration

$$Pt/Me, MeX_2//electrolyte \ (t_X = 1)//X_2(P = 0.01325 \ KPa)/Pt \qquad (A1)$$

where t_X the transport number of X^- anions in the electrolyte, gives a measure of the Gibbs free energy of formation $\Delta G^\circ_{MeX_2}$ of MeX_2

$$E = -\frac{RT}{2F}\Delta G^\circ_{MeX_2} \qquad (A2)$$

where E in volts is the EMF of the cell and F is the Faraday constant.

In this kind of measurement the electrolyte must be a suitable solid material presenting pure ionic (X^-) conductivity in the range of temperatures where $\Delta G^\circ_{MeX_2}$ should be measured, the left-hand electrode is a heterogeneous mixture of Me and MeX_2 suitably pressed in the form of a pellet and the right-hand electrode works as a reversible X_2 electrode. Alkaline earth fluorides, such as CaF_2, are good examples of F^- conducting electrolytes.

The metallic Pt electrodes allow the exchange of electrons involved in both half-cell reactions

$$Me + 2\,X^- \rightleftharpoons MeX_2 + 2e(Pt) \tag{A3}$$

$$X_2 + 2e(Pt) \rightleftharpoons 2X^- \tag{A4}$$

but could be substituted by other types of metals or electronic conductors able to satisfy the condition of electronic equilibrium between the metal electrode and the reactants in both the electrode compartments. The right-hand metallic electrode must behave as a reversible electrode, and this condition is not obvious for all types of X^-/X_2 redox couples.

Using the electrochemical potential definition given in Section 1.4.6 [7, 9] for the thermodynamic description of the processes occurring in the cell (Eq. A1), we can see that at the electrodes the following equilibria occur

$$2\eta_e^{Pt(1)} + \eta_{Me^{2+}}^{Me} = \mu_{Me} \tag{A5}$$

$$2\eta_{X^-}^{X_2} - \eta_e^{Pt(2)} = \mu_{X_2} \tag{A6}$$

$$\eta_{Me_2}^{Me} + \eta_{X^-}^{MeX_2} = \mu_{MX_2} \tag{A7}$$

At the different phase boundaries, instead, the following equilibria occur

$$\eta_e^{Pt(1)} = \eta_e^{Me} \tag{A8}$$

$$\eta_{Me^{2+}}^{Me} = \eta_{Me^{2+}}^{MeX_2} \tag{A9}$$

$$\eta_{X^-}^{MeX_2} = \eta_{X^-}^{Pt(2)} \tag{A10}$$

$$\eta_e^{X_2} = \eta_e^{Pt(2)} \tag{A11}$$

It is therefore possible to obtain a relationship between the difference in the internal electrical potentials $\Delta\varphi$ at the Pt electrodes and the cell reaction, by writing

$$\eta_e^{Pt(1)} - \eta_e^{Pt(2)} + \eta_e^{Me} - \eta_e^{X_2} = \eta_{Me^{2+}}^{Me} + \eta_{X^-}^{MeX_2} \tag{A12}$$

$$\mu_e^{Pt(1)} - \mu_e^{Pt(2)} = \mu_{Me^{2+}}^{Me} + \eta_{X^-}^{MeX_2} - \eta_e^{Me} - \eta_e^{X_2} = \mu_{MX_2} - (\mu_{Me} + \mu_{X_2})$$

$$= \Delta G^\circ_{MX_2} = -2F\Delta\varphi \tag{A13}$$

Therefore the EMF of the cell (Eq. (A2)) in thermodynamic equilibrium conditions (open circuit voltage) is

$$EMF = -\frac{\Delta G^\circ_{MX_2}}{2F} \tag{A14}$$

If MX_2 is MeO_2, the electrolyte consists of an yttria-doped ZrO_2 solid solution, presenting pure O^{2-} conductivity and the right-hand electrode works as an oxygen electrode at a pressure P = 0.101325 kPa, the overall cell reaction may be written as

$$Me + O_2(P = 0.101325 \text{ kPa}) \rightleftharpoons MeO_2 \tag{A15}$$

Therefore, under the assumption that both electrode reactions are reversible and that the ionic conductivity of the electrolyte is sufficiently high to allow fast equilibration, the EMF measured at the electrodes is proportional to the standard Gibbs free energy of formation of MeO_2, $\Delta G°(MeO_2)$ according to the following equation

$$E = -\frac{\Delta G^o_{MeO_2}}{4F} \tag{A16}$$

If at the left-hand electrode we have, instead, a non-stoichiometric oxide, formally considered as a solution of MeO in MO_2 and at the right-hand electrode we have pure MO_2, the measured EMF is proportional to the activity of MO_2 in the MeO_{2-x} solution at the left-hand electrode compartment

$$E = -\frac{RT}{4F} \ln a_{MeO_2} = -\frac{RT}{4F} \ln \gamma x_{MO_2} \tag{A17}$$

thus allowing determination of the activity $a = \gamma x$ and then the chemical potential of MO_2 in solution and of the interaction coefficient Ω for that particular solid solution system.

Provided a suitable electrolyte is available, the thermodynamic functions of a variety of complex solids may be investigated, as is the case for the potassium–graphite lamellar compounds, which have been studied [258] in the temperature range 200–350 °C with a solid-state EMF technique, using as the electrolyte a glass reversible to potassium ions and liquid potassium as the reference electrode.

The cell used was the following:

$$K_l // \text{glass} (t_{K+} = 1) // K_{\text{graphite}} \tag{A18}$$

using Ni wires as the electrodes. The EMF of the cell is proportional to the activity of K in the lamellar compound through the equation

$$E = -\frac{RT}{F} \ln a_{K(\text{graphite})} \tag{A19}$$

as the activity of K in liquid potassium is unity.

This technique has been rarely applied to semiconductors, as the challenge of EMF measurements for the determination of thermodynamic functions of semiconductors is the availability of suitable electrolytes. Oxides, fluorides, chlorides electrolytes are available, as well as pure H^+, Ag^+, Li^+ or Al^{3+} (β-Al_2O_3 or aluminium tungstate $Al_2(WO_4)_3$ electrolytes [259], which are, however, of little interest for most semiconductors of common use, with the potential[18] exception of Al-based compound semiconductors or hydrogenated semiconductors.

Therefore, as an alternative route, the possibility of equilibrating the semiconductor with an intermediate buffer phase which, in turn, should undergo fast equilibrium conditions with an oxide- or chalcogenide-type of electrolyte has been studied.

---

[18] To the author's knowledge, EMF measurements with Al^{3+} or H^+ ionic conductors have been never carried out on semiconductors.

In this case the problem could be the slow rate of the equilibration processes between the semiconductor and the intermediate buffer phase (which must also exhibit electronic conduction to satisfy electrochemical potentials equilibria) due to slow ion exchange reactions at the semiconductor/intermediate material interface.

This technique was originally applied [260] for the study of the structure relationships in the $CaO - ZrO_2$ system in the 900–1100 °C temperature range, using CaF_2 as the electrolyte, which is a pure F^- conducting electrolyte, in the following cell

$$Pt \ /(O_2)P = 0.101KPa//CaO(+CaF_2)//CaF_2(t_{F^-} = 1)//Zr_{1-x}Ca_xO_{2-x}$$

$$\times (+CaF_2)//(O_2)P = 0.101KPa/Pt \qquad (A20)$$

for which the EMF is proportional to the activity of CaO in the $CaO_xZrO_{2(1-x)}$ system

$$E = -\frac{RT}{2F}\ln a_{CaO(CaO_xZrO_{2(1-x)})} \qquad (A21)$$

In this case, a fast equilibrium between the oxide and CaF_2 phases in the left-hand and right-hand compartments was experimentally demonstrated to occur at the high temperatures of the experiment.

The same approach was recently followed [261] for the study of the thermodynamic functions of the ternary Zn_xCd_{1-x} Te alloys and binary Zn_xTe_{1-x} alloys using ZnO as the intermediate phase, and an yttria-doped ZrO_2 electrolyte.

The cell used to measure the activity of Zn in the alloy was:

$$Zn, ZnO \ //ZrO_2-Y_2O_3//Zn_xCd_{1-x}Te/ZnO \qquad (A22)$$

with a Zn/ZnO reference electrode which keeps the activity of Zn at unity.

The EMF of this cell

$$E = -\frac{RT}{2F} \ln a_{Zn \ (Zn_x \ Cd_{1-x} \ Te)} \qquad (A23)$$

is therefore directly proportional to the activity of Zn in the Zn_xCd_{1-x} Te alloy.

The results of the measurements fit well with the calculated ones, as can be seen in Figure A.1.

A $CdCl_2$-$CuCl$ electrolyte, presenting a pure Cu^+ conductivity, was instead used to study the properties of $Cu_{2\pm\delta}$ Se semiconductors [262] and a $Cu_4RbCl_3I_2$ was used as a Cu^+ ion conducting solid electrolyte for the study of the Cu–As–S system at temperatures from 300 to 370 K using EMF measurements [263]. From the EMF data the standard thermodynamic functions of formation and standard entropies of the ternary compounds were obtained, showing the significant potential of the EMF technique as a thermochemical tool in the field of compound semiconductors. EMF measurements were also used to study the oxygen activity in liquid silicon [264] with an oxygen sensor directly dipped in the melt of a CZ puller, using a zirconia-based electrolyte and a V/V_2O_3 reference electrode.

The EMF of the cell

$$C//O_{Si}/ZrO_2-Y_2O_2//V, V_2O_3//Pt-Rh \qquad (A24)$$

gives the relative activity of the oxygen in liquid silicon directly

$$E = \frac{RT}{2F} \ln \frac{p''_{O_2}}{p'_{O_2}} = \frac{RT}{2F} \ln a_O \qquad (A25)$$

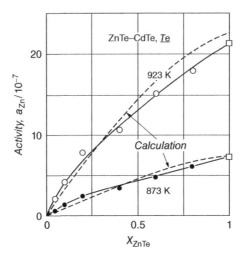

Figure A.1 *Activity of Zn (in 10^{-7} units) in the Te-rich region of the ternary $Zn_xCd_{1-x}Te$ alloy* [261]

where $p''(O_2)$ is the equilibrium partial pressure of oxygen of the oxygen-liquid silicon solution and $p'(O_2)$ is the equilibrium oxygen pressure of the V/V_2O_3 system.

As the oxygen content in a silicon melt at the temperature of the measurements (slightly above the melting temperature of silicon) is in the range 0.001–0.01%, the activity coefficient was assumed equal to 1. The results of the EMF measurements were compared to the results of oxygen content measured by FTIR measurements[19] and showed encouraging potential of this technique as a routine analytical tool.

Last but not least, the Gibbs free energies of formation of 3C- and 6H-silicon carbides were determined between 1623 and 1898 K by measuring the silicon solubility in liquid gold in equilibrium with silicon carbide and graphite [177]. The solution behaviour of the Au–Si system provides a good fit with the small difference in the activity of silicon exhibited by the two polytypes, thus offering greater experimental precision over previous similar methods. For 6H-SiC, the Gibbs free energy of its formation, ΔG (J mol^{-1}) $= -116\,900(\pm 7.2) + 38.2(\pm 4.1)T$ (where T is in K) relative to liquid silicon and graphite standard states, is in excellent agreement with other investigations, indicating that the method is capable of determining the Gibbs energy of formation with a precision of about 0.7 kJ mol^{-1}. The measurements on 3C-SiC exhibited greater experimental scatter, which may have been due to variations in the density of non-equilibrium defects of the crystals used, although experimental problems encountered with the complex crystal shapes may have also contributed. The precision of the technique is, however, sufficient to get information on polytype stability and perhaps on the Gibbs energy contributions of non-equilibrium defects.

---

[19] Fourier Transform IR measurements are routinely used for the determination of the concentration of oxygen in solid silicon samples.

References

1. Swalin, R.A. (ed) (1961) *Thermodynamics of Solids*, John Wiley & Sons, Inc., New York.
2. Rao, C.N.R. and Gopalakrishnan, J. (1986) *New Directions in Solid State Chemistry*, Cambridge University Press, Cambridge.
3. West, A.R. (1992) *Solid State Chemistry*, John Wiley & Sons, Ltd, Chichester.
4. Elliot, S. (1998) *The Physics and Chemistry of Solids*, John Wiley & Sons, Ltd, Chichester.
5. Le Roy, E. and O'Keeffe, M. (1970) *The Chemistry of Extended Defects in Non-Metallic Solids*, North Holland, Amsterdam.
6. Barron, A. *Chemistry of Electronic Materials Open Educational Resource*, http://cnx.org/content/col10719/latest/ (accessed 22 December 2014) (2013).
7. Guggenheim, E.A. (1952) *Thermodynamics*, North Holland, Amsterdam.
8. Kittel, C. (2005) *Introduction to Solid State Physics*, John Wiley & Sons, Inc., New York.
9. Ashcroft, N.W. and Mermin, N.D. (1988) *Solid State Physics HRW International Editions*, Sounders College, Philadelphia.
10. Cool, T., Bartol, A., Kasenga, M. *et al.* (2010) Gibbs: phase equilibria and symbolic computation of thermodynamic properties. *Calphad*, **34**, 393–404.
11. Santhanaman, P., Gray, D.J. Jr., and Ram, R.J. (2012) Thermoelectrical pumped light emitting diodes operating above unit efficiency. *Phys. Rev. Lett.*, **108**, 097403-1–097403-4.
12. Shockley, W. and Queisser, H.J. (1961) Detailed balance limit of efficiency of p-n junction solar cells. *J. Appl. Phys.*, **32**, 510–519. doi: 10.1063/1.1736034
13. Prigogine, I. (1967) *Thermodynamics of Irreversible Processes*, 3rd edn, Interscience, John Wiley & Sons, Inc., New York.
14. de Panfilis, S. and Filipponi, A. (1997) Short-range order in liquid tellurium probed by X-ray absorption spectroscopy. *Europhys. Lett.*, **37**, 397–402.
15. DeCicco, A., Rosoleny, M.J., Marassiz, R. *et al.* (1996) Short range order in liquid and solid KBr probed by EXAFS. *J. Phys. Condens. Matter*, **8**, 10779–10797.
16. Pizzini, S., Binetti, S., Calcina, D. *et al.* (2000) Local structure of erbium-oxygen complexes in erbium-doped silicon and its correlations with the optical activity of erbium. *Mater. Sci. Eng.*, **B 72**, 173–176.
17. Inui, M. (2000) EXAFS studies of liquid semiconductors. *J. Phys. Chem. Solids*, **61**, 2007–2012.
18. Lebedev, I., Michurin, A.V., Sluchinskaya, I.A. *et al.* (2000) EXAFS and electrical studies of new narrow-gap semiconductors: $inTe_{1-x}Se_x$ and $In_{1-x}Ga_xTe$. *J. Phys. Chem. Solids*, **61**, 2007–2012.
19. Byrne, A.P., Ridgway, M.C., Glover, C.J. and Bezakova, E. (2004) Comparative studies using EXAFS and PAC of lattice damage in semiconductors. *Hyperfine Interact.*, **158**, 245–231.
20. Kazmerski, L.L. (1998) Photovoltaics characterization: a survey of diagnostic measurements. *J. Mater. Res.*, **13**, 2684–2708.

21. Adams, F., Van Vaeck, L. and Barrett, R. (2005) Advanced analytical techniques: platform for nano materials science. *Spectrochim. Acta, Part B*, **60**, 13–26.
22. Razeghi, M. (2006) Semiconductor characterization techniques, in *Fundamentals in Solid State Engineering*, Springer, pp. 521–549.
23. Hockett, R.S. (2012) Advanced analytical techniques for solar grade feedstock, in *Advanced Materials for Photovoltaic Applications* (ed S. Pizzini), John Wiley & Sons, Inc., pp. 215–234.
24. Kumar, A., Gencarelli, F., Kambham, A.K. *et al.* (2012) Study of Sn migration during relaxation of Ge(1-x)Sn(x) layers using atom probe tomography. *Phys. Status Solidi C*, **9**, 1924–1930.
25. Miller, M.K. (1999) *Atom Probe Tomography*, Kluger Academic, Plenum Press, New York.
26. Lauhon, L.J., Adsumilli, P., Rosenheim, P. *et al.* (2009) Atom probe tomography of semiconductor materials and device structures. *MRS Bull.*, **34**, 738–743.
27. Hudelson, S., Newman, B.K., Bernardis, S. *et al.* (2010) Retrograde melting and internal liquid gettering in silicon. *Adv. Mater.*, **22**, 3948–3953.
28. Buonassisi, T., Istratov, A.A., Pickett, M.D. *et al.* (2006) Chemical natures and distributions of metal impurities in multicrystalline silicon materials. *Prog. Photovoltaics Res. Appl.*, **14**, 513–531.
29. Buonassisi, T., Marcus, M.A., Istratov, A. *et al.* (2005) Analysis of copper-rich precipitates in silicon: chemical state, gettering, and impact on multicrystalline silicon solar cell material. *J. Appl. Phys.*, **97**, 063503.
30. Buonassisi, T., Heuer, M., Istratov, A.A. *et al.* (2007) Transition metal co-precipitation mechanisms in silicon. *Acta Mater.*, **55**, 6119–6126.
31. Bertoni, M.I., Fenning, D.P., Rinio, M. *et al.* (2011) Nanoprobe X-ray fluorescence characterization of defects in large-area solar cells. *Energy Environ. Sci.*, **4**, 45–52.
32. Sarma, B.N., Prasad, S.S., Vijayvergiva, S. *et al.* (2003) Existence domains for invariant reactions in binary regular solution phase diagrams exhibiting two phases. *Bull. Mater. Sci.*, **26**, 423–430.
33. Sher, A., van Schilfgaarde, M., Chen, A.-B. and Chen, W. (1987) Quasichemical approximation in binary alloys. *Phys. Rev. B*, **36**, 4279–4295.
34. Wei, S.-H., Ferreira, L.G. and Zunger, A. (1990) First principle calculations of temperature-composition phase diagrams of semiconductors alloys. *Phys. Rev. B*, **41**, 8240–8269.
35. Vegard, L. (1921) Die Konstitution der Mischkristallen und die Raumfüllung der Atome. *Z. Phys.*, **5**, 17–26.
36. Crawford, J.H. Jr. and Slifkin, L.M. (1972) *Point Defects in Solids*, vol. **1** and **2**, Plenum Press, New York.
37. Auguillo-Lopez, F., Catlow, C.R.A. and Towsend, P.D. (1988) *Point Defects in Materials*, Academic Press, London.
38. Pizzini, S. (2002) *Defect Interaction and Clustering in Semiconductors*, Scitech Publications, Ueticon-Zürich.
39. Pichler, P. (2004) *Intrinsic Point Defects, Impurities and Their Diffusion in Silicon*, Springer.

40. Kröger, F.A. and Vink, H.J. (1956) Relations between the concentration of imperfections in crystalline solids, in *Solid State Physics*, vol. **3** (eds F. Seitz and D. Turnbull), Academic Press, New York, pp. 307–435.

41. *CRC Handbook of Chemistry and Physics*, 88th edn, (2007–2008) CRC Press, Taylor & Francis Group, Boca Raton, FL, London, and New York.

42. Soma, T. and Matsuo, H. (1982) Lindeman's Melting Law and the effect of pressure on the melting point of Si and Ge. *Phys. Status Solidi B*, **109**, 387–391.

43. Hänström, A. and Lazor, P. (2000) High pressure melting and equation of state of aluminium. *J. Alloys. Compd.*, **305**, 209–315.

44. Bundy, F.P. (1964) Phase diagrams of silicon and germanium to 200kBar at 1000°C. *J. Chem. Phys.*, **41**, 3809–3814.

45. Kailer, A., Gogotsi, Y.G. and Nickel, K.G. (2002) Phase transformations of silicon caused by contact loading. *J. Appl. Phys.*, **81**, 3057–3064, and references therein.

46. Besson, J.M., Itie, J.P., Polian, A. *et al.* (1991) High pressure phase transitions and phase diagram of gallium arsenide. *Phys. Rev. B*, **44**, 4214–4236.

47. Guinier, A., Bokij, G.B., Boll-Dornberger, K. *et al.* (1984) Nomenclature of polytype structures. *Acta Crystallogr., Sect. A*, **40**, 399–404.

48. Luo, F., Chen, Y.-R., Cai, L.-C. and Jing, F.-Q. (2011) The melting curves and the entropy of iron under high pressure. *J. Chem. Eng. Data*, **56**, 2063–2070.

49. Yang, C.C., Li, G. and Jiang, Q. (2003) Effect of pressure on melting temperature of silicon. *J. Phys. Condens. Matter*, **15**, 4961–4965.

50. Glazov, V.M. and Shchelikov, O.D. (2000) Volume changes during melting and heating of silicon and germanium melts. *High Temp.*, **38**, 405–412.

51. Lindeman, F. (1910) The calculation of molecular vibration frequencies. *Z. Phys.*, **11**, 609.

52. Knudson, M.D., Desjarlais, M.P. and Dolan, D.H. (2008) Phases of carbon: shock-wave exploration of the high pressure. *Science*, **322**, 1822–1825.

53. Soma, T., Itoh, T. and Kagaya, H.M. (1984) Pressure dependence of melting point of Al. *Phys. Status Solidi A*, **126**, 661–669.

54. Kagaya, H.-M., Imazawa, K., Sato, M. and Soma, T. (1998) Lindeman's melting law and solidus curve under pressure of Al-Si and Al-Ge solid solutions. *J. Mater. Sci.*, **33**, 2595–2599.

55. Yang, C.C., Li, J.C. and Jiang, Q. (2003) Effect of pressure on melting temperature of silicon determined by Clapeyron equation. *Chem. Phys. Lett.*, **372**, 156–159.

56. Garai, J. and Chen, J. (2008) Pressure Effect on the Melting Temperature, NSF Grant #0711321. Physics Archive arXiv:0805.0249.

57. Li, C., Wang, F. and Zhang, W. (2004) Pyrolysis effect of group V vapor sources on the composition ranges for metal-organic vapor phase epitaxy growth of III-V semiconductors. *J. Phase Equilib. Diffus.*, **25**, 53–58.

58. Kattner, U.R. (1997) Thermodynamic modeling of multicomponent phase equilibria. *J. Manage.*, **49**, 14–19.

59. Pelton, A.D. (1991) Thermodynamics and phase diagrams of materials, in *Materials Science and Technology* (eds R.W. Cahn, R. Haasen and E.J. Kramer), Wiley-VCH Verlag GmbH, Weinheim, pp. 3–73.

60. Stolen, S., Grande, T. and Allan, N.L. (2004) *Chemical Thermodynamics of Materials*, John Wiley & Sons, Ldt, Chichester.
61. Mbaye, A.A., Ferreira, L.G. and Zunger, A. (1987) First principle calculation of semiconductor alloy phase diagrams. *Phys. Rev. Lett.*, **58**, 49–52.
62. Ansara, I. (1998) Thermodynamic modelling of solution and ordered phases. *Pure Appl. Chem.*, **70**, 449–459.
63. Austin Chang, Y., Chen, S., Zhang, F. *et al.* (2004) Phase diagram calculation: past, present and future. *Prog. Mater. Sci.*, **49**, 313–345.
64. F. Bechsedt Nitrides as seen by a theorist, in *Low Dimensional Nitride Semiconductors*, B. Gil Ed, pp, 11–49, Oxford. Oxford University Press 2002.
65. van Laar, J.J. (1908) Melting-point and freezing-point curves in binary systems, when the solid phase is a mixture (amorphous solid solution or mixed crystals) of both components. *Z. Phys. Chem.*, **63**, 216.
66. van Laar, J.J. (1908) Die Schmelz- oder Erstarrungskurven bei biLaren Systemen, wenn die feste Phase ein Gemisch (amorphe feste Losung oder Mischkristalle) der beiden Komponenten ist. *Z. Phys. Chem.*, **64**, 257.
67. Hildebrand, J.H. (1929) Solubility. XII. Regular solutions. *J. Am. Chem. Soc.*, **51**, 66.
68. Fowler, R.H. and Guggenheim, E.A. (1939) *Statistical Thermodynamics*, Cambridge University Press, Cambridge, pp. 350–366.
69. Guggenheim, E.A. (1952) *Mixtures*, Oxford University Press.
70. Baldereschi, A. and Peressi, M. (1993) Atomic scale structure of ionic and semiconducting solid solutions I. *J. Phys. Condens. Matter*, **5**, 837–848.
71. A.D Pelton, Thermodynamics and phase diagrams of materials, in *Phase Transformations in Materials*. G. Kostorz Ed, 3–80 Wiley-VCH Verlag GmbH, Weinheim (2001)
72. Desay, P.T. (1986) The thermodynamic properties of iron and silicon. *J. Phys. Chem. Ref. Data*, **16**, 967–980.
73. Zhigilei , L.V. Phase Diagrams and Kinetics, (2015) University of Virginia, Department of Materials Science and Engineering **MSE 3050: Thermodynamics and Kinetics of Materials**
74. Pelton, A.D. and Thompson, W.T. (1975) Phase diagrams. *Prog. Solid State Chem.*, **10**, 119–155.
75. Oonk, H.A. (1981) *Phase Theory: The Thermodynamics of Heterogeneous Equilibria*, Elsevier, Amsterdam.
76. DeHoff, R.T. (1993) *Thermodynamics in Materials Science*, McGraw-Hill, New York.
77. Chang, Y.A., Chen, S.L., Zhang, F. *et al.* (2004) Phase diagram calculation: past, present and future. *Progr.Mater.Sci.*, **49**, 313–345.
78. Weber, E. (1983) Transition metals in silicon. *Appl. Phys. A*, **30**, 1–22.
79. Swain, S.K., Jones, K.A., Datta, A. and Lynn, K.G. (2011) Study of different cool down schemes during the crystal growth of detector grade CdZnTe. *IEEE Trans. Nucl. Sci.*, **58**, 2341–2345.
80. Fistul, V.I. (2005) *Impurities in Semiconductors: Solubility, Migration and Interactions*, CRC Press.
81. Istratov, A.A., Zhang, P., McDonald, R.J. *et al.* (2005) Nickel solubility in intrinsic and doped silicon. *J. Appl. Phys.*, **9**, 0235051.

82. McKelvey, A.L. (1996) Retrograde solubility in semiconductors. *Metall. Mater. Trans. A*, **27A**, 2704–2707.
83. Varamban, S.V. and Jacob, K.T. (1996) Discussion of retrograde solubility in semiconductors. *Metall. Mater. Trans. A*, **29A**, 1525–1526.
84. Wagner, R. and Kampmann, R. (1991) Homogeneous second phase precipitation, in *Materials Science and Technology*, Phase Transformations in Materials, vol. **5** (ed P. Haasen), Wiley-VCH Verlag GmbH, Weinheim, pp. 213–303.
85. Binder, K. (1991) in *Spinodal Decomposition in Materials Science and Technology*, Phase Transformations in Materials, vol. **5** (ed P. Haasen), Wiley-VCH Verlag GmbH, Weinheim, pp. 405–471.
86. Lubarda, V.A. (2003) On the effective lattice parameter of binary alloys. *Mech. Mater.*, **35**, 53–68.
87. Jeffs, N.J., Blant, A.V. and Cheng, T.S. (1998) EXAFS studies of group III nitrides. *MRS Proc.*, **512**, 519–524.
88. Miyano, K.E., Woicik, J.C., Robins, L.H. *et al.* (1997) Extended X rays absorption fine structure study of Al_xGa_{1-x} N films. *Appl. Phys. Lett.*, **70**, 2108–2110.
89. Jacob, T., Raj, S. and Rannesh, L. (2007) Vegard's Law: a fundamental law or an approximation? *Int. J. Mater. Res.*, **9**, 776–779.
90. Okamoto, H., Chakrabarti, D.J., Laughlin, D.E. and Massalki, T.B. (1987) The Au-Cu (gold-copper) system. *Bull. Alloy Phase Diagrams*, **8**, 454–473.
91. Bessiere, M., Lefebvre, M. and Calvayrac, Y. (1983) X ray diffraction study of short range order in a disordered Au_3Cu alloy. *Acta Crystallogr.*, **39B**, 145–153.
92. Okamoto, H. and Massalki, T.B. (1983) The Ag-Au (silver-gold) system. *J. Phase Equilib.*, **4**, 30–38.
93. Jian-Jun, G., Ji-Xian, Y. and Dong, D. (2007) First principle calculation on Au-Ag clusters. *Commun. Theor. Phys. (Bejing)*, **48**, 348–352.
94. Morse, M.D. (1986) Clusters of transition metal atoms. *Chem. Rev.*, **86**, 1049–1109.
95. Kobayashi, K., Kurita, N., Kumahora, H. and Tago, K. (1991) Bond energy calculations of Ag_2, Au_2 and CuAg with the general gradient approximation. *Phys. Rev. A*, **43**, 5810–5813.
96. Flanagan, W.F. and Averbach, B.L. (1956) Hall effect in solid solutions. *Phys. Rev.*, **10**, 11441–11442.
97. Weinberger, P., Drchal, V., Szunyogh, L. *et al.* (1994) Electronic and structural properties of copper-gold alloys. *Phys. Rev. B*, **49**, 13366–13372.
98. Szoldos, L. (1965) The determination of bond energies in the alloy system Au-Cu. *Phys. Status Solidi A*, **11**, 667–671.
99. Glazov, V.M. and Shchelikov, O.D. (2000) Volume changes during melting and heating of silicon and germanium melts. *High Temp.*, **38**, 405–412.
100. Kaczmarski, M., Bedoya-Martinez, O.N. and Hernadez, E.R. (2005) Phase diagrams of silicon by atomistic simulations. *Phys. Rev. Lett.*, **94**, 095701-1–095701-4.
101. Bernasconi, M., Chiarotti, G.L., Focher, P. *et al.* (1995) First-principle constant pressure molecular dynamics. *J. Phys. Chem. Solids*, **56**, 501–505.
102. Yang, C.C., Li, J.C. and Jiang, Q. (2004) Temperature–pressure phase diagram of silicon determined by Clapeyron equation. *Solid State Commun.*, **129**, 437–441.
103. Yang, C.C. and Yang, Q. (2004) Temperature–pressure phase diagram of germanium determined by Clapeyron equation. *Scr. Mater.*, **51**, 1081–1085.

104. *Properties of Silicon*, EMIS Data Review Series, vol. **4**, INSPEC, London (1988).

105. Hu, Z., Merkle, L.D., Manoni, C.S. and Spain, I.L. (1986) Crystal data for high-pressure phases of silicon. *Phys. Rev. B*, **34**, 4679–4684.

106. Piltz, R.O., Maclean, J.R., Clark, S.J. *et al.* (1995) Structure and properties of silicon XII: a complex tetrahedrally bonded phase. *Phys. Rev. B*, **52**, 4072–4084, and reference therein.

107. Bundy, F.P. (1964) Phase diagrams of silicon and germanium to 200 kbar, 1000°C. *J. Chem. Phys.*, **41**, 3809–3814.

108. Car, R. and Parrinello, M. (1985) Unified approach for molecular dynamics and density functional Theory. *Phys. Rev. Lett.*, **55**, 2471–2474.

109. Behler, J., Martonak, R., Donadio, D. and Parrinello, M. (2008) Pressure-induced phase transformations in silicon studied by neural network – based metadynamic simulations. *Phys. Status Solidi B*, **12**, 2618–2629.

110. Štich, I., Car, R. and Parrinello, M. (1991) Structural, bonding, dynamical and electronic properties of liquid silicon: an *ab initio* molecular-dynamics study. *Phys. Rev. B*, **44**, 4262–4274.

111. Waseda, W. and Suzuki, K. (1975) Structure of molten silicon and germanium by X-Ray diffraction. *Z. Phys. B*, **20**, 339–343.

112. Sato, Y., Nishizuka, T., Hara, K. *et al.* (2000) Density measurements of molten silicon by a pycnometric method. *Int. J. Thermophys.*, **21**, 1463–1471.

113. Stich, I., Car, R. and Parrinello, M. (1989) Bonding and disorder in liquid silicon. *Phys. Rev. Lett.*, **63**, 2240–2243.

114. Chelikowsky, J.R., Derby, J., Godlevsky, V. *et al.* (2001) *Ab initio* simulations of liquid semiconductors using the pseudopotential–density functional method. *J. Phys. Condens. Matter*, **13**, R817–R854.

115. Okada, J.T., Sit, P.H.-L., Watanabe, Y. *et al.* (2012) Persistence of covalent bonding in liquid silicon probed by inelastic x-ray scattering. *Phys. Rev. Lett.*, **108**, 067402.

116. Kulkarni, R.V., Aulbur, W.G. and Stroud, D. (1997) Ab initio molecular-dynamics study of the structural and transport properties of liquid germanium. *Phys. Rev. B*, **55**, 6896–6904.

117. Koga, T., Okumura, H., Nishio, K. *et al.* (2002) Simulation study of liquid germanium under pressure. *J. Non-Cryst. Solids*, **312–314**, 95–98.

118. Chaoui, N., Siegel, J., Wiggins, S.M. and Solis, J. (2005) Pressure-induced transient structural change of liquid germanium induced by high-energy picosecond laser pulses. *Appl. Phys. Lett.*, **86**, 221901–221903.

119. Olesinski, R.W. and Abbashian, G.I. (1984) The Ge-Si system. *Bull. Alloys Phase Diagrams*, **5**, 180–183.

120. Lee, S.M., Kim, E., Lee, Y.H. and Kim, N. (1998) Ge(Si) ordering on double layer stepped Si(Ge) (001) surface. *J. Korean Phys. Soc.*, **33**, 684–688.

121. Jivani, A.R., Trivedi, H.J., Gajjar, P.N. and Jani, A.R. (2005) Some physical properties of $S_{i1-x}Ge_x$ solid solutions using a pseudo alloy atom model. *Semicond. Phys. Quantum Electron. Optoelectron.*, **8**, 14–17.

122. Soma, T. (1979) The electronic theory of Si-Ge solid solution. *Phys. Status Solidi B*, **95**, 427–431.

123. Dunweg, B. and Landau, D.P. (1993) Phase diagram and critical behavior of the Si-Ge unmixing transition. *Phys. Rev. B*, **48**, 14182–14197.

124. Kajiyama, H., Muramatsu, S., Shimada, T. and Nishino, Y. (1992) Bond length relaxation in crystalline $Si_{1-x}Ge_x$ alloys: an extended X-ray absorption fine study. *Phys. Rev. B*, **45**, 14005–14010.

125. Shen, S., Zhang, D. and Fan, X. (1995) Tight-binding studies of crystalline $Si_{1-x}Ge_x$ alloys. *J. Phys. Condens. Matter*, **7**, 3529–3538.

126. Claeys, C. and Simoen, E. (2007) *Germanium Based Technologies: From Materials to Devices*, Elsevier.

127. Shao, L., Zhang, J., Chen, J. *et al.* (2004) Enhancement of boron solid solubility in Si by point-defect engineering. *Appl. Phys. Lett.*, **84**, 3325–3327.

128. Okamoto, H. (2005) B-Si (Boron-Silicon). *J. Phase Equilib. Diffus.*, **26**, 396.

129. Tang, K., Ovrelid, E. J., Tranell, G., Tangstad, M. (2009) *Thermochemical and Kinetic Databases for the Solar Cell Silicon Materials in Crystal Growth of Si for Solar Cells*, K. Nakajima, N. Usami Ed. Springer pp. 219–250 .

130. Scace, R.I. and Slack, G.A. (1959) Solubility of carbon in silicon and germanium. *J. Chem. Phys.*, **30**, 1551–1555.

131. Hu, B., Du, Y., Xu, H. *et al.* (2010) Thermodynamic description of the C-Ge and C-Mg systems. *J. Min. Metall. Sect. B*, **46**, 97–103.

132. Olesinski, R.W. and Abbaschian, G.J. (1984) The C-Si system. *Bull. Alloy Phase Diagrams*, **5**, 486–489.

133. Brazier, C.R. and Ruiz, J.I. (2011) The first spectroscopic observation of germanium carbide. *J. Mol. Spectrosc.*, **270**, 26–32.

134. Mahmood, A. and Sansores, L.E. (2005) Band structure and bulk modulus calculations of germanium carbide. *J. Mater. Res.*, **20**, 1101–1106.

135. Chroneos, A. (2008) Stability of impurity–vacancy pairs in germanium carbide. *J. Mater. Sci: Mater. Electron.*, **19**, 25–28.

136. Trumbmore, F.A., Porbanski, E.M. and Tartaglia, A.A. (1959) Solid soubilities of Aluminum and Gallium in germanium. *Phys. Chem.*, **11**, 239–245.

137. Yoshikawa, T. and Morita, K. (2003) Solid solubilities and thermodynamic properties of aluminum in solid silicon. *J. Electrochem. Soc.*, **150**, G465–G468.

138. Degtyareva, V.F., Chipenko, G.V., Belash, I.T. *et al.* (1985) F.C.C. solid solutions in Al-Ge and Al-Si Alloys under high pressure. *Phys. Status Solidi A*, **89**, 127–128.

139. Kagaya, H.-M., Imazawa, K., Sato, M. and Soma, T. (1998) Phase diagrams of Al–Si and Al–Ge systems. *Phys. B: Condens. Matter*, **245**, 252–255.

140. Ceccaroli, B. and Pizzini, S. (2012) Processes, in *Advanced Silicon Materials for Photovoltaic Applications* (ed S. Pizzini), John Wiley & Sons, Ltd, Chichester, pp. 21–78.

141. Rodot, M., Bourèe, J.E., Mesli, A. *et al.* (1987) Al-related recombination center in polycrystalline Si. *J. Appl. Phys.*, **62**, 2556–2558.

142. Marchand, R.L. and Shah, C.T. (1977) Study of thermally induced deep levels in Al doped Si. *J. Appl. Phys.*, **48**, 336–341.

143. Yoshikawa, T. and Morita, K. (2005) Refining of Si by the solidification of Si–Al melt with electromagnetic force. *ISIJ Int.*, **45**, 967–971.

144. Kotval, P.S. and Strock, H.B. (1980) *Process for producing of refined MG silicon*. US Patent 4,195,067, March 25, 1980.

145. He, G. and Atwater, H.A. (1997) Intraband transitions in Sn_xGe_{1-x} alloys. *Phys. Rev. Lett.*, **79**, 1937–1940.

146. Costa, R.D., Cook, C.S., Birdwell, A.G., Littler, C.L., Canonico, M., Zollner, S., Kouvetakis, J., Menendez, J. (2006) Optical critical points of thin-film $Ge_{1-y}Sn_y$ alloys: a comparative $Ge_{1-y}Sn_y/Ge_{1-x}Si_x$ study. *Phys. Rev. B*, **73**, 125207

147. Moontragoon, P., Ikonic, Z. and Harrison, P. (2007) Band structure calculations of Si–Ge–Sn alloys: achieving direct band gap materials. *Semicond. Sci. Technol.*, **22**, 742–748.

148. Mathews, J., Beeler, R.T., Tolle, J. *et al.* (2010) Direct-gap photoluminescence with tunable emission wavelength in $Ge_{1-y}Sn_y$ alloys on silicon. *Appl. Phys. Lett.*, **97**, 221912.

149. Lindl, J.D. (1995) Development of the indirect-drive approach to inertial confinement fusion and the target physics basis for ignition and gain. *Phys. Plasmas*, **2**, 3933–4024.

150. Bundy, F.P. (1963) Melting of graphite at very high pressure. *J. Chem. Phys.*, **38**, 618–631.

151. Bundy, F.P. (1963) Direct conversion of graphite to diamond in static pressure apparatus. *J. Chem. Phys.*, **38**, 631–643.

152. Bundy, F.F.P. (1989) Pressure-temperature diagram of elemental carbon. *Physica A*, **156**, 169–178.

153. Bundy, F.P., Bassett, W.A., Weathers, M.S. *et al.* (1996) The pressure-temperature phase and transformation diagram for carbon; updated through 1994. *Carbon*, **34**, 141–153.

154. Togaya, M. (1997) Pressure dependences of the melting temperature of graphite and the electrical resistivity of liquid carbon. *Phys. Rev. Lett.*, **79**, 2474.

155. Savvatimski, A.I. (2003) Melting point of graphite and liquid carbon. *Phys. Usp.*, **46**, 1295–1303.

156. Galli, G., Martin, R.M., Car, R. and Parrinello, M. (1990) Ab initio calculation of properties of carbon in amorphous and liquid state. *Science*, **250**, 1547.

157. Davydov, V.A., Rakhmanina, A.V., Rols, S. *et al.* (2007) Size-dependent phase transition of diamond to graphite at high pressures. *J. Phys. Chem. C*, **111**, 12918–12925.

158. Khaliullin, R.Z., Eshet, H., Kühne, T.D. *et al.* (2011) Nucleation mechanism for the direct graphite-to-diamond phase transition. *Nat. Mater.*, **10**, 693–697.

159. Wang, X., Scandolo, S. and Car, R. (2005) Carbon phase diagram from *Ab initio* molecular dynamics. *Phys. Rev. Lett.*, **95**, 185701–185704.

160. Bundy, F.P., Bovenwerk, H.P., Strong, H.M. and Wentorf, J.R.H. (1961) Diamond-graphite equilibrium line from growth and graphitization of diamond. *J. Chem. Phys.*, **35**, 383–391.

161. Ghiringhelli, L., Los, J.H., Mejier, E.J. *et al.* (2005) Modeling the phase diagram of carbon. *Phys. Rev. Lett.*, **94**, 145701–145704.

162. Grumbach, M.P. and Martin, R.M. (1996) Phase diagram of carbon at high pressures and temperatures. *Phys. Rev. B*, **54**, 15730–15741.

163. Dmitriev, V.P., Rochal, S.B., Gufan, Y.M. and Toledano, P. (1989) Reconstructive transitions between ordered phases: the martensitic fcc-hcp and the graphite-diamond transitions. *Phys. Rev. Lett.*, **62**, 2495–2498.

164. Correa, A.A., Bonev, S.A. and Galli, G. (2006) Carbon under extreme conditions: phase boundaries and electronic properties from first-principle calculations. *Proc. Natl. Acad. Sci. U.S.A.*, **103**, 1204–1208.

165. Steinbeck, J., Dresselhaus, G. and Dresselhaus, M.S. (1990) The properties of liquid carbon. *Int. J. Thermophys.*, **11**, 789–796.
166. Steinbeck, J., Braunstein, G., Dresselhaus, M.S. *et al.* (1985) A model for pulsed laser melting of graphite. *J. Appl. Phys.*, **58**, 4374–4381.
167. Malvezzi, A.M., Bloembergen, N. and Huang, C.Y. (1986) Time-resolved picosecond optical measurements of laser-excited graphite. *Phys. Rev. Lett.*, **57**, 146.
168. Savvatimskiy, A.I. (2008) Liquid carbon density and resistivity. *J. Phys. Condens. Matter*, **20**, 114112–114117.
169. Rao, K.J. and Mohan, R. (1981) Chemical Bond approach to determining conductivity band gaps in amorphous chalcogenides and pnictides. *Solid State Commun.*, **39**, 1065–1068.
170. Choyke, W.J., Mastromami, H. and Pensl, G. (1997) *Silicon Carbide: A Review of Fundamental Questions and Applications to Current Device Technology*, John Wiley & Sons, Ltd, Chichester.
171. Choyke, W.J., Matsunami, H. and Pensl, G. (2004) *Silicon Carbide: Recent Major Advances*, Springer-Verlag, Berlin.
172. Friedrichs, P., Kimoto, T., Ley, L. and Pensl, G. (2009) *Silicon Carbide: Growth, Defects, and Novel Applications*, vol. **1**, John Wiley & Sons, Inc., New York.
173. Friedrichs, P., Kimoto, T., Ley, L. and Pensl, G. (2009) *Silicon Carbide, Power Devices and Sensors*, vol. **2**, John Wiley & Sons, Inc., New York.
174. Sakwe, S.A., Stockmeier, M., Hens, P. *et al.* (2009) Bulk growth of SiC – Review on advances of SiC vapor growth for improved doping and systematic study on dislocation evolution, in *Silicon Carbide: Growth, Defects, and Novel Applications*, vol. 1 (eds P. Friedrichs, T. Kimoto, L. Ley and G. Pensl), John Wiley & Sons, Inc., pp. 1–31.
175. Madar, R. (2004) Materials science: silicon carbide in contention. *Nature*, **430**, 974–975.
176. Fissel, A. (2001) Relationship between growth conditions, thermodynamic properties and crystal structure of SiC. *Int. J. Inorg. Mater.*, **3**, 1273–1275.
177. Sambasivan, S., Capobianco, C. and Petuskey, W.T. (1993) Measurement of the free energy of formation of silicon carbide using liquid gold as a silicon potentiometer. *J. Am. Ceram. Soc.*, **76**, 397–400.
178. Pattanaik, A.K., Robi, P.S. and Srinivasan, A. (2003) Microhardness of Pb-modified calchogenide glasses. *J. Optoelectron. Adv. Mater.*, **5**, 35–38.
179. Sanderson, R.T. (1976) *Chemical Bonds and Bond Energy*, 2nd edn, Academic Press, London.
180. Singh, K., Saxena, N.S., Srivastava, O.N. *et al.* (2006) Energy gap of selenium chalcogenide glasses. *Chalcogenide Lett.*, **3**, 33–36.
181. Silva, L.A. and Cutler, M. (1990) Optical properties of liquid Se-Te alloys. *Phys. Rev. B*, **42**, 7103–7113.
182. Kumar, A., Barman, P.B. and Sharma, R. (2010) Study of the physical properties with compositional dependence of Bi content in Bi-Se-Te glasses. *Adv. Appl. Sci. Res.*, **1**, 47–57.
183. Grison, E. (1951) Studies on tellurium-selenium alloys. *J. Chem. Phys.*, **19**, 1109–1113.

184. Fthenakis, V. and Alsema, E. (2006) Photovoltaics energy payback times, greenhouse gas emissions and external costs: 2004– early 2005 status. *Prog. Photovoltaics Res. Appl.*, **14**, 275–280.
185. Raugei, M., Bargigli, S. and Ulgiati, S. (2007) Life cycle assessment and energy pay-back time of advanced photovoltaic modules: CdTe and CIS compared to poly-Si. *Energy*, **32**, 1310–1318.
186. Guskov, N., Greenberg, J.H., Fiederle, M. and Benz, K.-W. (2004) Vapour pressure investigation of CdZnTe. *J. Alloys Compd.*, **371**, 118–121.
187. Cavallini, A. and Fraboni, B. (2013) Point defects in cadmium zinc telluride. *J. Cryst. Growth*, **379**, 41–54.
188. Capper, P. (1994) *Properties of Narrow Gap Cadmium Based Compounds*, EMIS Review Series, vol. **10**, INSPEC.
189. Franc, J., Grill, R., Hl'ıdek, P. *et al.* (2001) The influence of growth conditions on the quality of CdZnTe single crystals. *Semicond. Sci. Technol.*, **16**, 514–520.
190. Szeles, C. and Driver, M.C. (1998) Growth and properties of semi-insulating CdZnTe for radiation detector applications. *Proc. SPIE*, **3446**, 1–8.
191. Greenberg, J.H., Guskov, V.N., Lazarev, V.B. and Shebershneva, O.V. (1992) Vapor pressure scanning of nonstoichiometry in CdTe. *J. Solid State Chem.*, **102**, 382–389.
192. Greenberg, J.H. (1996) $P - T - X$ phase equilibrium and vapor pressure scanning of non-stoichiometry in CdTe. *J. Cryst. Growth*, **161**, 1–4.
193. Feltgen, T., Greenberg, J.H., Guskov, V.N. *et al.* (2001) P–T–X phase equilibrium studies in Zn–Te for crystal growth by the Markov method. *Int. J. Inorg. Mater.*, **3**, 1241–1249.
194. Guskov, V.N., Greenberg, J.H., Alikhanyan, A.S. *et al.* (2002) P–T–X phase equilibrium in the Zn–Te System V.N. *Phys. Status. Solidi B*, **229**, 137–140.
195. Weltner, W. (1969) On polytypism and internal rotation. *J. Chem. Phys.*, **51**, 2469–2484.
196. McMahon, M.I., Nelmes, R.J., Wright, N.G. and Allan, D.R. (1994) Crystal structure studies of II-VI semiconductors using angle dispersive diffraction techniques with an image plate detector. *AIP. Conf. Proc.*, **309**, 633. doi: 10.1063/1.46413.
197. Laugier, A. (1973) Thermodynamics and phase diagram calculation in II-VI and IV-VI ternary systems using an associated solution model. *Rev. Phys. Appl.*, **8**, 259–270.
198. Moskvin, P., Olchowik, G., Olchowik, J.M. *et al.* (2011) New approach to the determination of phase equilibrium in the ZnTe system. *Task Q.*, **15**, 209–217, http://www.bop.com.pl (accessed 20 December 2014).
199. Brebrick, R.F. (2010) The Cd–Te phase diagram. *Calphad*, **34**, 434–440.
200. Adachi, S. (2009) *Properties of Semiconductor Alloys: Group IV, III-V and II-VI Semiconductors*, John Wiley & Sons, Ltd, Chichester.
201. Rudolph, P. (2003) Non-stoichiometry related defects at the melt growth of semiconductor compound crystals: a review. *Cryst. Res. Technol.*, **38**, 542–554, and references therein.
202. Sharma, R.C. and Chang, Y.A. (1989) The Cd-Te system. *J. Phase Equilib.*, **10**, 334.
203. J. Greenberg, *Thermodynamic Basis of Crystal Growth* Springer (2002), ISBN: 3642074529, ISBN: 13: 9783642074523.

204. Sharma, R.C. and Chang, Y.A. (1966) The Cd-Se system. *J. Phase Equilib.*, **17**, 140–145.

205. Sharma, R.C. and Chang, Y.A. (1987) The Te-Zn system. *J. Phase Equilib.*, **85**, 417.

206. Moskvin, P., Olchowik, G., Olchowik, J.M. *et al.* (2011) New approach to the determination of phase equilibrium in the ZnTe system. *Task Q.*, **15**, 209–217.

207. Gavrichev, K.S., Sharpataya, G.A., Guskov, V.N. *et al.* (2002) High-temperature heat capacity and thermodynamic functions of zinc telluride. *Thermochim. Acta*, **381**, 133–138.

208. Gavrichev, K.S., Sharpataya, G.A., Guskov, V.N. *et al.* (2002) Thermodynamic properties of ZnTe in the temperature range 15–925 K. *Phys. Status Solidi*, **229**, 133–135.

209. Stukes, A.D. and Farrel, G. (1964) Electrical and thermal properties of alloys of CdTe and CdSe. *J. Phys. Chem. Solids*, **25**, 477–482.

210. Zabdyr, L.A. (1984) Thermodynamics and phase diagram of the pseudobinary ZnTe-CdTe system. *J. Electrochem. Soc.*, **131**, 2157–2160.

211. Haloui, A., Feutelais, Y. and Legendre, B. (1997) Experimental study of the ternary system Cd-Te-Zn. *J. Alloys Compd.*, **260**, 179–192.

212. Alikhanian, A.S., Guskov, V.N., Natarovsky, A.M. *et al.* (2002) Mass spectrometric study of the CdTe–ZnTe system. *J. Cryst. Growth*, **240**, 73–79.

213. Patrick, S., Chen, A.B., Sher, A. and Berding, M.A. (1988) Phase diagrams and microscopic structures of (Hg,Cd)Te, (Hg,Zn)Te and (Cd,Zn)Te alloys. *J. Vac. Sci. Technol., A*, **6**, 2643–2649.

214. Williams, D.J. (1994) Densities and lattice parameters of CdTe,CdZnTe and CdTeSe, in *Properties of Narrow Gap Cd Based Compounds*, Emis Data Review Series, vol. **10** (ed P. Capper), INSPEC, p. 399.

215. Ohata, K., Saraie, J. and Tanaka, T. (1973) Phase diagram of the CdS-CdTe pseudobinary system. *Jpn. J. Appl. Phys.*, **12**, 1198–1204.

216. Lane, D.W., Conibeer, G.J., Wood, D.A. *et al.* (1999) Sulphur diffusion in CdTe and the phase diagram of the CdS–CdTe pseudo-binary alloy. *J. Cryst. Growth*, **197**, 743–748.

217. Tmar, M., Gabriel, A., Chatillon, C. and Ansara, I. (1984) Critical analysis of the thermodynamic properties and phase diagrams in the III-V compounds: the InP-In and Ga-P systems. *J. Cryst. Growth*, **68**, 557–580.

218. Mizuno, O. (1975) Vapor growth of InP. *Jpn. J. Appl. Phys.*, **14**, 451–457.

219. Haitz, R. and Tsao, J.T. (2011) Solid-state lighting: "The Case" 10 years after and future prospects. *Phys. Status Solidi A*, **208**, 17–29.

220. Nakamura, S., Senoh, M., Nagahama, S.-I. *et al.* (1996) InGaN-based multi-quantum-well-structure laser diodes. *Jpn. J. Appl. Phys.*, **35**, L74–L76.

221. Kuykendall, T., Ulrich, P., Alon, S. and Yang, P. (2007) Complete composition tunability of InGaN nanowires using a combinatorial approach. *Nat. Mater.*, **6**, 951–956.

222. Li, C., Li, J. and Zhang, W. (2000) A thermodynamic assessment of the GaInP system. *J. Phase Equilib.*, **21**, 357–363.

223. Marzari, N., de Gironcoli, S. and Baroni, S. (1994) Structure and phase stability of $Ga_x In_{1-x}P$ solid solutions from computational Alchemy. *Phys. Rev. Lett.*, **72**, 4001–4004.

224. Stringfellow, G.B. (1974) Calculation of ternary and quaternary III-V phase diagrams. *J. Cryst. Growth*, **27**, 21–34.

225. Jenichen, A. and Engler, C. (2010) Stability and band gaps of InGaP, BGaP, BInGaP alloys: density functional supercell calculations. *Phys. Status Solidi B*, **247**, 59–66.

226. Ansara, I., Chatillon, C., Lukas, H.L. *et al.* (1994) A binary data base for III-V compound semiconductor systems. *Calphad*, **18**, 177–222.

227. Scofield, D., Davison, J.E., Smith, S.R. (1991) Calculation of Phase Diagrams for Metal-GaAs System. Interim Report for period January 1990 January 1991, project WL-TR-91-2097.

228. Charykov, N.A., Litvak, A.M., Mikhailova, M.P. *et al.* (1997) Solid solution $In_xGa_{1-x}As_ySb_zP_{1-y-z}$: a new material for infrared optoelectronics. I. Thermodynamic analysis of the conditions for obtaining solid solutions, isoperiodic to InAs and GaSb substrates, by liquid-phase epitaxy. *Semiconductors*, **31**, 344–349.

229. Quiao, H., Shen, J., Xi, G. and Chatillon, C. (1994) Calculation of phase diagram for pseudobinary GaAs-InAs. *Trans. Nonferrous Met. Soc.*, **4**, 25–28.

230. Ohtani, H., Kobayashi, K. and Ishida, K. (2001) Thermodynamic study of phase equilibria in strained III-V alloy semiconductors. *J. Phase Equilib.*, **22**, 276–286.

231. Dinsdale, A.T. (1991) SGTE data for pure elements. *Calphad*, **15**, 317–425.

232. Okada, T., Weatherly, G.C. and McComb, D.W. (1997) Growth of strained InGaAs layers on InP substrates. *J. Appl. Phys.*, **81**, 2185–2196.

233. Behbehani, M.K., Piner, E.L., Liu, S.X. *et al.* (1999) Phase separation and ordering coexisting in $In_xGa_{1-x}N$ grown by metal organic chemical vapor deposition. *Appl. Phys. Lett.*, **75**, 2202–2204.

234. Ruterana, P., Nouet, G., der Stricht, W.V. *et al.* (1998) Chemical ordering in wurtzite $In_xGa_{1-x}N$ layers grown on (0001) sapphire by metalorganic vapor phase epitaxy. *Appl. Phys. Lett.*, **72**, 1742–1745.

235. Yamaguchi, S., Kariya, M., Nitta, S. *et al.* (1998) Structural and optical properties of AlInN and AlGaInN on GaN grown by metalorganic vapor phase epitaxy. *J. Cryst. Growth*, **195**, 309–313.

236. Teles, L.K., Marques, M., Scolfaro, L.M.R. *et al.* (2002) Phase separation suppression in InGaN epitaxial layers due to biaxial strain. *Appl. Phys. Lett.*, **80**, 769–771.

237. Silveira, E., Tabata, A., Leite, J.R. *et al.* (1999) Evidence of phase separation in cubic $In_xGa_{1-x}N$ epitaxial layers by resonant Raman scattering. *Appl. Phys. Lett.*, **75**, 3602–3604.

238. Teles, L.K., Scolfaro, L.M., Leite, J.R. *et al.* (2002) Phase diagram, chemical bond and gap bowing in cubic $In_xAl_{1-x}N$ alloys: ab initio calculations. *J. Appl. Phys.*, **92**, 7109–7113.

239. Teles, L.K., Marques, M., Scolfaro, L.M.R. *et al.* (2004) Phase separation and ordering in group-III nitride alloys. *Braz. J. Phys.*, **34**, 593–597.

240. Gan, C.K., Feng, Y.P. and Srolovitz, D.J. (2006) First principle calculations of the thermodynamics of $In_xGa_{1-x}N$ alloys: effect of lattice vibrations. *Phys. Rev. B*, **73**, 235214.

241. Zhong Zhang, J. (2009) *Optical Properties and Spectroscopy of Nanomaterials*, World Scientific, Hackensack, NJ, pp. 127–133.

242. Wise, F.W. (2000) Lead salt quantum dots: the limit of strong quantum confinement. *Acc. Chem. Res.*, **33**, 773–780.

243. Stucky, G.D. and McDougall, J.E. (1990) Quantum confinement and host/guest chemistry: probing a new dimension. *Science New Ser.*, **247**, 669–678.
244. Zhang, L.Z., Sun, W. and Chen, P. (2003) Spectroscopic and theoretical studies of quantum and electronic confinement effects in nanostructured materials. *Molecules*, **8**, 207–222.
245. Schmidt, V., Wittemann, J.V. and Gösele, U. (2010) Growth, thermodynamics, and electrical properties of silicon nanowires. *Chem. Rev.*, **110**, 361–388.
246. Sun, J. and Simon, S.L. (2007) The melting behavior of aluminum nanoparticles. *Thermochim. Acta*, **463**, 32–40.
247. Zhao, M., Zhou, X.-H. and Jiang, Q. (2001) Comparing of different models for melting point changes of metallic nanocrystals. *J. Mater. Res.*, **16**, 3304–3308.
248. Jiang, Q., Li, J.C. and Chi, B.Q. (2002) Size-dependent cohesive energy of nanocrystals. *Chem. Phys. Lett.*, **366**, 551–554.
249. Liang, L.H., Liu, D. and Jang, Q. (2003) Size-dependent continuous binary solution phase diagram. *Nanotechnology*, **14**, 438–442.
250. Sun, C.Q., Tay, B.K., Lau, S.P. *et al.* (2001) Bond contraction and lone pair interaction at nitride surfaces. *J. Appl. Phys.*, **90**, 2615–2617.
251. Sheu, H.-S., Jeng, U.-S., Shih, W.-J. *et al.* (2008) Phase separation inside the CdTe-CdSe type II quantum dots revealed by synchrotron x-ray diffraction and scattering. *J. Phys. Chem. C*, **112**, 9617–9622.
252. Davydov, V.A., Rakhnìmanina, A.V., Rols, S. *et al.* (2007) Size dependent phase transition of diamond to graphite at high pressures. *J. Phys. Chem. C*, **111**, 12918–12925.
253. Tolbert, S., Herhold, A.B., Brus, L.E. and Alivisatos, A.P. (1996) Pressure-induced structural transformations in Si nanocrystals: surface and shape effects. *Phys. Rev. Lett.*, **76**, 4384–4387.
254. van Gool, W. (ed) (1973) *Fast Ion Transport in Solids*, North-Holland/American Elsevier.
255. Bruce, P. (ed) (1997) *Solid State Electrochemistry*, Cambridge University Press.
256. Maier, J. (2009) Fundamentals, applications, and perspectives of solid-state electrochemistry: a synopsis, in *Solid State Electrochemistry I* (ed V.V. Kharton), Wiley-VCH Verlag GmbH, Weinheim, pp. 1–13.
257. Mari, C.M., Pizzini, S., Manes, L. and Toci, F. (1977) A novel approach to the oxygen microdetermination of oxides by EMF measurements. *J. Electrochem. Soc.*, **124**, 1831–1835.
258. Aronson, S., Salzano, F.J. and Bellafiore, D. (1968) Thermodynamic properties of the potassium–graphite lamellar compounds from solid-state emf measurements. *J. Chem. Phys.*, **49**, 434.
259. Kobayashi, Y., Egawa, T., Tamura, S. *et al.* (1997) Trivalent Al^{3+} ion conduction in aluminum tungstate. *Solid Chem. Mater.*, **9**, 1649–1654, 1649.
260. Pizzini, S. and Morlotti, R. (1972) E.M.F. measurements with solid electrolyte galvanic cells on the calcium oxide + zirconia system: determination of the phase relationships. *J. Chem. Soc., Faraday Trans.*, **68**, 1601–1610.
261. Katayama, I., Tanaka, T. and Iida, T. (2003) Thermodynamic study of semiconducting related materials by use of EMF method with solid electrolyte. *J. Min. Metall.*, **39**, 453–464.

262. Leushina, A.P., Makhanova, E.V. and Zlomanov, V.P. (2010) Influence of the nature and density of defects on the thermodynamic and electrical properties of semiconductor materials. *Inorg. Mater.*, **46**, 269–275.

263. Babanly, M.B., Gasanov, Z.T., Mashadiev, L.F. *et al.* (2012) Thermodynamic study of the Cu–As–S system by EMF measurements with $Cu_4RbCl_3I_2$ as a solid electrolyte. *Inorg. Mater.*, **48**, 225–228.

264. Müller, G., Mühe, A., Backofen, R. *et al.* (1999) Study of oxygen transport in Czochralski growth of silicon. *Microelectron. Eng.*, **1**, 13–147.

265. *Phase Transformations in Metals and Alloys*, by D. A. Porter, K. E. Easterling and M. Y. Sherif, CRC Press, 3rd edn (2009) p. 309.

2

Point Defects in Semiconductors

2.1 Introduction

Defect is the name given to a variety of imperfections of different dimensionality present in a crystalline lattice. Vacancies, self-interstitials, antistructural defects and substitutional or interstitial impurities are 0D (point) defects, dislocations are 1D defects, surfaces and internal interfaces (grain boundaries) are 2D defects and voids and precipitates, are 3D defects, all of which violate the conditions of local order in the lattice [1].

Although it is now possible to image point defects directly using high resolution transmission electron microscopy (HRTEM) [2], atomic resolution scanning tunnelling microscopy (STM) and atom probe tomography (APT) (and luminescence spectroscopy for optically active defects in diamond [3] and compound semiconductors), their presence has been traditionally deduced in the past from properties of solids that were otherwise difficult to explain.

Point defects are treated like true chemical (neutral or charged) species, subjected to conditions of thermodynamic equilibrium and mass action law. Considering that thermodynamic equilibrium can be achieved only if the defect generation/recombination processes are fast and the defects are mobile and randomly distributed in the lattice, equilibrium is generally a high-temperature condition, achieved above temperatures roughly corresponding to $T_m/2$ (K), T_m being the melting temperature of the particular solid considered.

Like every chemical species, defects interact with each other and with impurities. Defect interaction in solids is a particularly sensitive issue, even at relatively moderate temperatures where defects very seldom survive as individual species but react with other defects or impurities giving rise to defect complexes. Only at high temperatures do these complexes dissociate, allowing native defects to be studied as such.

It is via chemical interaction processes that defects are actually 'visible', though in a modified configurational or energy state.

Extended defects are non-equilibrium species and their formation may result from a number of causes of thermal, mechanical, structural and chemical origin, including the

Physical Chemistry of Semiconductor Materials and Processes, First Edition. Sergio Pizzini.
© 2015 John Wiley & Sons, Ltd. Published 2015 by John Wiley & Sons, Ltd.

coalescence (condensation) of supersaturated point defects and impurities, the anomalous stacking of lattice planes and crystallographic shear (CS) [4].[1]

While point defects in ionic crystals mostly influence and control the transport of matter and charge and the properties that stem from this, point defects in semiconductors influence both the transport of matter and the vibrational, electronic and optical properties of the host matrix. Different to bulk vibrational modes, defects and impurities introduce localized vibrational modes (LVMs) that are localized in the space, that is are localized close to the defects [5]. The role played by defects (point-like and extended) and impurities, and by their mutual interaction, on the properties of ionic solids and of elemental and compound semiconductors is an excellent example of how thermodynamics, kinetics, strain and diffusion dominate most of the physical properties of the solid phases over a wide range of temperatures.

Defects in solids have been the topic of tens of thousands of papers and of a large number of books and textbooks in the distant [6–8] and recent past [9–13]. Still today there is much interest in the field of advanced experimental and theoretical modelling studies, which are the basis of the deep understanding we have of defects in elemental and compound semiconductors [14–17].

The following sections will, therefore, be mostly limited to the physico-chemical aspects of point defect formation and of defect–impurity interactions in semiconductors, on which there is less emphasis in the literature, although they are of crucial interest for a better theoretical understanding of the problems encountered in the growth (see Chapter 4) and processing of semiconductor materials (see Chapter 5).

As a general introduction to the subject, but of direct interest also for compound semiconductors, in Section 2.2 the properties of defects in ionic solids will be brieflyly considered. They were the first to be studied in the first decades of the last century, using arguments and strategies of exceptional originality for the time, which were of large influence on the later investigations on defects in semiconductors. The great advantage in the investigation on solids that are purely ionic in character, such as the alkali and alkaline earth halides, is the practical absence of electronic conduction contributions. This unique feature allowed the almost exclusive use of doping, migration (electrical conductivity) and tracer diffusion experiments to study the physico-chemical properties of point defects and of their interaction. Quite different is the case of mixed valence materials, such as transition metal oxides, where electronic and ionic conduction coexist and defects induce, also, the set-up of shallow and/or deep levels in the forbidden gap.

It is important to mention here that defect studies on ionic solids culminated with the discovery of fast ion transport, that allowed a number of important technological applications in the field of electrochemical sensors, medium- and high-temperature fuel cells and electrical power storage, of which lithium batteries are a good example [18].

The third section of this chapter addresses atomic defects and impurities in Group IV semiconductors, not only for their technological importance, but also because defects in elemental semiconductors present a well understood behaviour, in comparison with that of compound semiconductors. In these materials vacancies and interstitials are thermodynamically stable as individual species, interact with themselves, with impurities and with

---

[1] CS is typical of non-stoichiometric oxides with the ReO_3 structure.

extended defects (dislocation and grain boundaries) and induce the presence of shallow and deep levels in the gap, with a massive influence on their bulk properties and on their technological applications.

The final section of this chapter is devoted to point defects and non-stoichiometry in compound semiconductors, whose impact on the development of modern microelectronics, optoelectronics and radiation detection is substantial.

2.2   Point Defects in Ionic Solids: Modelling the Electrical Conductivity of Ionic Solids by Point Defects-Mediated Charge Transfer

For thousands of years, solid materials were engineered to their final utilization employing the initial composition, the processing temperature and the applied stress as the process variables. For these materials, consisting of metallic alloys, glasses and ceramics, the manufacturing process was almost entirely dedicated to the fabrication of mechanical tools or containers.

More recently, when structural and chemical analysis techniques (X-ray diffraction, transmission electron microscopy (TEM), extended X-ray absorption fine structure (EXAFS), scanning electron microscopy (SEM), secondary ion mass spectroscopy (SIMS) and atomic force microscopy (AFM) became available, macro- and microstructure and material composition became additional process variables,leading to major improvements in the final product.

At the beginning of the last century, Tuband's discovery of ionic transport in ionic solids and their high temperature electrical conductivity and the work of Wagner gave birth to a new branch of physics, the physics of defects in crystalline solids, and the idea, a few decades later, for the development of the defect engineering of semiconductors.

The results of the early studies demonstrated that the electrical conductivity of ionic solids is thermally activated and typically shows a high-temperature region with an activation energy E_1 and a low-temperature region with an activation energy $E_2 < E_1$, given by the slopes of the Arrhenius plots of their conductivity displayed in Figure 2.1a, for the case of NaCl. It was also shown that the isothermal electrical conductivity values of the low-temperature regions are almost linearly dependent on the concentration n_{MX2} of deliberately added aliovalent impurities, as seen in Figure 2.1b.

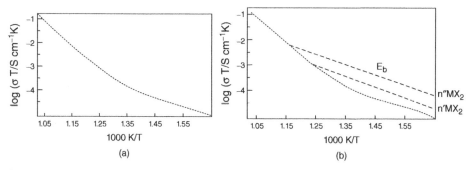

Figure 2.1   *(a) Electrical conductivity of NaCl (b) electrical conductivity of a doped alkali halide*

These features could be explained assuming the occurrence of a thermally activated migration process associated with the presence of equilibrium point defects [2, 19, 20] which, in the high-temperature region, are *intrinsic* to the particular solid and whose concentration depends, in the low-temperature region, on doping with aliovalent impurities.

In the absence of any experimental, direct proof of the existence of point defects, now available using, for example, APT, see Figure 1.2 of Chapter 1, the physical arguments supporting this model were originally based on the assumption that a direct exchange process among neighbour atoms in a perfect solid lattice is an energetically forbidden process, in comparison with that occurring via a jump to an empty position.

This assumption has been shown to hold for the case of semiconductor silicon using first principle calculations [21], but is supported also by thermodynamic arguments, for which a perfect solid is necessarily less stable than a defective solid, as defects (and impurities) add a configurational entropy contribution $\Delta S_{conf} = \sum_j x_j \ln x_j$ to the Gibbs energy ΔG° of the perfect solid.

From these arguments, and from the exponential dependence on temperature of the ionic conductivity, came the hypothesis that intrinsic point defects are stable, quasi-chemical species, whose formation follows a thermodynamic equilibrium process. Defects should be charged species in ionic solids and thus defect formation requires also the fulfillment of electrical neutrality conditions. For this reason, the elementary process of defect formation in ionic solids necessarily involves the birth of pairs of oppositely charged defects.

The concentration of defect pairs depends on the temperature with an exponential law

$$x^2 = \exp -\frac{\Delta G_f}{kT} \tag{2.1}$$

where x is the atomic fraction of each individual defect, ΔG_f is the Gibbs free energy of formation of a defect pair in the perfect crystal, occurring by a formal transfer of lattice atoms to the surface or by the transfer of a lattice atom to an interstitial position and k is the Boltzmann constant.

Under these conditions, two different models for point defect formation were originally suggested, known in the literature by the names of the proposers. The Schottky model envisages the transfer of a couple of ions of opposite charge from their stable lattice positions to surface sites with the consequent formation of a pair of vacant lattice sites

$$M_M X_X \rightleftharpoons M_{surf} + X_{surf} + V_M^- + V_X^+ \tag{2.2}$$

using for the defects (not for the charges of the defects[2]) the symbols proposed by Kröger and Vink [22].

The equilibrium concentration of Schottky pairs is given by the following equation

$$K_S = [V_M][V_X] = \exp -\frac{\Delta G_S}{kT} = \exp \frac{\Delta S_S}{k} \exp -\frac{\Delta H_S}{kT} = A \exp -\frac{\Delta H_S}{kT} \tag{2.3}$$

where K_S is the equilibrium constant and $\Delta G_S, \Delta H_S$ and ΔS_S are the Gibbs free energy, the enthalpy and the entropy of formation of Schottky pairs, respectively.

---

[2] In the Kröger and Vink notation a dot is used for a negative charge, a ball for a positive charge and a cross for a neutral species, here and in what follows the plus and minus sign will be instead used for immediate understanding.

Due to the very low equilibrium concentration of defects, concentrations of isolated, non-interacting defects can be expressed directly in terms of their atomic fractions instead of their thermodynamic activities.

Therefore, the $[V_M]$ and $[V_x]$ terms in Eq. (2.3) are expressions of true concentrations.

It can be seen immediately that in the absence of impurities which could influence the concentration of the defects (see later), the following condition holds

$$[V_M] = [V_X] = K_S^{1/2} \tag{2.4}$$

Frenkel defects are an interstitial type of defects, formed thanks to another elementary process that involves a single sublattice and which can be described by the following equations

$$M_M \rightleftharpoons M_i + V_M \tag{2.5}$$

and

$$X_X \rightleftharpoons X_i + V_X \tag{2.6}$$

where M_i is a metallic (cationic) species in an interstitial position and X_i is an anionic species in an interstitial position.

The equilibrium constants and the equilibrium concentration of the Frenkel pairs are given by the following equations

$$K_{M,F} = [M_i][V_M] = B \exp -\frac{\Delta H_F'}{kT} \tag{2.7}$$

$$K_{X,F} = [X_i][V_X] = C \exp -\frac{\Delta H_F''}{kT} \tag{2.8}$$

where $\Delta H_F'$ and $\Delta H_F''$ are the enthalpies of formation of the two Frenkel defect pairs.

Under full equilibrium conditions, the total concentration of vacancies and interstitials depends on the simultaneous occurrence of both the Schottky and the Frenkel processes.

This classical scheme, however, does not fully evidence the interplay of extended defects (external or internal surfaces, dislocations) as heterogeneous sources and sinks for both vacancies and self-interstitials, in all the processes involving atomic species in a solid lattice, as is well known [23] for the case of semiconductor silicon.

In fact, the presence of extended defects may allow the restoration of point defect equilibrium conditions when a defect excess is generated by temperature fluctuations or by interaction of the lattice host with energetic ($hv \gg \Delta H_f$) radiations or particles.

Excess defect trapping or recombination at extended defects in ionic solids would require defect pairs annihilation, to allow mass (and charge) conservation, as shown with the following equation

$$M_i + V_M \underset{\text{surf}}{\overset{\text{surf}}{\rightleftharpoons}} M_M \tag{2.9}$$

for the case of a system with dominant Frenkel equilibrium in the cationic sublattice and the surface as the defects' source and sink.

Nevertheless, since surfaces and line defects bring an excess charge, the electroneutrality constraints at extended defects may be relaxed, with the formation of a space charge region.

Table 2.1 *Formation enthalpies of Schottky and Frenkel defects and migration activation enthalpies for selected ionic compounds. The last column reports some values of the formation enthalpies of vacancy complexes with divalent cations*

	Compound	Schottky defect formation ΔH_f (eV)	Cation migration enthalpy ΔH_c (eV)	Anion migration enthalpy ΔH_a (eV)	Divalent cation impurity association enthalpy ΔH_p (eV)
Schottky defects	LiF	2.34	0.7–0.73	—	—
	LiCl	2.12	—	—	—
	LiBr	1.80	—	—	—
	NaCl (*)	2.45 ± 0.05	0.63 ± 0.01	0.86 ± 0.1	0.55 ± 0.05 (Sr) 0.31 (Ca)
	KCl (*)	2.52 ± 0.05	0.74 ± 0.02	0.9 ± 0.01	0.58 ± 0.01 (Sr)
	KBr (*)	2.37 ± 0.06	0.667 ± 0.02	0.92 ± 0.04	0.57 ± 0.003 (Ca)
	RbCl	2.04	0.54	1.45	0.64 (Sr)
	KI (*)	2.21	0.63 ± 0.03	1.29	0.54 ± 0.03 (Sr)
Frenkel defects		Frenkel defect formation enthalpy ΔH_f (eV)	Anion vacancy migration enthalpy $\Delta H_{m,V}$ (eV)	Anion interstitial migration enthalpy $\Delta H_{m,I}$ (eV)	—
	CaF$_2$	2.3–2.80	0.5–0.9	1.7	—
	SrF$_2$	2.3	—	1	—
	BaF$_2$	1.5–1.9	0.56–0.85	0.78	—
	—	Frenkel defect formation enthalpy ΔH_f (eV)	Cation vacancy migration enthalpy $\Delta H_{m,V}$ (eV)	Cation interstitial migration enthalpy $\Delta H_{m,I}$ (eV)	—
	AgCl	1.25–1.55	0.008–0.16	0.27–0.35	—
	AgBr	1.06–1.27	0.06–0.15	0.30–0.34	—
	β-AgI	0.70	—	—	—

Crawford & Slifkin, 1972, [6]. Reproduced with permission from Springer Science and Business Media. (*) doped material

As can be seen from Table 2.1 [6], alkali halides are dominated by Schottky defects, with comparable values of their formation enthalpies ΔH_f, while in alkaline earth halides and silver halides Frenkel defects prevail.

Isolated defects of opposite charge, such as Schottky defects here below, might interact and associate to neutral and non-interacting defect pairs, the bonding being essentially of electrostatic nature.

$$V_M^- + V_X^+ \rightleftharpoons [V_M^- - V_X^+] \tag{2.10}$$

Considering as a first approximation only nearest neighbours pairs, in the framework of the chemical equilibrium, the ratio of defect pairs x^P to the total number of defects is given by:

$$\frac{x^P}{x^2} = Z \exp -\frac{\Delta G^P}{kT} \qquad (2.11)$$

where Z is the number of indistinguishable orientations of each pair and ΔG^P is the Gibbs energy of formation of the defect pair.

In the case of an AB solid with prevailing vacancy defects, Eq. (2.11) would be written

$$\frac{x^P}{x_V^A x_V^B} = Z \exp -\frac{\Delta G^P}{kT} \qquad (2.12)$$

where x^P is the fraction of lattice sites occupied by defects that are paired, x_V^A and x_V^B are the fraction of A and B sites occupied by unpaired defects, here cation and anion vacancies.

As defects can be approximated to point-like charged species in ionic crystals, their interaction is long range and of Coulombic nature, screened by the lattice charge. If the crystal is treated as a continuum, the Coulomb interaction energy between a pair of oppositely charged defects at the distance R of closest approach (which could be considered the formation energy of the pair) is given by the equation

$$\Delta G^P = k\frac{z^2 q^2}{\varepsilon n R} \qquad (2.13)$$

where $k = 2.31 \cdot 10^{-19}$ J $\cdot$ nm, ε is the dielectric constant and zq is the charge. For NaCl, $\varepsilon \approx 6$, $zq = 1$, $R = 0.5$ nm as the closest approach distance and the interaction energy $\Delta G^P = 42.21$ kJ mol^{-1} = 0.479 eV.

The calculated concentration of pairs for two extreme values of interaction energy (60 and 300 kJ mol^{-1}) is reported in Table 2.2, which shows that for ionic solids at high temperatures the molar fraction of pairs should stay in the range of 1%.

As the interaction energy, according to Eq. (2.13), decreases slowly with increase in the interatomic distance, it is possible to assume that pairs separated by an interatomic distance multiple of R would also be stable as unique and identifiable chemical species, provided $\Delta G^P = k\frac{z^2 q^2}{\varepsilon n R} \gg k_B T$, where nR is an integer multiple of the interatomic distance and k_B is Boltzmann's constant.[3]

Table 2.2 *Concentration of Schottky defect pairs (in molar fractions) for different values of the interaction energy ΔG_p*

T (K)	$\Delta G_p = 60$ kJ mol^{-1}	$\Delta G_p = 300$ kJ mol^{-1}
300	$5.72 \cdot 10^{-6}$	$6.12 \cdot 10^{-27}$
1000	$2.65 \cdot 10^{-2}$	$1.27 \cdot 10^{-8}$

---

[3] In the remainder of the chapter Boltzmann's constant will be written as k when obvious from the context.

Using the Debye–Hückel theory, developed and applied for aqueous ionic solutions, the thermodynamic activities of defects may by calculated by multiplying the atomic fraction of each unassociated defect species i in Eq. (2.12) by an activity coefficient γ

$$\ln \gamma_i = -\frac{q_i^2 K_{DB}}{2\varepsilon kT(1 + KR)} \qquad (2.14)$$

where q is the charge, ε the dielectric constant, R the distance of closest approach and K_{DB} is the reciprocal of the Debye–Hückel length λ_D or the 'effective' coulombic radius of the defect, given by the ratio

$$K_{DB} = (\lambda_D)^{-1} = \left[\frac{4\pi \left[\sum_i q_i^2 x_1 / \Omega_i \right]}{\varepsilon kT} \right]^{1/2} \qquad (2.15)$$

where Ω_i is the atomic volume of the ith species in the host lattice.

For the Debye–Hückel theory the value of R is undetermined, but it may be taken equal to the next nearest neighbour ions distance, as the 'effective' coulombic radius of the defect includes at least the nearest neighbour ions.

Doping ionic solids with heterovalent impurities, in the range of their solid state solubility, leads to a change in the concentration of point defects. As an example, doping NaCl, or any other alkali metal chloride with $CaCl_2$ [24], would lead to the formation of a substitutional Ca_{Na} species sitting in a cationic lattice position and of an equivalent number of sodium vacancies.[4]

$$CaCl_2 \xrightarrow{NaCl} Ca_{Na}^+ + 2Cl_{Cl} + V_{Na}^- \qquad (2.16)$$

The consequence is the shift of the concentration of sodium vacancies from their equilibrium value to that of calcium in solution

$$[V_{Na}] = c_{CaCl_2} = [Ca_{Na}] \qquad (2.17)$$

while the concentration of chlorine vacancies takes, according to Eq. (2.3) a new equilibrium value

$$[V_{Cl}] = \frac{K_S}{[Ca_{Na}]} \qquad (2.18)$$

where K_S is the Schottky equilibrium constant. Aliovalent impurities dissolved in the host lattice are charged species and interact with point defects. As an example, see Eq. (2.16), a Ca^{2+} ion in a substitutional Na position bears a unit (excess) positive charge and may interact with a sodium vacancy, which bears a negative charge, leading to formation of a neutral pair:

$$Ca_{Na}^+ + V_{Na}^- \rightleftharpoons [Ca_{Na}^+ - V_{Na}^-]^x \qquad (2.19)$$

using for the zero charge of the neutral pair (x) the Kröger–Vink notation [22].

---

[4] Reaction (2.16) is not an equilibrium reaction.

The concentration of pairs (actually, their thermodynamic activity) depends on the enthalpy of formation ΔH^p of the pair (actually, on the Gibbs energy of formation ΔG^p), assuming the validity of the mass action law also in this case.

$$\frac{[Ca^+_{Na} - V^-_{Na}]^x}{[Ca^+_{Na}] \, [V^-_{Na}]} = K_p = A \exp -\frac{\Delta H^p}{kT} \tag{2.20}$$

The entropy contribution to the Gibbs energy of formation, considered independent of the temperature, is included in the pre-exponential term A.

Table 2.1 reports some values of the enthalpy of formation ΔH^p of divalent impurities-vacancy complexes in alkali halides, from which one can see that, in the point charge approximation, the interaction energies are close to the calculated coulombic interaction energies of point defect pairs in NaCl at the distance of closest approach.

Ions are the actual charge carriers and the most energetically favourable charge- and mass-transport process is a jump of an ion from an occupied lattice site to a neighbour empty one or to an interstitial position above a potential barrier. Defects diffuse with the same rate as the charge carrier, as a defect substitutes an ion having left its lattice position with a jump. Only the ionic species maintain, however, their own identity in a migration or diffusion process.

Diffusion and migration processes[5] have been systematically used to study the properties of defects in ionic solids, but in this section only migration processes will be considered, as they give sufficient information concerning the nature and the properties of defects in ionic solids.

Diffusion and migration processes are driven by a gradient of concentration or of electrical potential, as shown by the following empirical equations, of which the first is Fick's law

$$J_i = D_i \mathrm{grad}\, c_i \tag{2.21}$$

$$J_i = \Lambda_i \mathrm{grad}\, \varphi \tag{2.22}$$

where D_i is the diffusion coefficient, Λ is the conductance and φ is the electrical potential.

At high temperatures, when ions and defects are mobile, the carrier migration process in isothermal conditions is described by the equation

$$\sigma = \sum_i \mu_i n_i \tag{2.23}$$

where σ is the electrical conductivity, μ_i is the thermally activated mobility of the ith charge carrier and n_i is its temperature-dependent concentration.

In turn, the mobility of the ith mobile species

$$\mu_i = \frac{S_r q_1 \, a^2 \Gamma_i}{kT} \tag{2.24}$$

depends on the number of equivalent sites S at a distance a into which the carrier can jump and on the jump frequency Γ_i of the ion above a barrier of height E_b

$$\Gamma_i = v \exp -\frac{E_b}{kT} \tag{2.25}$$

---

[5] Diffusion processes are driven by a concentration gradient while migration processes are driven by an electric field.

where v is an effective frequency associated with the vibration of the defect in the direction of the saddle point [25].

In solids exhibiting Schottky-type disorder, cations and anions are the charge carriers and their individual role in the migration process entirely depends on their mobilities, which differ amongst the alkali halide family. In solids exhibiting Frenkel-type disorder interstitial- and vacancy-supported migration may occur, but cation migration is predominant in silver halide crystals and anion migration is predominant in alkaline earth halides crystallizing with the fluorite structure, as seen in Table 2.1.

In the case of undoped alkali halides, where both cation- and anion- vacancies are available, the conductivity is supported both by cations and anions and the intrinsic conductivity σ is given by

$$\sigma = \frac{1}{T}\sum_i^2 A_i \exp-\frac{\Delta H_D}{kT}\exp-\frac{E_{b,i}}{kT} = \frac{1}{T}\sum_i^2 A_i \exp-\frac{E_i^*}{kT} \tag{2.26}$$

where the $1/T$ term comes from the temperature dependence of the mobility according to Eq. (2.24).[6] A curvature of the Arrhenius plot of the σT product is therefore expected if cation and anion vacancies exhibit different mobilities, as expected from Eq. (2.24) and Eq. (2.25) and shown in Table 2.1.

The absence of any appreciable curvature in the high-temperature, intrinsic conductivity region of Figure 2.1 relative to undoped NaCl has been debated for years until it was demonstrated that for NaCl, pure and SrCl$_2$-doped, cations and anions have similar activation enthalpies for the mobility (0.626 and 0.744 eV, respectively [26]. In other halides, such as LiF, the ionic conductivity is actually the sum of the cation and anion contributions and the Arrhenius plot shows a slight, continuous curvature.

In the case of a prevailing role of a single defect over a wide range of temperatures, the Arrhenius plot of the intrinsic electrical conductivity of doped or undoped alkali halide would therefore be given by the equation

$$\sigma T = A\exp-\frac{\Delta H_D}{kT}\exp-\frac{E_b}{kT} = A\exp-\frac{E^*}{kT} \tag{2.27}$$

where $E^* = \Delta H_D + E_b$ is the slope of the high-temperature curve of Figure 2.1 and A is a constant.

The low-temperature conductivity of a virtually undoped sample is shown, instead, to depend on the residual presence of aliovalent impurity which contaminates the sample (see Figure 2.1), or on the concentration n_{MX2} of aliovalent dopants intentionally added to fix the concentration of carriers.

In this case, that is the general case of an AX solid doped with aliovalent impurities

$$MX_2 \overset{AX}{\rightleftharpoons} M_A + 2X_X + V_A \tag{2.28}$$

the low-temperature conductivity is given by the equation

$$\sigma T = n_{MX_2} A^* \exp-\frac{E_b}{kT} \tag{2.29}$$

---

[6] To take into account the temperature mobility dependence the σT products are plotted in an Arrhenius plot.

where n_{MX_2} is the concentration of the dopant, A^* is a constant and E_b is the activation energy needed for a carrier to jump above a barrier of energy E_b.

It is therefore possible to obtain, from the linearized slope of the Arrhenius plots of the low-temperature conductivity, the activation enthalpy for the elemental transport process of cation vacancies. Linearization works well in the doping concentration region where the intrinsic concentration of carriers is much lower than that arising from doping.

Furthermore, at the experimental temperature where the intrinsic and extrinsic conductivities take the same value, given by the intersection points of Figure 2.1b, we have

$$K \exp-\frac{\Delta H_f}{kT} \exp-\frac{E_b}{kT} = n_{MX_2} A' \exp-\frac{E_b}{kT} \tag{2.30}$$

where the pre-exponential K is the product of formation and mobility entropy terms, A' is the mobility entropy term and

$$n_{MX_2} = A \exp-\frac{\Delta H_S}{kT} = k_S^{1/2} \tag{2.31}$$

It would therefore be possible to determine, with a reasonable statistical accuracy, both the defect concentration and its temperature dependence, provided a proper number of electrical conductivity measurements are carried out at different doping levels.

It should, however, be observed that the nominal values of the slopes of the conductivity curves of a doped halide and of their intersections with the intrinsic conductivity curve are necessarily influenced by the occurrence of impurity-defect pairing interaction.

Therefore, these measurements would give only approximate values of the migration and defect formation enthalpies.

2.3 Point Defects and Impurities in Elemental Semiconductors

2.3.1 Introduction

As in the case of ionic solids, mass transport in semiconductors occurs thanks to the presence of point defects. Lattice disorder in semiconductors is associated with a variety of defect species, ranging from point defects to extended defects, with impurities playing an even more powerful role.

In this section, native atomic defects (vacancies and self-interstitials) in elemental semiconductors will be discussed, together with light and metallic impurities, in order to present their individual properties, their influence on the properties of the host in which they are embedded and the physico-chemical aspects of their mutual interaction.

The interaction of impurities and defects with extended defects will be discussed in Chapter 3, as this argument requires a different and more addressed treatment.

Defects (and impurities) in semiconductors introduce vibrational modes which are spatially localized in the vicinity of the defect, distinct from bulk modes and shallow- or deep-states in the band gap of the semiconductor. These properties allow one to carry out a study of defects using spectroscopic measurements such as Raman and IR absorption spectroscopies as well as electron spin resonance (ESR), photoluminescence (PL) and deep level transient spectroscopy (DLTS).

As in the case of ionic solids, vacancies and self-interstitials in elemental semiconductors are equilibrium species, whose generation, using silicon as a typical example, is considered to occur via a Schottky-type of process

$$Si_{Si} \rightleftharpoons Si_{surf} + V_{Si} \tag{2.32}$$

or via a Frenkel mechanism

$$Si_{Si} \rightleftharpoons Si_i + V_{Si} \tag{2.33}$$

where Si_{Si} is a silicon atom in the silicon lattice, Si_{surf} is a silicon atom at the surface, V_{Si} is a silicon vacancy and Si_i is a silicon atom sitting in an interstitial position, that is a self-interstitial. In what follows vacancies will also be indicated with the symbol V and the self-interstitials with the symbol I, when their identification as specific defects of a particular host is not necessary.

Assuming the validity of the law of mass action, the defect concentration depends on the temperature with an Arrhenius type of law

$$[V_{Si}] = K_S = k_S \exp{-\frac{\Delta H_S}{kT}} \tag{2.34}$$

$$[V_{Si}][Si_i] = K_F = k_F \exp{-\frac{\Delta H_F}{kT}} \tag{2.35}$$

where ΔH_S and ΔH_F are the Schottky and Frenkel enthalpies of formation of defects. The entropy terms enter into the pre-exponential terms k_S and k_F of Eqs. (2.34) and (2.35).

Defects in semiconductors may be stable in different charge states. Therefore, the enthalpy of formation of a defect, or its formation energy E_f, is not simply a constant at constant temperature, but depends on the Fermi level ε_F of the host, that is on doping.

As an example, the formation energy for a vacancy of charge q in germanium $E_f^{V_{Ge}^q}$ (but obviously this condition could be extended to any other semiconductor) might be given by the following expression

$$E_f^{V_{Ge}^q} = E_{tot}^{Ge+V_{Ge}^q} - E_{tot}^{Ge} + \mu_{Ge} + \varepsilon_F \tag{2.36}$$

where the difference $E_{tot}^{Ge+V_{Ge}^q} - E_{tot}^{Ge}$ is the contribution of the vacancy of charge q to the total energy of the system, μ_{Ge} is the chemical potential of bulk germanium and the Fermi level ε_F is refers to the maximum of the valence band [27].

In the absence of external sinks or sources for either vacancies or self-interstitials,[7] from Eq. (2.35) we have for the intrinsic concentration of Frenkel defects in silicon

$$[V_{Si}] = [Si_i] = K_F^{1/2} = k_F^{1/2} \exp{-\frac{\Delta H_F}{2kT}} \tag{2.37}$$

Very rarely the intrinsic concentration of defects is, however, preserved, because a number of defect sinks (traps) and sources exist, consisting of impurities, dislocations, external and internal surfaces, precipitates, which actually determine their effective concentration.

---

[7] Homogeneous and heterogeneous sinks or sources of vacancies and self-interstitials may contribute to the defect population generated by the intrinsic defect formation processes. A Schottky process is a typical heterogeneous process.

An example of a defect trap might be an impurity M that interacts with the native defect, either a vacancy or an interstitial, and here labelled D, with the formation of a stable D–M pair

$$D + M \rightleftharpoons DM \tag{2.38}$$

whose equilibrium concentration is given by the equation

$$K_{eq} = \frac{[DM]}{[D][M]} \tag{2.39}$$

where again the equilibrium constant K_{eq} depends exponentially on the temperature

$$K_{eq} = Z \exp{-\frac{E_b^{DM}}{kT}} \tag{2.40}$$

and E_b^{DM} is the pair binding energy while Z is the number of possible orientations of the D–M pair which maintain the same symmetry [28].

The actual concentration of defect complexes, and therefore also of native defects, at a specific temperature, may, however, be close or far from the equilibrium one, depending on the reaction rate of reaction (2.38), with significant differences in the physical properties of the semiconductor where these interaction processes occur. As for conventional chemical reactions, the overall rate of a defect–impurity complex formation depends on the diffusivities of the reaction partners and on the true reaction rate

$$\frac{dC_{MD}}{dt} = k_r \, (C_M, C_D, T) \tag{2.41}$$

which is a function of the temperature and of the concentration C of the reactants, via a kinetic constant k_r.

At moderate temperatures, when the diffusion processes are slow, the rate determining step of the impurity–defect interaction reaction is the diffusion of the reaction partners at the reaction site. Only at process temperatures where diffusion processes may be fast, will the true reaction rate be rate determining. A quantitative analysis of the interaction processes dominated by diffusion is given in Section 3.4.6.

Both ionic and covalent bonds may be involved in the formation of the D–M pairs, which could be treated, using chemical language, as true complexes dissolved in the bulk solid lattice which is like the solvent in a conventional liquid solution. They could also behave as the precursors of a new phase, which segregates when the solubility limit is reached.

As in the case of ionic compounds, mass transport processes in semiconductors are directly linked to the presence of defects. Diffusion measurements have been, therefore, the main tools used for their study, although spectroscopic techniques were and are also used extensively, as anticipated in the previous section. Obviously, electrical conductivity measurements should be excluded, as electronic conduction via electrons and holes dominates the electrical conductivity of undoped (intrinsic) and doped (extrinsic) semiconductors.

2.3.2 Vacancies and Self-Interstitials in Semiconductors with the Diamond Structure: an Attempt at a Critical Discussion of Their Thermodynamic and Transport Properties

Despite the fact that vacancies and self-interstitials are hardly detectable as such, it is a matter of experience and the result of theoretical modelling that vacancies and self-interstitials should be taken into account to explain mass transfer and chemical equilibrium processes occurring in semiconductors that otherwise are not understandable.

The diffusion of dopants (P, B, Sb) and of interstitial metallic (3d) impurities, the formation of impurity–defect pairs, the segregation of impurities from their solutions, the nucleation and growth of dislocations and the segregation of vacancy voids and of hydrogen bubbles in silicon are typical examples of point defect-assisted or point defect-coupled processes. Knowledge of their properties is, therefore, a prerequisite to understanding their unique role in such a variety of processes.

It has already been shown in the previous section that that the generation of vacancies and self-interstitials in semiconductors may be formally depicted as Schottky- and/or Frenkel-type defect generation/recombination processes (see Eqs. (2.32) and (2.33)). It is, however, well known that non-equilibrium concentrations of point defects can also be generated by high-energy particles irradiation, a technique which was originally used to demonstrate their presence in semiconductors.

Under thermal equilibrium conditions, the intrinsic concentration of defects depends on the occurrence of both defect generation and recombination processes. The intrinsic defect generation/recombination processes, however, occur systematically in competition with heterogeneous defect generation/recombination/trapping reactions at internal or external surfaces or with homogeneous interaction reactions with impurities in the bulk or at the surface, which determine the *effective* concentration of vacancies and self-interstitials in a semiconductor phase.

Vacancies and self-interstitials are, therefore, generally far from their thermal equilibrium concentration [29].

Different from ionic solids, defects in covalent solids have a specific spatial extension, that depends on their bonding configuration inside the host lattice and on their charge states.

In addition, the point symmetry of atomic defects in diamond, silicon and germanium is often lower than cubic, as they can present different unrelaxed and relaxed crystallographic configurations [30]. As an example, the schematic configuration of an unrelaxed silicon neutral vacancy, with four dangling bonds pointing towards the missing silicon atom, is displayed in Figure 2.2.

Computational studies show, however, that the equilibrium configuration of an isolated vacancy is reached through an inward relaxation of the four silicon atoms neighbouring the tetrahedral vacant site, accompanied by important bond distortions [31, 32].

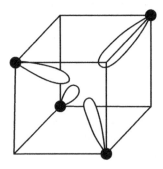

Figure 2.2   *Dangling bonds at an unrelaxed neutral silicon vacancy*

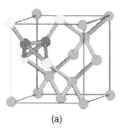 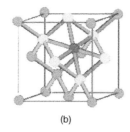 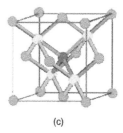

(a)   (b)   (c)

Figure 2.3   *Configuration of the different self-interstitials in silicon (a) split <100>, (b) hexagonal and (c) tetrahedral. Leung* et al., *1999, [34]. Reproduced with permission from American Physical Society*

Computational studies show also that the silicon vacancy can exist in five charge states, from V^{2-} to V^{2+}, while in Ge two charge states are probably stable, V^{2-} and V° [33]. Computational studies show that the free (unreacted) vacancy in silicon is a fast moving species with an activation energy for the mobility of only 0.1 eV.

In semiconductors having the diamond structure, three different types of self-interstitials can be foreseen, the tetrahedral (T), the hexagonal (H) and the split-interstitial (with a dumbbell configuration) one (see Figure 2.3), where the Si-Si bond is displaced either along the <110> or the <100> direction [34]. The dumbbell interstitial can be conceived as a pair of silicon atoms that occupy a single lattice position [31].

It is worth noting that in germanium the split <110> configuration is preferred for the neutral self-interstitial [33], while the split <110> and a configuration close to the hexagonal one is preferred for the positive or double positive ones. In the case of silicon the split <100> and a configuration close to the hexagonal are the most stable for the neutral state [34]. In diamond the split interstitial configuration has also been found to be the stable one [35].

As anticipated above, both vacancies and self-interstitials are extended in the space, at least in a region including their covalent bonds with the nearest neighbours and cannot be considered as true point-like defects.[8]

As the equilibrium concentration of native defects in Group IV semiconductors is so small as to be undetectable using conventional methodologies, most of the early studies on these defects in silicon and germanium, and the recent studies on diamond, were carried out by deliberately enhancing their concentration by irradiation with high energy electrons or particles. Then, various spectroscopies were used to detect them in the irradiated samples and to get information on their properties.

As an example, after 1–3 MeV electron irradiation of silicon at cryogenic temperatures, the presence of isolated vacancies in two charge states (V^+ and V^-) was inferred by ESR [36–38] measurements.

In diamond, high energy electron or neutron irradiation leads to the formation of vacancies that become mobile at temperatures higher than 550 °C and interact with nitrogen, which is the most common impurity in diamond [39]. Nitrogen-vacancy (NV) centres at the surface of diamond are particularly interesting for the potential development of quantum processors, owing to the possibility of their optical read-out [3] as most of the defects in diamond are colour centres.

---

[8] The comparison here is with point defects in purely ionic solids.

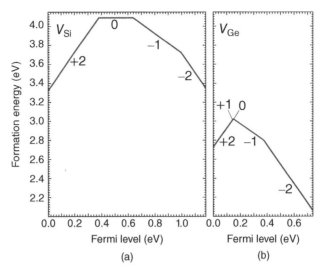

Figure 2.4   *Calculated formation energies of vacancies in silicon (a) and Ge (b). Weber et al., 2013, [27]. Reproduced with permission from American Physical Society*

Due to their optical activity, vacancy characterization in diamond can be carried out by luminescence spectroscopy, but Raman spectroscopy may also be used for both vacancies and self-interstitials, with results which fit well with those obtained with molecular dynamics (MD) simulations [35].

Negative and neutral vacancies well in excess of their equilibrium concentration may be detected in silicon by positron annihilation spectroscopy (PAS), with reasonable sensitivity ($\approx 10^{15}$ cm^{-3} for negative vacancies and $\approx 10^{16}$ cm^{-3} for neutral vacancies [40, 41].

Also DLTS on irradiated silicon has been able to detect a vacancy level at $E_V + 0.13$ eV in the forbidden gap [37].

Information about the properties of vacancies and self-interstitials in electron irradiated Ge has been obtained by perturbed angular correlation (PAC) spectroscopy [42], revealing the presence of an acceptor state for the vacancy at $E_V + 0.20$ eV and, tentatively, a donor state for the interstitial, close to the conduction band ($E_C - 0.040$ eV).

For germanium, in addition to irradiation, quenching from high temperatures followed by annealing at lower temperatures and plastic deformation at high temperatures have been important tools to study intrinsic point defect properties [43, 44].

The evidence that the concentration of vacancies in Ge is much higher than in silicon was one of the important results of these studies, which fits with their predicted lower energy of formation [27], as shown in Figure 2.4.

Self-interstitials are more elusive than vacancies to be caught.

The predicted self-interstitial presence in silicon has been, in fact, only indirectly inferred from the ESR spectra of their trapped configurations with Al, B and carbon [45–46], from the enhancement of the dopants diffusion in silicon by surface oxidation [47][9] and from the profile of the gold diffusion after gettering [48].

---

[9] The explanation given for this process is that the strain induced by the volume misfit of silica (SiO_2) in the silicon lattice makes an oxidized silicon interface the source of interstitial type defects, which are the mobile species assisting the dopant diffusion.

The elemental mechanism of this fast diffusion process is the formation of a dopant–self-interstitial pair $D_{Si} - Si_i$, which dissociates to a mobile dopant in an interstitial position, according to the following equation

$$D_{Si} - Si_i \rightarrow D_i + Si_{Si} \tag{2.42}$$

where D_{Si} is a generic dopant (B, Ga, P, As, Sb) which normally sits in a substitutional position.

All the aforementioned studies give information on the *effective* concentration of native defects, when the semiconductor is or was submitted to an external stress, rather than on their intrinsic equilibrium concentrations. To obtain these concentrations, measurements should be carried out after having annealed the solid at a temperature at which the isolated defects are stable and interact neither with impurities nor with heterogeneous trapping/recombination centres.

For silicon and germanium, temperatures slightly below their melting point would be necessary, ruling out the direct use of ESR and DLTS for defect identification and characterization. Measurements could be carried out, however, on samples fast cooled to room temperature.

X-ray diffraction techniques[10] are, in principle, limited by the very low values of the equilibrium concentrations of defects, although used to detect their presence on electron irradiated germanium at 2 K, with a concentration of defects larger than 10^{19} cm^{-3} [43].

In addition to ESR on fast cooled samples, self-diffusion measurements have been the main tools to obtain information on defect concentration and mobilities [49].

For a single species in one single charge state, the self-diffusion coefficient D^{SD} is given by

$$D^{SD} = D_0 \exp -\frac{Q^{SD}}{kT} \tag{2.43}$$

where Q^{SD} is a migration enthalpy term.

When the mass transport is simultaneously mediated by vacancies and self-interstitials, the self-diffusion coefficient is the sum of all the different contributions arising from them, in all their stable charge states, and of an additional term (D_{exch}), which accounts for the contribution of a direct exchange of neighbouring lattice atoms

$$D^{SD} = \sum_j f_j \, x_j D_j + D_{exch} \tag{2.44}$$

where x_j is the equilibrium atomic fraction of the j defect normalized to the atomic density of silicon, D_j is the diffusivity of the j defect and the f_j terms are correlation factors,[11] given by the ratio of the mean square displacement of the species moving in the correlated fashion to that of the same species moving randomly $f = \frac{\langle x^2 \rangle_{corr}}{\langle x^2 \rangle_{random}}$.

Consequently, the Arrhenius plot of the self-diffusion coefficient should deviate from linearity when different neutral and charged species contribute simultaneously to the diffusion process, giving a first signature of a process mediated by different defect species.

---

[10] The density of defects in a crystal may be determined by coupling a gravimetric density measurement to a lattice constant determination using XR diffraction, this last giving the theoretical density of a defect-free crystal.

[11] In the case of diffusion in solids, if a vacancy or interstitial mechanism holds, a correlation occurs between successive jumps of the diffusing species, even if the defect steps are uncorrelated

However, the main problem with self-diffusion measurements is the impossibility to directly infer the values of the individual contributions of the different defects, when they coexist and are multiply charged [50], even if the direct exchange D_{exch} term may be neglected, being generally much smaller than the other terms of Eq. (2.44) [21, 51].

Further, self-diffusion is commonly assumed to take place under thermal equilibrium conditions, that is assuming that the concentrations of self-interstitials and vacancies at the beginning of the self-diffusion measurement are equal to their thermodynamic equilibrium values.

Moreover, it is supposed that the defect concentrations are not influenced by the self-diffusion process itself or by heterogeneous sinks such as surfaces and dislocations [52, 53]. Therefore, self-diffusion is taken as a direct proof of the existence of point defects and its measurement remains a practical tool for the identification of their presence and for their preliminary characterization.

Silicon self-diffusion measurements on undoped and doped samples were historically carried out[12] using the radioactive isotope Si^{31}. In the last 20 years self-diffusion measurements in Si, Ge, GaAs and GaP, were, instead, carried out using isotopically enriched heterostructures and measuring the diffusion profiles with SIMS [54–57].

As a typical example, self-diffusion measurements in the 800–1100 °C range were carried out [51, 57] using a Si sample consisting of a surface layer, approximately 90–210 nm thick, with the three stable Si isotopes in their natural relative abundances and of a 5 μm thick buried layer heavily depleted in Si^{29} and Si^{30}. The Si^{30} isotope was employed as the tracer and the self-diffusion profile was obtained by mass spectrometry measurements on thin sections of the buried layer.

In an analogous manner, self-diffusion measurements in germanium were carried out in the 400–600 °C range using Ge^{70}/Ge^{nat} isotope multilayer structures [58] or by carrying out dopant diffusion measurements [59].

All measurements on Si and Ge fit well with an activation energy value of the self-diffusion coefficient close to 5 and 3 eV, respectively (see Table 2.3) [56, 58]. Further, the linear shape of the Arrhenius plot of their self-diffusion coefficients in Si and Ge (see Figure 2.5) [54, 56] is taken as a demonstration that a single defect species controls the self-diffusion over the whole measurement temperature interval.

As anticipated above, it is, however, unrealistic to extract from self-diffusion measurements the diffusivity and the concentration of each defect in each charge state, at least in the most general case.

Direct information about the defect concentration in germanium and silicon has been, instead, obtained using diffusivity measurements of Zn [50], Cu [60], Pt [61, 62], Au [63] and Ir [64].[13]

A common property of all these metals is that they dissolve in Si or Ge as substitutional and interstitial species, with an equilibrium concentration of the substitutional species much larger than that of the interstitial species, which are, however, the most mobile.

---

[12] For a review, see Ref. [53].

[13] Metal diffusion experiments are carried out by depositing the metal by vacuum evaporation or sputtering on the surface of a semiconductor. The metal is subsequently diffused into the sample by heating it at a defined temperature for a defined time. Depth profiles of the diffused metals are eventually measured at room temperature on the quenched samples using DLTS. Details may be found in Refs. [50, 60, 63, 64].

Table 2.3 *Self-diffusivity data for silicon and germanium*

	Measurement range ($T°C$)	$D°$ ($cm^2 s^{-1}$)	Q^D (eV)	References
Si	1100–1400	560	4.76	[56]
Ge	429–904	25.4	3.13	[58]

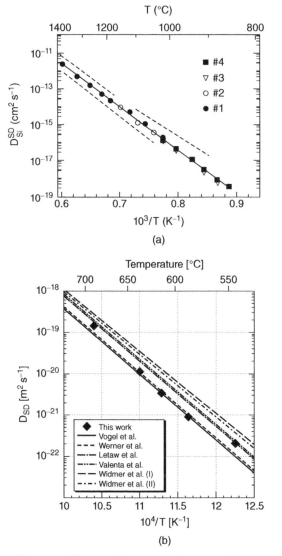

(a)

(b)

Figure 2.5 *(a) Self-diffusion coefficients of n-type and p-type silicon samples. Bracht et al., 1998, [56]. Reproduced with permission from American Physical Society. (b) Self-diffusion coefficients of germanium samples) Fuchs et al., 1995, [54]. Reproduced with permission from American Physical Society*

The interchange processes that are supposed to occur at the beginning of a long-range diffusion process may be described by the following equilibria

$$M_s \rightleftharpoons M_i + V \tag{2.45}$$

$$M_s + I \rightleftharpoons M_i \tag{2.46}$$

where the subscripts s and i label substitutional and interstitial positions in the host lattice.

Reaction (2.45) running from left to right is assumed to be sufficiently fast to establish a local equilibrium among any concentration of M_i almost instantaneously [63], being independent of any external vacancy source, while the concentration of M_i according to reaction (2.46) running from left to right, depends on self-interstitial supply from local sources or sinks. In the presence of a high density of interstitial sources (or sinks), the concentration c_I of self-interstitials corresponds to its equilibrium concentration c_I^{eq}, as vacancies and self-interstitials are maintained at their equilibrium concentration by Eq. (2.47)

$$V + I \rightleftharpoons O \tag{2.47}$$

where O stands for the unperturbed lattice, independently of the amount of metal dissolved, provided the kinetics of the processes working as sources or sinks for defects are very fast.

The process of Eq. (2.45) is known as the dissociative or Frank and Turnbull (FT) mechanism, whereas that under Eq. (2.46) is known as the kick-out or Gösele mechanism.

In its original formulation [60], the FT mechanism postulates for the long-range transport of Cu in *dislocated* Ge,[14] a sequence of diffusion steps of interstitial Cu_i, correlated to the interchange of substitutional (s) and interstitial (i) species mediated by vacancies which, as in the case of metals, are the predominant defects in Ge.

In the original work [60] it is assumed that the equilibrium concentration of substitutional copper Cu_s is higher than that of the interstitial species. It is assumed also that the dissolution of Cu in Ge from an external Cu source occurs with the formation of an excess of interstitial copper Cu_i, which then reaches its equilibrium concentration with respect to substitutional Cu_s

$$Cu_i + V \rightleftharpoons Cu_s \tag{2.48}$$

up to the concentration ratio $\alpha = \frac{[Cu_s]}{[Cu_i]} \approx 100$ at 710 °C, leading to a transient undersaturation of vacancies.

Local equilibrium is re-established thanks to the supply of vacancies from the reaction interface, which is an infinite source of vacancies, though not necessarily infinitely fast, even if the reaction is carried out at sufficiently high temperatures.

It should be noted that this reaction step is common to every case of substitutional solid solution formation from solid reactants, but not explicitly mentioned when discussing in Chapter 1 the thermodynamics of solid solution formation.

As the Cu solubility in Ge, taken as that of the substitutional species Cu_s,[15] is constant at constant temperature and the concentration ratio $\alpha = \frac{[Cu_s]}{[Cu_i]}$ is determined by the equilibrium constant

$$K_V = \frac{[V_{Ge}][Cu_i]}{[Cu_s]} = \alpha^{-1}[V_{Ge}] \tag{2.49}$$

---

[14] Dislocations behave as powerful vacancy sources and sinks.

[15] The concentration of the interstitial species is negligible in comparison with that of the substitutional species.

the concentration of interstitial copper $[Cu_i] = K'_V [V_{Ge}]^{-1}$ depends on the reciprocal of the vacancy concentration.

In most general terms, as soon as the system reaches equilibrium conditions, that is at time zero of the diffusion process, the following relation holds between the initial concentration of vacancies C^0_V, their equilibrium concentration C^{eq}_V, the equilibrium concentration of the substitutional metal C^{eq}_s and the actual concentration of the substitutional metal C^b_s in the bulk of the diffusion source[16] [62].

$$C^0_V(t = 0) = C^b_s \left(1 + \frac{C^{eq}_V}{C^{eq}_s} \right) \qquad (2.50)$$

Equation (2.50) states the starting conditions of the diffusion process and shows that the minimum value of the initial concentration of vacancies is that of the metal dissolved in a substitutional position, for values of the metal solubilities higher than the equilibrium concentration of vacancies. Values of the initial vacancy concentration close to their equilibrium concentration may only be obtained for diffusing metals having solubilities close to the equilibrium concentration of vacancies.

This condition discriminates between experiments capable of leading to the determination of the *effective* or *equilibrium* concentration of vacancies, in the absence of additional thermodynamic constraints, such as those due to vacancy complex formation, which will be illustrated in the next section.

The kick-out mechanism, proposed by Gösele *et al.* [63] for gold diffusion in silicon, postulates that self-interstitials I mediate the interchange of the gold between substitutional Au_{Si} and interstitial Au_i positions

$$Au_i + I \rightleftharpoons Au_{Si} \qquad (2.51)$$

Different from the case of Ge, self-interstitials in silicon are the dominant defects but, as for Cu in Ge, the Au concentration is higher in substitutional positions whereas the mobility of Au_i interstitials is higher than that of substitutional Au_s.

In full analogy with the former process, the local equilibrium conditions between the concentrations of Au_i, Au_{Si} and Si self-interstitials involved in the process are described by an equilibrium constant

$$K_i = \frac{[Au_i]}{[Au_{Si}][I]} \qquad (2.52)$$

and the dissolution of Au from the metal source[17] generates a supersaturation of self-interstitials.

To get information on the equilibrium concentration of vacancies and self-interstitials from metal diffusion measurements, the condition that the defect species at their equilibrium concentration effectively control the process rate should be necessarily and preliminarily fulfilled.

Under this hypothesis and assuming that the defect generation/recombination reactions are given by Eqs. (2.32) and (2.33), the description of the diffusion process leads to a system

---

[16] That is the layer of diffused Cu at the beginning of the experiment.

[17] Au from a metal source dissolves in Si as interstitial Au_i but the equilibrium is reached for Au in substitutional positions, with the contemporaneous emission of an excess of self-interstitials.

of partial differential equations [63, 65] involving substitutional and interstitial impurities, silicon self-interstitials and vacancies.

The following set of equations holds for the dissociative, Frank–Turnbull mechanism

$$\frac{\delta C_s}{\delta t} = v_s[k_V C_i (C_V - C_s)] \tag{2.53}$$

$$\frac{\delta C_i}{\delta t} = D_I \frac{\delta^2 C_i}{\delta x^2} - \frac{\delta C_s}{\delta t} \tag{2.54}$$

$$\frac{\delta C_V}{\delta t} = D_V \frac{\delta^2 D_V}{\delta x^2} - \frac{\delta C_s}{\delta t} + A_V \left(1 - \frac{C_V}{C_V^{eq}} \right) \tag{2.55}$$

and the following ones for the kick-out process

$$\frac{\delta C_s}{\delta t} = -v_i[k_i C_i (C_s - C_i)] \tag{2.56}$$

$$\frac{\delta C_i}{\delta t} = D_I \frac{\delta^2 C_i}{\delta x^2} - \frac{\delta C_s}{\delta t} \tag{2.57}$$

$$\frac{\delta C_I}{\delta t} = D_I \frac{\delta^2 D_I}{\delta x^2} + \frac{\delta C_s}{\delta t} - A_I \left(\frac{C_I}{C_I^{eq}} - 1 \right) \tag{2.58}$$

Here v_s is the dissociation frequency of the substitutional impurity in an interstitial impurity and a vacancy, v_i is the frequency of dissociation of the interstitial impurity in a substitutional impurity and a self-interstitial, k_V is the equilibrium constant for the FT reaction (2.50) and k_i that for the kick-out mechanism (Eq. (2.51)). D_V and D_I are the diffusion coefficients of vacancies and self-interstitials and further symbols are the concentrations of vacancies (C_V), self-interstitials (C_I), substitutional (C_s) and interstitial (C_i) impurities.

Once the actual and equilibrium concentration of the diffusing metal is suitably determined or calculated, the solution of the system of differential equations can be obtained by numerical integration, for the appropriate initial and boundary conditions, yielding the D_I and C_I^{eq} or D_V and C_V^{eq} values, within the temperature interval where the dissociation or the kick-out process is followed.

For the FT mechanism two limiting conditions can hold [63], which simplify the solution of the system of partial differential equations.

It is possible to assume either a large or a small density of vacancy sources within the sample used in the measurement. In the case of a high density of vacancy sources

$$C_V = C_V^{eq} \tag{2.59}$$

and

$$D_{eff} = \frac{D_i}{1 + k_V C_V^{eq}} = D_i \frac{C_i^{eq}}{C_i^{eq} + C_s^{eq}} \tag{2.60}$$

where D_i is the diffusion coefficient of the metal impurity, k_V is the equilibrium constant of the FT reaction and D_{eff} is the *effective* diffusion coefficient of the metal diffusing impurity, which does not depend on the vacancy diffusivity as vacancies are available everywhere.

In the second case, vacancies are made available by the dissociation mechanism, the long-range metal transfer is due to their diffusivity, the metal concentration is constant

$$C_i = C_i^{eq} \tag{2.61}$$

and

$$D_{eff} = \frac{D_V}{1 + k_V C_i^{eq}} = D_V \frac{C_V^{eq}}{C_V^{eq} + C_s^{eq}} \tag{2.62}$$

when $C_s^{eq} \ll C_V^{eq}$.

For the kick-out process, the assumption of a large density of self-interstitial sinks brings their concentration locally uniform and makes it possible to assume that

$$C_I = C_I^{eq} \tag{2.63}$$

and

$$D_{eff} = \frac{D_i \, k_I C_I^{eq}}{1 + k_I C_I^{eq}} = D_i \frac{C_i^{eq}}{C_i^{eq} + C_s^{eq}} \tag{2.64}$$

which is equivalent to the solution (2.60). Therefore, in the case of a large density of self-interstitial sinks, the kick-out and the Frank–Turnbull mechanisms predict the same effective diffusivity.

Instead, for the condition valid for a high initial concentration of substitutional metal, which is generally satisfied for almost all metals used in diffusion measurements, the effective diffusion coefficient is directly proportional to the diffusion coefficient of the self-interstitials [63].

$$D_{eff} = D_I \frac{C_I^{eq} C_s^{eq}}{C_s^2} \tag{2.65}$$

This condition might not be fulfilled at temperatures lower than 850 °C, because the Frenkel process (Eq. (2.47)), which establishes the equilibrium bulk concentration of self-interstitials, becomes ineffective below this temperature [50, 65].

A significant example of the use of these procedures is the determination of the diffusivity and the equilibrium concentration of vacancies and self-interstitials using Si-self-diffusion and Zn-diffusion measurements. The case of silicon is particularly favourable because the measured self-diffusion values are shown to be almost independent of doping over a wide range of temperatures [56] and the direct exchange could be neglected.

It is, therefore, possible to apply Eq. (2.44) to a system populated only by neutral point defects, writing

$$D_{Si}^{SD} = f_I \, x_I^{eq} D_I + f_V \, x_V^{eq} D_V \tag{2.66}$$

where f_I and f_V are the calculated correlation coefficients for the interstitialcy and vacancy mechanism in the diamond lattice (0.74 and 0.5, respectively).

Using for the $x_I^{eq} D_I$ term the results obtained with Zn diffusion measurements [50]

$$x_I^{eq} D_I = 2980 \exp -\frac{4.95 \text{ eV}}{kT} \tag{2.67}$$

and for the self-diffusion coefficient the following equation, which fits its temperature dependence over the whole range of measurements [56], see Figure 2.5a

$$D_{Si}^{SD} = 530 \pm 250 \exp{-\frac{-4.75 \pm 0.04 \text{ eV}}{kT}} \tag{2.68}$$

the $x_V^{eq} D_V$ term is obtained

$$x_V^{eq} D_V = 0.92 \exp{-\frac{4.14 \text{ eV}}{kT}} \tag{2.69}$$

The equilibrium concentrations of point defects are then extracted from the solution of the system of partial differential Eqs. (2.54)–(2.59) [50, 56], together with the individual values of the diffusion coefficients

$$c_V^{eq} \approx 1.4 \cdot 10^{23} \exp{-\frac{2.0 \text{ eV}}{kT}} \ (\text{cm}^{-3}) \tag{2.70}$$

$$D_V \approx 3 \cdot 10^{-2} \exp{-\frac{1.8 \text{ eV}}{kT}} \ (\text{cm}^2\text{s}^{-1}) \tag{2.71}$$

$$c_I^{eq} \approx 2.9 \cdot 10^{24} \exp{-\frac{3.18 \text{ eV}}{kT}} \ (\text{cm}^{-3}) \tag{2.72}$$

$$D_I \approx 5.9 \cdot 10^{1} \exp{-\frac{1.77 \text{ eV}}{kT}} \ (\text{cm}^2\text{s}^{-1}) \tag{2.73}$$

It is possible to see, by displaying Eq. (2.71) and (2.73) in an Arrhenius plot, that the mass transfer process is mediated by vacancies at low temperatures and by self-interstitials above 890 °C, in good agreement with the results of computer simulation experiments [66], which show, see Figure 2.6, [66, 67], that interstitials are more mobile than vacancies at temperatures in excess of 900 °C.

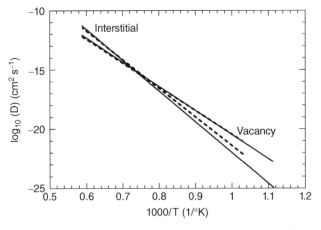

Figure 2.6 *Comparison between the calculated and experimental self-diffusion coefficients of vacancies and self-interstitials diffusivity in silicon. Solid curves, theoretical estimates, dashed curves, experimental data. Tang et al., 1997, [66]. Reproduced with permission from American Physical Society*

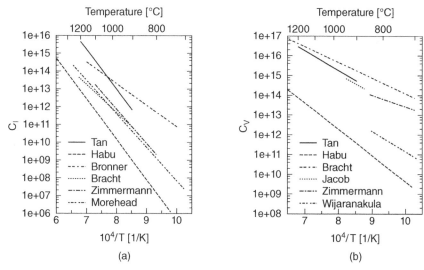

Figure 2.7 *Comparison of literature data for the equilibrium concentration (in atoms cm^{-3}) of silicon self-interstitials (a) and vacancies (b) in silicon, [23, 47, 50, 61, 65, 68–71]. Zimmermann, 1998, [68]. Reproduced with permission from Trans Tech Publications*

From a comparison of the cumulative literature data concerning interstitials and vacancy concentration from metal diffusion measurements in silicon [68], displayed in Figure 2.7 [23, 47, 50, 61, 65, 68–71], it can be recognized immediately, however, that the large spread of experimental data makes the evaluation of the true equilibrium concentration of defects, and especially that of vacancies, at least questionable.

As the problem concerning interstitials is less challenging, because the experimental results concerning their formation enthalpies correspond rather well with the calculated values of the <100> split interstitial, quite independently of the modelling approximation used, see Table 2.4, the main question involves vacancies.

As an attempt to account for this issue, it should be questioned, first, whether the problem lies in the experimental determination of the diffusion profiles or is intrinsic to the metal–defect interchange process itself.

Part of the problem may arise from inaccurate experimental measurements, but not for the results obtained using, as an example, DLTS, with its lower limit of trap concentration determination lying in the 10^{10}–10^{14} cm^{-3} range [68], well below the experimental vacancy concentrations.

Therefore, the problem could be intrinsic to the interchange process. In fact, the dissolution of a foreign atom stable both in substitutional and interstitial configurations generates a non-equilibrium concentration of point defects. This effect could be dramatic when the solubility of the diffusing impurity at the temperature of the experiment is much larger than the equilibrium concentration of the defects (vacancies or self-interstitials) [50]. Restoration of equilibrium conditions may occur only if the homogeneous defect generation/recombination process is sufficiently fast or external defect sources and sinks (surfaces and dislocations) [52] are available to allow fast defect re-equilibration.

Table 2.4 *Formation enthalpies of vacancies and self-interstitials in silicon*

Type of Si	Theory						Experimental			
Technique	EDIP	TBMD	LDA	LDA	FP-MD	GGA	FZ (undoped) Positron annealing	FZ, B-doped Ir-diffusion	FZ Au diffusion	FZ (undoped) Zn diffusion
Experimental temperature (°C)							Rt – 1175 °C	875–1050	1000–1100	870–1208
Vacancy	3.47	3.97	4.36	—	4.1	—	—	—	—	—
Relaxed	3.22	—	3.17	—	—	3.66 (LDA)	—	—	—	2.0
Experimental	—	—	—	—	—	—	3.6 ± 0.2	2.44 ± 0.15	—	—
Interstitial T	4.05	4.39	—	3.43	—	3.25 (LDA)	—	—	—	—
Interstitial H	4.16	4.93	—	3.31	—	3.47	—	—	—	—
Interstitial split	3.35	3.8	—	3.31	3.3	3.44	—	—	—	—
Experimental	—	—	—	—	—	—	—	—	3.8	3.18
References	[72]	[66]	[73]	[34]	[74]	[75]	[40]	[64]	[71]	[50]

EDIP = Eenvironmental dependent interatomic potential; TBMD = Tight binding molecular dynamics; FP – MA = first principle molecular dynamics; LDA = local density approximation and GGA = General gradient approximation.

A problem would arise again, however, at temperatures lower than 850 °C, when the defect generation/recombination rate through the homogenous Frenkel mechanism is too slow.

One further reason, of a quite different nature, has been postulated [76] when discussing the vacancy interaction/aggregation phenomena during the cooling stage of a Si ingot grown with the Czochralski (CZ) process (see Chapter 4, Section 4.2.1).

This topic will be discussed in depth at the end of Section 2.3.3, but it can be anticipated here that the vacancy concentration in CZ silicon could never correspond to the equilibrium concentration in a virtually uncontaminated material, even when measurements are carried out in full equilibrium conditions.

In fact, oxygen, the main contaminant of CZ grown silicon ($[O_i] \approx 10^{18}$ cm^{-3}) interacts with vacancies with the formation of the VO complex, the so-called A centre, well known from spectroscopic measurements [77]

$$V_{Si} + O_i \rightleftharpoons \text{V-O} \tag{2.74}$$

shifting the equilibrium of reaction (2.74) well to the right, at least up to temperatures of the order of 1000 °C, above which the complex dissociates.

Above this temperature, undissociated vacancies may aggregate in vacancy clusters that transform in voids, truly 3D defects, which are further vacancy sinks.

This objection could be significant also, even if at a lesser extent, for measurements carried out using float zone FZ silicon samples, as in FZ-Si the oxygen concentration is only 2 orders of magnitude lower ($[O_i] \approx 5 \cdot 10^{16} - 10^{17}$ cm^{-3}) than in CZ-Si (see Chapter 4), but close or above the equilibrium vacancy concentration at the temperature of the measurements.

Leaving open, for the moment, the question about the true nature of the vacancy species involved in metal diffusion processes, it is worth mentioning here that a large difference between the experimental and theoretical equilibrium vacancy concentration remains, even using the most appropriate tool to measure the concentration of equilibrium vacancies, which is the Ir in-diffusion [64].

The singularity of Ir diffusion in Si is that it might occur both via self-interstitials (kick-out mechanism) and vacancies (FT mechanism)

$$Ir_s + I \rightleftharpoons Ir_i \tag{2.75}$$

$$Ir_s \rightleftharpoons Ir_i + V \tag{2.76}$$

but the crucial advantage is that Ir, compared to other metals, such as Au or Pt, has a solubility two orders of magnitude lower, while it shares with other metals the larger mobility of its interstitial species Ir_i.

The equilibrium solubility of Ir is, therefore, very close to the Si-vacancy concentration in thermal equilibrium C_V^{eq}, which, as already seen, is the condition to avoid the shift of the vacancy concentration from the equilibrium value. As a consequence, Ir diffusion into Si should be directly sensitive to the actual equilibrium values of the silicon vacancy and self-interstitial concentration.

From the fit of the diffusion profile of Ir, accounting for both the kick-out and the dissociative mechanism, a value of the enthalpy of formation of vacancies of 1.99 ± 0.2 eV is obtained, with an Arrhenius prefactor of $2.63 \cdot 10^{22}$ cm^{-3}.

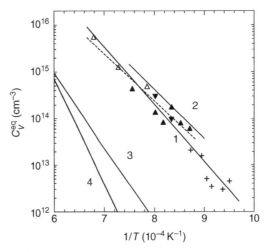

Figure 2.8 *Experimental values of vacancy concentration in Si deduced from literature.* Δ *from Ir diffusion in FZ silicon [79].* ▲▼ *from Ir diffusion in FZ Si [64]* + *from Si self-diffusion under proton irradiation [78]. Solid line 1 fits all data points (Eq. (2.77)). Solid line 2: fits Zn outdiffusion [80]. Solid lines (3) and (4): from studies of defect structures after crystal growth [76, 81]. Lerner & Stolwijk, 2005, [64]. Reproduced with permission AIP Publishing*

Combining this result with previous literature ones [78], the temperature dependence of the vacancy concentration is given by

$$C_V = 1.4\ 10^{24} \exp{-\frac{2.44(\text{eV})}{kT}}$$ (2.77)

from which one obtains a value of the equilibrium vacancy concentration $C_V^{eq} = 4 \cdot 10^{16}$ (cm^{-3}) at the melting point of Si. It is apparent that Ir measurements lead to vacancy concentration values not very different from those discussed previously, as can be seen in Figure 2.8 [64].

A question still remains open, therefore, concerning the difference between the results of experimental measurements carried out with metal diffusion experiments [64], those obtained by indirect structural analysis of defect formation during CZ silicon growth and those resulting from theoretical simulations carried out using quantum mechanical or first principle calculations.

Apparently, the large difference between results obtained with metal diffusion measurements [78–80] and from studies of defect structures during CZ-Si growth [76, 81], see again Figure 2.8, gives quantitative evidence that structural studies are unable to determine the equilibrium concentration of vacancies but only the *effective* vacancy concentrations involved in oxygen aggregation phenomena.

A preliminary suggestion about the origin of the difference between the results of metal diffusion measurements and those arising from simulation experiments may be obtained from an inspection of Table 2.4, which reports a critical selection of theoretical [34, 66, 72–75] and experimental results [40, 50, 64, 71], while a full compilation of data can be found in Refs. [82, 83].

It could be noted, first, the substantial agreement between the formation enthalpy of vacancies evaluated theoretically and the experimental values obtained with PAS, a technique which is selective for vacancies and voids in solid hosts and is frequently used for silicon and compound semiconductors [84–86].

The formation energies of vacancies obtained with metal diffusion measurements are, instead, systematically different to the theoretical values. The hypothesis could be advanced that the difference depends on the fact that the simulation experiments are carried out on perfect, impurity-free, virtual samples, whereas measurements are carried out, even using the best FZ samples, on materials containing native or unintentionally added impurities, which may interact with vacancies.

It will be shown in the next section that these deviations may be reconciled by considering the interaction of point defects with impurities, such as carbon, oxygen, hydrogen and dopants, leading to the formation of complex species which affect the thermodynamics and the kinetics of metal diffusion processes.

2.3.3   Effect of Defect–Defect Interactions on Diffusivity: Trap-and-Pairing Limited Diffusion Processes

One of the most straightforward evidences of defect–defect or defect–impurity interactions in solids was deduced from charge transport and/or diffusivity experiments in ionic solids, which was dealt with in Section 2.2. When interaction occurs, not only does the isothermal conductivity/concentration plot of an ionic solid doped with aliovalent impurities deviate from linearity, but the depth profiles of the diffusing impurity in any material also exhibit a variety of shapes, different from the standard complementary error function [87].

In the most general case, common to ionic solids and semiconductors, one could observe a change of slope in the Arrhenius plot of the diffusion coefficient, a drastic difference in the diffusivity or mobility of a specific impurity in a pure and doped semiconductor and some scattered behaviour in the case of samples of the same material but of different origin, as would be the case of CZ-Si and FZ-Si.

However, the most direct indication of the possible occurrence of defect–impurity interactions could be deduced from the strong deviations of the experimental values of the defect formation enthalpies and of defect concentrations from the computed ones, which have been observed in the previous section and which are directly related to the thermodynamics of interaction.

Limiting the attention, here, to the effects of interactions between defects and impurities on their mobilities, it could be supposed that their diffusivities are limited by pairing- or trap-limited processes, consequent to the formation of stable, almost immobile complexes, which would result in a delay of diffusion (shallow traps) or in the formation of a diffusion barrier (deep traps).

The lattice defect at which the mobile species are trapped may be another atomic defect, an impurity, a dislocation or an external or internal interface (the surface of a precipitate, of an inclusion or a grain boundary). In what follows, only point defect–impurity interactions in the bulk of a semiconductor will be taken into account, because point defect interactions with extended defects will be discussed in the next chapter.

Typical examples of defect–impurity interactions are given by vacancies in Si, which tend to form pairs with many species, such as interstitial oxygen, substitutional donors

(P, As, Sb) and acceptors (B, Al). Self-interstitials-impurity interactions also occur, as is the case of their interaction with carbon [87, 88], which leads to an effective self-interstitial undersaturation consequent to the formation of an interstitial carbon species C_i

$$C_s + I \rightleftharpoons C_i + Si_{Si} \tag{2.78}$$

The interstitial carbon C_i is a fast-diffuser, but its diffusivity is slowed by a pairing reaction

$$C_s + C_i \rightleftharpoons [C_s - C_i] \tag{2.79}$$

where the $[C_s - C_i]$ complex is the immobile species.

The C–I interaction is also the origin of the formation of a C–I complex [88], which might be supposed to be the embryo of the SiC phase which segregates in supersaturated solutions of C in Si, which will be discussed in Section 2.3.4.2.

The simplest approach to deal with these pairing reactions would be the trap model of Wert and Frank (WF) [89], aimed at describing the diffusion of a mobile species A in a lattice presenting a distribution of immobile species B capable of interacting with A to form a low mobility, complex species AB.

The basic features of the model are illustrated schematically in Figure 2.9, which shows the configurational potential energy diagram of a system where the depth of a regular lattice well is E_M, while that of the trap is $E_M + E_B$, with E_B as the binding (i.e. formation) energy of the AB complex. Therefore, whereas the transfer rates $\omega^0 \approx v^0 \exp -\frac{E_M}{kT}$ for a jump of an atom from a normal lattice position in the trap or in an empty regular site, where v^0 is the frequency factor at a normal lattice site, are ruled by the same activation energy E_M, the activation energy for the re-emission from the trap is, instead, given by the sum $E_M + E_B$. Consequently, the transfer rate for the un-trapping process is $\omega^t \approx v^t \exp -\frac{E_M + E_B}{kT}$, where v^t is the frequency factor at a trap site. In turn, the binding energy E_B might have a coulombic, covalent or mixed character.

The macroscopic modelling of trapping and un-trapping processes leads to an effective diffusion coefficient D_{eff}, given by the following equation[18]

$$D_{eff} = \frac{D^0}{1 - 2x_T + 2x_T(v^0/v^t)\exp\frac{E_B}{kT}} \tag{2.80}$$

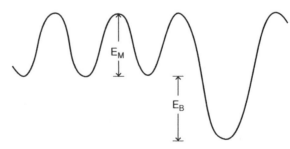

Figure 2.9   *Potential energy configurational diagram for a diffusion, trapping and un-trapping process*

---

[18] See Ref. [89] for details.

where D° is the diffusion coefficient in the absence of traps and x_T is the atomic fraction of traps.

The calculated influence of the trap binding energy E_B and the concentration of traps on the diffusion coefficient is displayed in Figure 2.10 [89].

The values of D_{eff} coincide with D° at a temperature T_{diss} at which the traps could be thermally emptied and T_{diss} depends on both the trap depth and trap concentration

$$T_{diss} = \frac{E_B}{k} \ln \frac{2x_T(v^\circ/v^t)}{1-2x_T} \tag{2.81}$$

Pairing reactions between impurities were also discussed in the early 1950s of by Reiss, Fuller and Morin (RFM) [89], dealing with the interactions of the positively charged interstitial lithium with a family of acceptors in germanium.[19] The theoretical treatment of pairing processes between oppositely charged species was carried out applying the mass action law under the assumption that the statistics satisfied by electrons and holes (which are necessarily reaction partners) also remain classical.

It can be demonstrated [90] that the effective diffusion coefficient D_{eff}, for $N_d \ll N_a$ takes the form

$$D_{eff} = \frac{D^\circ}{1 + K_{eq}N_a} \tag{2.82}$$

where N_d is the donor concentration, N_a the acceptor concentration and K_{eq} the equilibrium constant for the pairing reaction, given by

$$K_{eq} = \frac{[P]}{[N\text{-}P]^2} = 4\pi \int_a^b r^2 \exp \frac{V(r)}{kT} dr \tag{2.83}$$

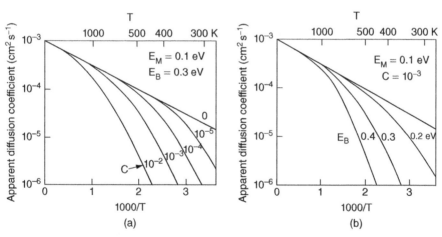

Figure 2.10   *Effect of trap concentration C on the temperature dependence of the effective diffusion coefficient, for a specific value of the trap concentration (a) effect of trap binding energy E_B on the effective diffusion coefficient, for a typical value of the activation energy E_M of the intrinsic diffusivity. The trap concentration is given in molar fractions (b). Pizzini, 2001, [89]. Reproduced with permission from Trans Tech Publications Ltd*

---

[19] Reiss *et al.* [90] were the first to show that the mass action law could be applied to chemical reactions occurring in semiconductor hosts, assuming that semiconductors, particularly silicon and germanium, provide a reaction medium similar to water.

where [P] is the concentration of pairs, [N-P] is the concentration of unpaired species of one sign, a is the distance of closest approach and b is the capture radius of the pair interaction potential $V(r)$, assumed to be given by a purely coulombic term $V(r) = \frac{z^2 q^2}{R_C}$, with $R_C = a$.

It has been demonstrated [91] that the results of WF and RFM models coincide for deep trapping conditions, using for K_{eq} in Eq. (2.84) the ratio

$$K_{eq} = \frac{\tau_c^{-1}}{N_a \tau_{diss}^{-1}} = \frac{D^\circ R_C}{\tau_{diss}^{-1}} \tag{2.84}$$

with τ_c and τ_{diss} being the temperature-dependent time constants of the association and dissociation processes and R_C the capture radius of the mobile donor by the immobile acceptor.

To show the ability of trapping models in the analysis of diffusion processes, one could consider, as a first example, the results relative to the interaction between lithium, which behaves as a donor, and various acceptors in germanium [90]. In this case one would expect that the effect of pairing on lithium diffusivity would be independent of the specific chemical nature of the interacting acceptor species, if only electrostatic forces were involved in the pairing process.

Figure 2.11 [90] confirms the expected onset of deviation of the diffusivity of Li in acceptor-doped Ge samples from the intrinsic values (dotted curve) but shows also an appreciable difference between the effective diffusivities of Li in Ga- and Zn-doped Ge. In fact, in spite of a factor of ten higher concentration of Ga, even normalized for the different charge states of Ga ($z = 1-$) and Zn ($z = 2-$), the deviations associated with Zn-pairing are much more relevant, indicating that short-range covalent binding prevails on electrostatic binding.

The case of Cu diffusion in B-doped Si, where Cu_i also behaves as a donor, is different [91]. Here large deviations are observed from the intrinsic behaviour over the whole range of B-doping (see Figure 2.12) [91], but electrostatic binding seems to prevail, in spite of the at least partially covalent character of the metal–acceptor bonds of the copper complexes (CuB, CuAl, CuGa, CuIn) [93], for which the binding energy increases from 0.61 eV for CuB to 0.7 eV for CuIn.

The experimental points [91, 92] fit well over a very wide range of B-concentrations (N_a from 10^{14} to 10^{20} cm^{-3}) with the values calculated (solid lines in Figure 2.12) assuming interaction purely due to a Coulomb screened potential, with an electrostatic capture radius $R_C = \frac{z^2 q^2}{4\pi\varepsilon\varepsilon_0 kT}$ amounting to several nanometres and, therefore, much larger than the covalent capture radius.

Another case of a trap-limited diffusion process is that involving hydrogen in B-doped Si and Ge. In B-doped Si and Ge, H forms stable B-H bonds and a common consensus exists on the modelling features which could be adopted [94–98] to account for bonding.

A result which can be deduced from these studies is that the effective diffusion coefficient, in the presence of a random distribution of unsaturated (or partially filled) point-like traps, consisting of BH complexes, is given by a variant of Eq. (2.82)

$$D_{eff} = \frac{D^\circ}{1 + rx_B \exp\frac{E_B}{kT}} \tag{2.85}$$

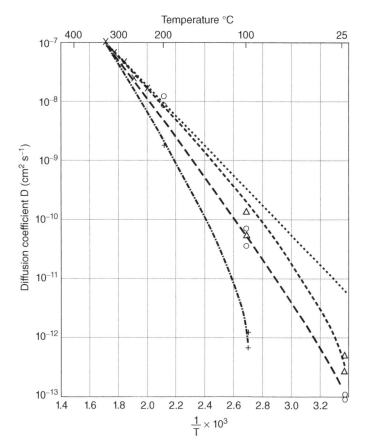

Figure 2.11 *Effect of In- and Ga- doping on the lithium diffusivity in Ge:* ○ *In-doped Ge,* △ *Ga-doped Ge, + Zn-doped Ge (the dotted curve displays the Li diffusivity in undoped Ge). Reiss* et al., *1956, [90]. Reproduced with permission from John Wiley & Sons*

where r is a factor which depends on the changes in entropy (configurational and vibrational) associated with the trapping of an H atom at an acceptor site, x_B is the atomic fraction of boron in silicon and the bonding energy E_B is $\approx 0.6 \, eV$.

This conclusion is confirmed by the results of experiments carried out on heavily B-doped ($10^{19} \, cm^{-3}$) Si samples, using deuterium instead of hydrogen [95], which show (Figure 2.13) [95] that trapping occurs and that the binding energy of the BH complex, is $0.61 \pm 0.1 \, eV$, in good agreement with *ab initio* studies.

The diffusion of vacancies in CZ silicon is analysed as a final example, to complete the discussion carried out in the last section concerning the difference existing between the theoretical and experimental values of the formation enthalpy of vacancies. It has already been shown that vacancies interact with oxygen with the formation of a stable VO complex, see Eq. (2.74), with a calculated binding energy E_B ranging between 1.4 and 1.8 eV [77, 99–102] and a dissociation enthalpy $\Delta H_{VO}^{diss} = 1.85 \, eV$ [101].[20]

---

[20] Binding and dissociation enthalpy are homologous terms, but differently calculated.

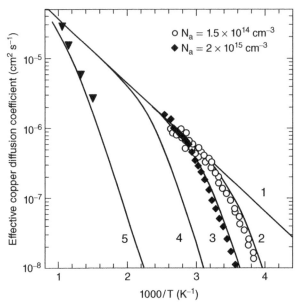

Figure 2.12   *Comparison between the calculated values of the effective diffusion coefficient of copper in silicon at different boron doping levels N_a (solid lines of which line 1 refers to undoped Si), and the experimental data:* $\bigcirc N_a = 1.5 \cdot 10^{14} \ cm^{-3}$ *[91],* $\blacklozenge N_a = 2 \cdot 10^{15} \ cm^{-3}$ *[91],* $\blacktriangledown N_a = 5 \cdot 10^{20} \ cm^{-3}$ *[92]. Istratov, et al., 1998, [91]. Reproduced with permission from American Physical Society*

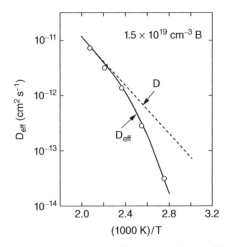

Figure 2.13   *Temperature dependence of the effective hydrogen diffusion coefficient in heavily B-doped silicon (----); undoped Si (O), experimental points (solid line calculated values). Herrero et al., 1990, [95]. Reproduced with permission from American Physical Society*

As a consequence of the VO complex formation, the density of *free* vacancies is orders of magnitude lower than that of the bound vacancies in CZ silicon, where the concentration of oxygen is around 10^{18} cm^{-3}. The same conclusion holds for FZ silicon, where the concentration of oxygen ranges between $5 \cdot 10^{15}$ and 10^{16} cm^{-3} and the formation of VO complexes has been demonstrated by IR measurements on transmutation doped[21] FZ silicon [103].

The hypothesis was advanced [101] that the diffusion of vacancies involves two mass transport mechanisms, involving interstitial oxygen and VO complexes

$$O_i(a) + V \rightleftharpoons VO \rightleftharpoons V + O_i(b) \tag{2.86}$$

$$VO(a) \rightleftharpoons O_i + V(a) \rightleftharpoons V + O_i(b) \rightleftharpoons VO(b) \tag{2.87}$$

where (a) and (b) are the initial and final positions of a species involved in the transport process in a space coordinates diagram. It is also supposed that in the VO complex the vacancy is trapped close to the interstitial oxygen forming a bond-centre structure.

The formation of an isolated VO species leads in the case of the mechanism of Eq. (2.86) to the diffusion of a free vacancy from (a) to (b) with O_i transfer, whereas the dissociation of VO in the case of the mechanism of Eq. (2.87), leads to the diffusion of a free vacancy from (a) to (b), without O_i transfer. As the (calculated) activation energies for oxygen and VO diffusion, are very close ($H^m_{O_i} = 2.02$ eV and $H^m_{VO} = 1.98$ eV), both processes occur simultaneously.

A mechanism assuming the diffusion of trapped vacancies alone, however, does not account for the experimental values of the diffusivities of vacancies in CZ silicon. It is, in fact, shown in Figure 2.14 [100] that a concentration of 10^{20} cm^{-3} of trapped vacancies would be needed to account for the experimental diffusivities (solid lines in the figure), much in excess of the actual oxygen concentration in CZ Si [100].

To set the question, it is first worth remembering that free vacancies are much more mobile than bound vacancies (see Table 2.5) [66, 82, 101, 104] and that the enthalpy of formation of VO complexes amounts to about 1.8–2.0 eV[22] in both Si and Ge (see Table 2.6 [99]).

It is, therefore, quite reasonable to suppose that almost all the vacancies, at least in CZ silicon, are bound as VO complexes, not to assume that bound vacancies are the diffusing species.

Let us then consider that, in the presence of oxygen impurities, the diffusion environment contains both bound vacancies VO and ZnO complexes.

It is, however, known that ZnO complexes dissociate at $T > 700\,°C$ [105], whereas the VO complexes are stable in Si (and Ge) up to an estimated dissociation temperature around $1050\,°C$ [106], with a calculated dissociation enthalpy ΔH^{diss}_{VO} of 2 eV [99–101].

Above $700\,°C$ and up to $900\,°C$,[23] the FT process should, therefore, occur with the simultaneous diffusion of Zn_i and vacancies originated from the dissociation of the VO complex. Consequently, the enthalpy term in the exponential of Eq. (2.70) is not the

---

[21] Neutron transmutation doping is a current technology used for P-doping of silicon at dedicated neutron reactors when very uniform, when relatively low doping levels on samples of relatively large size is required. The silicon isotope involved is $^{30}_{14}Si$ and the transmutation reaction is $^{30}_{14}Si \xrightarrow{n,\gamma} {}^{31}_{14}Si \xrightarrow{2.62h} {}^{31}_{15}P$.

[22] The calculated formation enthalpy of VO given in Ref. [99] has been subtracted for the formation enthalpy of interstitial oxygen.

[23] Figure 2.6 shows that above $900\,°C$ self-interstitials are the mobile species.

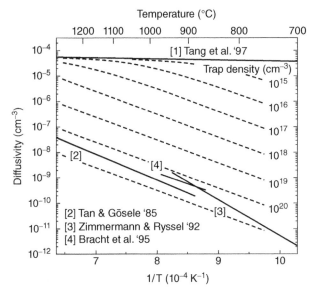

Figure 2.14   *Comparison of the calculated values of the diffusivity of trapped vacancies (dotted lines) with the experimental values of vacancy diffusivities using metals: [2] Ref. [3, 23] Ref. [4, 65] Ref. [8, 50] Ref. [66]. Casali et al., 2001, [100]. Reproduced with permission from AIP Publishing*

Table 2.5   *Migration enthalpy H_V^m values for free- and bound-vacancies in silicon*

H^m (eV)	Charge state	References
0.273	V^{2+}	[104]
0.32 ± 0.02	V^{2+}	[82]
0.190	V^{+}	[104]
0.45 ± 0.04	V°	[82]
0.355	V°	[104]
0.035	V^{-}	[104]
0.18 ± 0.02	V^{2-}	[82]
0.151	V^{2-}	[104]
≈ 0.1	V	[66]
1.98	VO	[101]

Table 2.6   *Calculated formation energies (in eV) of VO centres in silicon and germanium*

	VO	References
Si	2.04	[99]
Ge	1.84	[99]
Si	1.7	[100]
Si	1.85	[101]

formation enthalpy of free vacancies but the dissociation enthalpy of the VO complex to form free vacancies

$$VO \rightarrow V + O_i \qquad (2.88)$$

Possibly, however, what discussed here so far is not the conclusive explanation of problems involving mobility and formation enthalpies of vacancies. The question has been, in fact, recently reexamined by V.V. Voronkov,[24] who shows that in Pt diffusion experiments and vacancy-assisted oxygen precipitation three different vacancy species coexist, characterized by different mobilities. Further details on this issue have been also discussed recently by Bracht [107].

2.3.4 Light Impurities in Group IV Semiconductors: Hydrogen, Carbon, Nitrogen, Oxygen and Their Reactivity

These impurities represent the major, and almost unavoidable, contaminants of most crucible grown semiconductor crystals, as they arise from partial crucible dissolution or corrosion (oxygen from quartz crucibles, carbon from graphite crucibles), from gas phase contamination (nitrogen) and transport (carbon) and from trace amounts of water vapour or by the use of hydrogen as cover gas in the case of Ge growth. For thin epitaxial layers of Si-Ge alloys and for germanium and silicon microcrystalline and nanocrystalline films the hydrogen contamination originates from the plasma phase of silane (SiH_4) or germane (GeH_4), used as silicon or germanium precursors.

Hydrogen is, instead, deliberately used as a passivating agent of donor and acceptor dopants, metal impurities and point and extended defects, as will be shown in Chapter 5.

The basic properties of these impurities are already very well known, at least as contaminants of most semiconductors of technological interest, including compound semiconductors (GaAs, InP). For this reason, we will reserve attention in what follows to some relevant aspects of their chemical reactivity only.

2.3.4.1 *Hydrogen*

Atomic hydrogen is a common impurity in silicon and germanium [108–113]: it is a fast diffuser and easily interacts with impurities and native defects in both Ge and Si with a similar physics, although some differences are evident in its chemical behaviour.

As hydrogen is present in most of the chemicals used to clean, store, process and encapsulate semiconductor materials and devices, it is a universal contaminant.

For its reactivity and fast diffusivity hydrogen tends to occupy lattice positions where it can be trapped by a nearby impurity or a defect. Therefore, due to the formation of stable H-B and H-P pairs or H-O and H-C complexes [114], the equilibrium solubility of hydrogen should be experimentally determined in undoped, low oxygen and carbon silicon or germanium.

The temperature dependence of hydrogen diffusivity D_H and solubility S_H in silicon and germanium has been obtained from hydrogen permeation measurements [113] between

---

[24] Voronkov, V.V., Falster R. (2014) Multiple structural forms of a vacancy as evidenced by vacancy profiles produced by RTA *Phys.Status Solidi (B)* **251** 2179–2184.

1090 and 1200 °C for silicon

$$D_{\mathrm{H}}^{\mathrm{Si}}(\mathrm{cm}^2\mathrm{s}^{-1}) = 9.4 \cdot 10^{-3} \exp - \frac{0.48(\mathrm{eV})}{kT}; \; S_{\mathrm{H}}^{\mathrm{Si}}(\mathrm{cm}^{-3}) = 4.8 \; 10^{24} \exp - \frac{1.88(\mathrm{eV})}{kT} \quad (2.89)$$

and over the temperature range 800–910 °C for germanium

$$D_{\mathrm{H}}^{\mathrm{Ge}}(\mathrm{cm}^2\mathrm{s}^{-1}) = 2.7 \cdot 10^{-3} \exp - \frac{0.38(\mathrm{eV})}{kT}; \; S_{\mathrm{H}}^{\mathrm{Si}}(\mathrm{cm}^{-3}) = 3.2 \cdot 10^{24} \exp - \frac{2.3(\mathrm{eV})}{kT} \quad (2.90)$$

(see Figure 2.15).

It is apparent that the equilibrium solubility of hydrogen in Si and Ge is almost negligible, but it could be introduced at higher concentrations using ion implantation, with the potential formation of supersaturated solutions and second phase formation (molecular hydrogen, hydrogen bubbles or hydrogen platelets).

Once in-diffused or implanted, hydrogen outward diffusion and desorption is limited by a surface barrier [115, 116], whose calculated height is at least 1.4 eV. This surface barrier is due to the formation a surface layer of hydrogen dimers, stable in the 150–600 K temperature range. Surface-limited hydrogen out-diffusion is also at the origin of the establishment of a sub-surface hydrogen accumulation layer, which could be detected using SIMS measurements on deuterated[25] samples.

Surface barrier limited out-diffusion of hydrogen occurs also in the case of homoepitaxial diamond films [117], evidencing that the formation of a surface barrier to the out-diffusion of hydrogen is a phenomenon common to all members of Group IV semiconductors.

In the atomic form, hydrogen dissolves in Si as an interstitial species and is stable as a neutral (H_i°) deep-donor level or as a positively and negatively charged ($H_i^{\pm 1}$) interstitial,

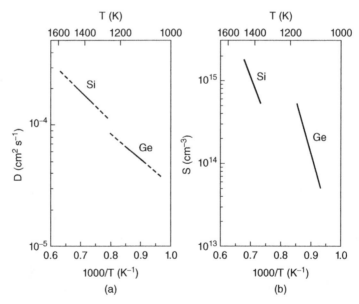

Figure 2.15 *Diffusivity (D) and solubility (S) of hydrogen in silicon and germanium. Courtesy of M. Stavola, Lehigh University, USA*

---

[25] SIMS analysis of hydrogen suffers from a high hydrogen background originating from vacuum oils which limits the the precision of its analysis: for this reason, deuterium is often used as an alternative to hydrogen for quantitative measurements of hydrogenation effects

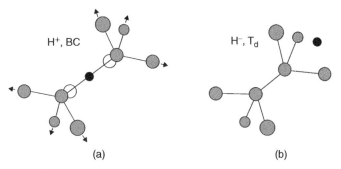

H⁺, BC

H⁻, T_d

(a)   (b)

Figure 2.16 *Isolated H_i in (a) a BC and (b) a tetrahedral T_d site. Courtesy of M. Stavola, Lehigh University, USA*

depending on the position of the corresponding gap levels with respect to the Fermi level, and thus on doping.

In p-type silicon, hydrogen takes a positive charge (H_i^{+1}) and in n-type silicon, a negative charge (H_i^{-1}).

Most of the theoretical calculations [118] indicate that the equilibrium conditions neutral H_{BC}^o and positively charged H_{BC}^+ in silicon sit in a bond centred (BC) position, with a symmetric outward relaxation of the Si atoms, getting a Si-H bond length around 1.6 Å. For the negatively charged H_T^- species, the antibonding (tetrahedral, T) position see Figure 2.16) is the most stable [113].

The activation energy for diffusion of H_{BC}^+ from BC to BC positions is 0.48 eV, while that of H_T^o from tetrahedral to tetrahedral sites is only 0.1 eV, making it extremely fast [119].

The formation of stable pseudomolecular hydrogen, via a dimerization reaction, can also occur. In the molecular form hydrogen occupies a T site, oriented along a <111> direction in both Si and Ge.

Hydrogen (and deuterium) may be introduced into silicon and germanium by several wet and dry techniques. Intentionally or unintentionally, it can be incorporated in the atomic form at the surface or in the bulk by immersion in water or HF solutions [118].

Reactions involving diluted aqueous solutions of HF are very important, both in silicon processing technologies and in the laboratory practice [120, 121], because they induce the compensation of acceptors and donors in a sub-surface layer and the metastable hydrophobicity of silicon surfaces associated with their hydrogen termination.

Hydrogen termination, which can be obtained both by exposure to HF-containing solutions or by annealing in hydrogen ambient, provides clean and reproducible silicon surfaces temporarily suitable for further processing.

Several dry techniques may be used to hydrogenate semiconductors, such as plasma hydrogenation and implantation, but hydrogen implantation is the most important one. It shares the advantages of ion implantation of dopants and metal impurities, as it allows to control with the ion beam energy and intensity $J_b = q\, n_{H+}\Delta\varphi$ the amount and the spatial localization of the material introduced into the crystal.

It results, however, in local damage of the crystals [122], which may even cause surface amorphization, unless operating at low implant energies (ca. $\leq$ 15 keV).

In the case of low implant energies, the resulting damage is essentially due to the formation of vacancy–interstitial pairs and of a local lattice distortion [123], which can be recovered by thermal annealing procedures. Hydrogen implantation or hydrogenation with a hydrogen plasma followed by neutron irradiation enhances the thermal donor formation (see Section 2.3.4.3) and induces the formation of shallow hydrogen donors (HDs) detectable with photothermal ionization spectroscopy (PTIS) and interpreted as complexes between radiation induced defects and hydrogen [124]. This process is a low thermal budget process which might enable the doping of silicon at low temperatures with shallow junction formation.

If hydrogen implantation creates a local hydrogen supersaturation, predetermined by selecting the proper ion beam intensity and time, the excess hydrogen segregates with the formation of a layer of hydrogen bubbles. This process condition is intentionally used in the Soitec's Smart Cut process™, a technology which makes use of the local brittleness induced by the presence of the bubbles to decouple a top semiconductor layer separated from the substrate by a layer of hydrogen bubbles.

The most important application of hydrogen in semiconductor technology is defects passivation (see Chapter 5), based on its capability to react with acceptor- and donor-dopants, shallow- and deep-level impurities and defects, including extended defects (external and internal surfaces, dislocations) [118], with the formation of hydrogen bonds.

It can be carried out by wet and dry processes and the consequence of hydrogen bonding is the electrical deactivation of donor or acceptor species, the suppression of the minority carriers recombination at native defects and at several metallic impurities and the saturation of dangling bonds of extended defects, which will be dealt with in Section 3.5.2 and in Chapter 5.

Having amphoteric behaviour (H^+/H^-), hydrogen in Si passivates ionized shallow acceptors (B^-) and donors (P^+). The oppositely charges species react with the formation of partially covalent bonds, forming neutral H-acceptor or H- donor complexes.

Most of the stable complexes of hydrogen with vacancies and metal impurities are IR active in the 1800–2300 cm^{-1} range or detectable with PTIS.

An important example is the VH_4 complex, which is IR active (2223 cm^{-1}) and forms without the need for a pre-existing vacancy with a mechanism that starts with the formation of a metastable VIH_2 complex ($\Delta H = 2.1$ eV), which then reacts with a second H_2 molecule ($\Delta H = 2.3$ eV)

$$VIH_2 + H_2 \rightarrow VH_4 + I \tag{2.91}$$

Eventually, the excess of self-interstitials formed by reaction (2.91) is spontaneously recovered ($\Delta H = +3$ eV) at existing heterogeneous sinks [125]. The overall process is spontaneous and exothermic (+0.47 eV), having also taken into account the heat of solution of H_2 (1.7 eV) [110].

Further information about the equilibrium concentration of vacancies as VH_4 complexes was obtained indirectly [126] on Si samples first submitted to a hydrogenation process at 1200–1390 °C. The evolution of the relative concentration of VH_x complexes was followed by IR measurements after a fast quench and an annealing step at 450 °C. The assumption is made that the VH_x complexes formed at high temperature do not decompose during the rapid cooling step and that the different VH_x complexes transform into the VH_4 complex after annealing at 450 °C. From the Arrhenius plot of the IR absorption intensities of VH_4 at 2223 cm^{-1} a formation enthalpy of about 4 eV for the VH_4 complex is determined.

In C-rich silicon ($8.8 \cdot 10^{17}$ cm^{-3}) high-temperature hydrogenation results in the formation of two different complexes ($VH_3 - HC$ and C_2H_2) [119] and in a VH_4 complex trapped by carbon [127].

V_nO_mH complexes were also found to be formed in the neutron transmutation doping of FZ silicon grown in a hydrogen atmosphere ($[O_i] \sim 10^{17}$ cm^{-3}) by Xiang-Ti [103]. In the irradiated material, vacancies are released by neutron irradiation- induced vacancy clusters and VP complexes and are captured to form VO complexes.

The formation of hydrogenated VO complexes occurs during thermal annealing at 200 °C of the neutron transmutation doped samples and can be followed by monitoring the intensity of the Si-H band at 2066 cm^{-1}, whose strength is controlled by oxygen. The results suggest that the formation of a VOH_2 complex might be due to the hydrogen decoration of the two residual Si dangling bond of the VO complex [77, 103].

Among the hydrogen–metal complexes, for which exists an extensive literature, the electrical signature of Au-H, evidenced using DLTS and photoinduced current transient spectroscopy (PICTS) [128] has been used to detect the presence of Au in silicon nanowires grown by CVD (chemical vapour deposition) with Au as the catalyst.

Knowledge of the physico-chemical properties of hydrogenated metal complexes might be, therefore, of extreme interest for the ongoing development of silicon or germanium nanowires, whose synthesis is generally carried out in hydrogenated atmospheres.

Different from Si, the hydrogen passivation of donors and acceptors in Ge is, instead, not yet totally clarified, although the low thermal stability of the H-acceptor complexes has already been ascertained [118].

2.3.4.2   Carbon

Although being iso-electronic with silicon and therefore electrically inactive, substitutional carbon was recognized already in the early 1960s as a harmful impurity when present in supersaturated conditions, because SiC precipitates, segregated in the silicon matrix during the ingot cooling process, behave as minority carriers recombination centres. Carbon, as oxygen, is an almost unavoidable impurity in silicon crystals grown from the melt, where the main sources of carbon contamination are the hot graphite parts in the CZ or directional solidification (DS) furnaces, as will be discussed in Chapter 4.

Graphite reacts with traces of oxygen contained in the argon gas used to ensure a protecting atmosphere. The reaction product is CO, which transports carbon to the silicon melt, where it will dissolve until equilibrium between the carbon sources and the melt is established. Carbon contamination is particularly severe in Solar Grade silicon produced by purification of metallurgical (MG) silicon, which is C-saturated, with important consequences for the practicability of the process itself [129].

Once dissolved in liquid Si, carbon may interact with the dissolved oxygen with the formation of C-O complexes and the carbon contamination of the melt is directly transferred to the crystal during solidification.

Thus, the average carbon concentration in the melt and in the solid ingot is the result of complex equilibria, which depend on how (and how long) the growth process is carried out, an issue which will also be discussed in Chapter 4.

The solubility of carbon in silicon has long been a matter of debate, in part because of the difficulty of being precisely measured by IR spectroscopy [130] but also because its concentration depends on the oxygen content, which is different in FZ and CZ silicon.

In commercial CZ silicon ingots the typical carbon content is below 0.05 ppma ($2.5 \cdot 10^{15}$ cm^{-3}) [131] which is, however, sufficiently large to influence the oxygen precipitation [132], to induce the formation of C-O defects, to react with hydrogen in wet hydrogenation processes [133] and, eventually, to contribute to the annihilation of thermal donors [134] (see Section 2.3.4.3).

A further effect is the reduction of implantation-induced self-interstitial supersaturation [135]. All these phenomena indicate the strong reactivity of carbon with native defects and impurities in silicon.

It is today commonly agreed that the equilibrium solubility of C in FZ silicon is $3.5 \pm 0.4 \cdot 10^{17}$ cm^{-3} and that in CZ silicon is $2.5 \cdot 10^{18}$ cm^{-3}, with an effective segregation coefficient $k_{eff} = 0.07$ in FZ silicon and 0.3 in CZ silicon [136, 137], in good agreement with more recent results [138], which give for the equilibrium solubility of carbon at the melting point of silicon a value of $2.8 \pm 1.5 \cdot 10^{18}$ at cm^{-3}.

The large difference between these results and some of the earliest ones, of which some values are presented in Figure 2.17 [139–141], may be explained by the fact that that equilibrium solubility measurements at the melting point of silicon should be carried out on a melt saturated with SiC, a condition probably not systematically fulfilled in the experiments of previous authors.

The same problem arises with the measurement of the solubility of carbon in solid silicon at temperatures below its melting point, complicated by the circumstance that the solid solubility of carbon depends on the oxygen content and that the segregation of SiC from saturated solutions of C in Si is a hostile process.

In fact, segregation of SiC from supersaturated C-Si solutions is reported to occur, even in the case of extremely high carbon contents, only in the presence of a large supersaturation

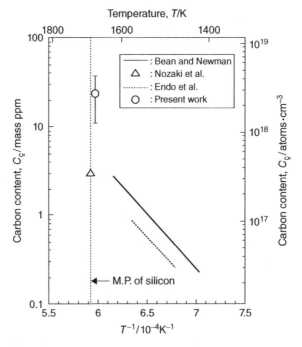

Figure 2.17 *Equilibrium solubility of carbon at the melting point of silicon Materials (References on the figure: Bean and Newman [139] Nozaki et al. [140] Endo et al. [141]). Narushima et al., 2002, [138]. Reproduced with permission from The Japan Institute of Metals and Materials*

of silicon self-interstitials (40 or more) or of high concentrations of oxygen [132], while the precipitation of silicon carbide is generally not observed in oxygen-free Si crystals.

The influence of oxygen on the solubility of carbon may be qualitatively understood (and quantitatively evaluated) considering, first, that carbon sits in substitutional positions and bonds covalently with silicon to form C-Si$_4$ tetrahedra, but the large difference in the Si-Si (0.235 nm) and Si-C bond lengths (0.189 nm) leads to a strong decrease in the lattice constant (see Figure 2.18) [142] and to a severe lattice relaxation [143].

In turn, the local lattice tensile strain which arises from the formation of Si-C bonds could be almost fully compensated by the compressive strain induced by interstitial oxygen–silicon bond formation, as can again be seen in Figure 2.18 [142].

The metastability of supersaturated solid solutions of C in Si was qualitatively explained [142] with simple volume considerations, arguing that a lattice contraction of about one atomic volume occurs for each carbon atom incorporated.

Therefore, the suggestion is given that in order to have carbon precipitation as SiC, the local volume deficit should be counterbalanced by the capture of one self-interstitial for each carbon atom incorporated.

Looking, however, at the lattice constants of Si and cubic 3C-SiC ($a_{Si} = 0.543$ nm, $a_{SiC} = 0.436$ nm) and their molar volumes (see Table 2.7) which fit neatly, there is not a volume deficit to comply. Actually, phase segregation occurs when clusters having the

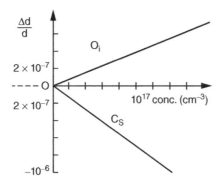

Figure 2.18 *Effect of carbon and oxygen incorporation on the lattice constant of silicon. Gösele, 1986, [142]. Reproduced with permission from Cambridge University Press*

Table 2.7 *Molar volumes of silicon, silicon carbide and of silica polymorphs*

Phase	Structure	Temperature stability range (°C)	Coexistence pressure (MPa)	Molar volume (cm^{3}mol^{-1})
Silicon	Cubic	<1412	—	12.508
SiC (3C)	Cubic	<2830	—	12.664
β-Cristobalite	Cubic	>1470	—	25.785
β-Tridymite	Hexagonal	>870	—	26.350
β-Quartz	Hexagonal	>570	—	23.747
α-Quartz	Rhombohedral	<570	—	22.671
Coesite	Monoclinic	≈800	>2000	20.505
Stishovite	Tetragonal	—	>8000	13.970

structure of the elemental cell of SiC nucleate in the silicon matrix with the incorporation of silicon self-interstitials.

$$C_{Si} + Si_i \overset{\delta V \simeq 0}{\rightleftharpoons} SiC(Si) \tag{2.92}$$

Segregation of carbon as SiC, therefore, occurs only in the active presence of self-interstitial sources. It will be seen in the next section that the oxygen precipitation process is a typical source of self-interstitials, thus explaining also why carbon segregation occurs only in the presence of high oxygen concentrations.

2.3.4.3   Oxygen

Oxygen is the silicon impurity most carefully studied [99, 144–149] for its ubiquitous presence in both FZ-and CZ-grown silicon crystals, influencing many of their physical properties.

Oxygen contamination of CZ-grown crystals is particularly severe and arises from the spontaneous, partial dissolution of the quartz crucible during the pulling process, as will be seen in Chapter 4. The oxygen contamination of FZ-grown silicon is about 2 orders of magnitude less and probably arises from the dissolution of SiO vapours in the molten silicon charge. In turn, SiO is the product of a vapour/solid phase reaction between Si subliming (or diffusing out) from the molten charge and the silica walls of the FZ furnace (see Chapter 4)

$$Si(v) + SiO_2(s) \rightarrow SiO(v) \tag{2.93}$$

The contamination of germanium with oxygen is less challenging, because Ge is currently grown using H_2 as cover gas and quartz presents a limited solubility in liquid Ge at its melting point because the reaction

$$Ge + SiO_2 \rightleftharpoons GeO_2 + Si \tag{2.94}$$

is strongly shifted to the left, with a Gibbs free energy of reaction of $+345.87$ kJ mol^{-1} at the melting point of Ge (1210.55 K). When needed, oxygen can be dissolved in Ge by growing it under an oxygen–argon mixture as cover gas, as the solubility of oxygen in germanium at its melting point appears to be $2 \cdot 10^{18}$ at cm^{-3}, comparable to that in silicon [150].

In spite of the vast amount of research on oxygen in silicon over the last 60 years, some of its properties were under debate until recently.

As an example, the solubility of oxygen in liquid and solid silicon has been studied by a number of authors [151–155], but frequently the accuracy of the results is affected by substantial deviations from thermodynamic equilibrium conditions occurring during its measurement. Equilibrium measurements would, in fact, require the saturation of the liquid or solid-phase with silicon oxide, according to

$$Si_{Si} + O_i \rightleftharpoons SiO_2(Si) \tag{2.95}$$

a condition not systematically accomplished.

When equilibrium conditions are experimentally satisfied [138, 154] a linear Arrhenius dependence is obtained for the oxygen solubility over a wide temperature range, as shown in Figure 2.19 [154].

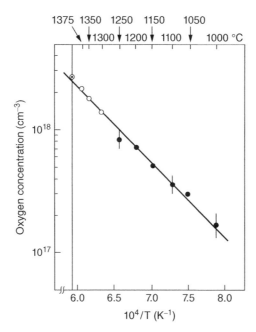

Figure 2.19 *Solubility of oxygen in silicon. Itoh and Nozaki, 1985, [154]. Reproduced with permission from The Japan Society of Applied Sciences*

There is experimental evidence, however, that the room temperature oxygen concentration in silicon samples directly cut from CZ- or FZ-grown ingots far exceeds its room temperature equilibrium concentration. It ranges from 10^{18} at cm^{-3} in CZ-grown to less than 10^{16} at cm^{-3} in FZ silicon, showing the metastability of supersaturated solutions, which cover, in the case of CZ silicon, the entire two-phase region of the diagram of Figure 2.20 [146], with supersaturation ratios close to 10^4 at 600 K.

From the earliest IR investigations [149] it was found that oxygen induces the presence of a number of local vibrational modes, which fit well with a configuration of oxygen occupying a staggered interstitial position close to the BC of a Si-Si bond (see Figure 2.21b). In this configuration the Si-O-Si bond angle should be 164 °, in order to allow for the Si-O bond a length of 0.156 nm, typical of Si-O bonds in quartz and silicates, and for the Si-Si distance the value of the Si-Si bond length 0. 235 nm in silicon.

More recently, theoretical investigations have shown, however, that the activation barrier of the process occurring in order to shift oxygen from the staggered position to the bond centre is only 1 meV. Such a small difference should allow oxygen to tunnel towards the very symmetrical C_{3d} configuration (see Figure 2.21a), with a bond angle of about 180° [99], which would imply a symmetrical outward relaxation of the silicon atoms.

The calculated energy of formation E_f of a BC oxygen interstitial was shown to be between 1.1 and 1.8 eV [99, 156], see also Table 2.6, in reasonable agreement with the experimental data from which a value of 1.4 eV could be inferred.

The outward relaxation of the silicon atoms when an interstitial O sits in a BC position is responsible for the increase in the lattice constant of solutions of O in Si displayed in Figure 2.20. It is also compatible with the large differences (around a factor of 2) between

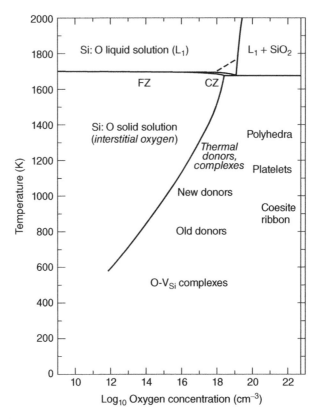

Figure 2.20   *Phase diagram of the Si–O system. Mikkelsen, 1986, [145]. Reproduced with permission from Cambridge University Press*

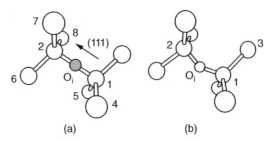

Figure 2.21   *(a) Symmetrical and (b) staggered configurations for interstitial oxygen. Coutinho et al., 2000, [99]. Reproduced with permission from APS*

the molar volumes of Si and of most of the SiO_2 polymorphs (see Table 2.7), which potentially coexist with saturated solutions of O in Si.

Precipitation of oxygen in supersaturated solutions,[26] therefore, generates a very large amount of strain energy, which is quoted to hold around 1–1.5 eV per silicon atom [144],

---

[26] In oxygen saturation conditions the Gibbs energy of formation of the SiO_2 phase $\Delta G_{SiO_2} = 0$.

while the chemical excess energy available from oxygen supersaturation

$$-\Delta G_{ss} = RT \, \ln \frac{c}{c^*} \qquad (2.96)$$

where c^* is the solubility, amounts to a maximum of 0.81 eV at 1000 K for a supersaturation ratio of 10^3.

Thus, the chemical energy is not sufficient to counterbalance the strain associated with the formation of the oxide phase, at least at temperatures well below the brittle–ductile transition of silicon. Therefore, the oxygen segregation process evolves with the formation, via a sequence of clustering steps leading to embryonic assemblies of Si and O atoms of increasing complexity, whose structure depends on the temperature at which the process occurs.

These oxygen-silicon clusters, called old donors, new donors, thermal donor complexes (see Figure 2.20) for their electronic properties which assign a role of donors to these complexes [157, 158], are the precursors of the oxide phase and act to partially relieve supersaturation conditions.

The rate of oxygen clustering becomes important at temperatures above 350–400 °C, while oxygen segregation in the form of oxide precipitates occurs only at temperatures in excess of 600 °C, when the oxide phase precipitation is assisted by the ejection of silicon self-interstitials

$$2O_i + xSi_{Si} \underset{\delta V \approx \Omega}{\rightleftharpoons} SiO_2(Si) + (x-1)Si_i \qquad (2.97)$$

or by the absorption of silicon vacancies

$$2O_i + xSi_{Si} + (1-x)V_{Si} \underset{\delta V \approx \Omega}{\rightleftharpoons} SiO_2(Si) \qquad (2.98)$$

with $x \approx 2$, as suggested by the exigent volume theory (EVT) of Tan [159] and Hu [144].

Vacancies absorption or self-interstitials ejection supplies the local volume excess (one atomic volume Ω of Si for all the possible low pressure coexisting phases) needed for an almost strain-free transformation.

The process described by Eq. (2.97) is a source of silicon self-interstitials, and for this reason has been compared to the surface oxidation of silicon:

$$(1+x) \, Si_{surf} + O_2(g) \rightarrow SiO_{2(surf)} + x \, Si_i \qquad (2.99)$$

which is a powerful source of silicon self-interstitials. In the case of surface oxidation of Si, however, the energetic cost of the self-interstitial formation (≥ 3 eV) is granted by the availability of the Gibbs energy ΔG_{SiO_2} of the reaction $Si + O_2 \rightarrow SiO_2$, which amounts, at 1000 K, to 730.25 kJ mol^{-1} or 7.57 eV, while in the case of SiO_2 segregation from a saturated solution the Gibbs energy available is that arising from supersaturation.

The segregation of the oxide is, therefore, a complex process, where the nucleation of embryos of the second phase, vacancies absorption and interstitials ejection simultaneously play a role and the thermodynamic barriers are overcome thanks to a very large oxygen supersaturation ratio. Segregation of oxygen should be accompanied by the emission of dislocations, if local self-interstitials supersaturation does occur.

A significant example of an oxide segregation process associated with dislocation emission is given in Figure 2.22, which displays the TEM picture of an oxide particle surrounded

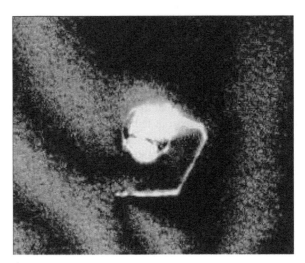

Figure 2.22 *TEM micrograph of an oxide precipitate associated to a dislocation (the size of the precipitate is about 0.8 μm). Pizzini et al., 2000, [160]. Reproduced with permission from IOP*

by a hexagonal dislocation half loop. The precipitation process consisted of an 8 h nucleation step at 650 °C and a precipitation annealing step at 1000 °C [160].

Although oxygen, as an isolated impurity, is electrically inactive both as dopant and as a minority carriers recombination centre, it determines a number of physical and chemical properties of silicon which have been of customary importance in very large scale integration (VLSI) technology.

At first, it behaves as a strong hardening agent. It develops an impurity atmosphere around stationary or slowly moving dislocations, behaving as a locking environment for the dislocation motion. This property is particularly important in microelectronic processes, where dislocation generation and motion should be avoided during the device fabrication processes. This property is also beneficial for detectors working in harsh radiation environments, minimizing the radiation damage.

It should be noted, however, that hardening does not withstand oxygen precipitation processes, and the superior mechanical stability of CZ silicon is, therefore, lost after oxygen precipitation.

Further, oxygen is a very reactive species. It tends to oxidize almost all metal impurities dissolved in silicon (and germanium), generating a number of oxygenated complexes and binary oxides. As an example, the standard Gibbs energy of binary metal oxide formation ranges from 550 kJ mol^{-1} at 1273 K for 3d metals and up to 1000 kJ mol^{-1} for Group II, III and IV metals including silicon.[27]

Among oxygenated defects, $B - O_x$ complexes are known as an interesting case of defect species formed in B-doped CZ silicon by high energy electron irradiation or by light exposure ator slightly above room temperature.

The indirect evidence of the formation of B-O complexes came from the severe, but reversible, light induced degradation (LID) of the efficiency of terrestrial and space solar

---

[27] Interactive Ellingham diagrams www.doitpoms.ac.uk.

cells manufactured with B-doped CZ wafers, which is associated with the generation of a deep recombination centre with an energy level lying between $E_v + 0.35$ and $E_c - 0.45$ eV [161–164].

LID is attributed to the formation of B-O complexes, because it occurs only in B-doped CZ -Si, not in Ga-doped silicon solar cells. B-O complexes are relatively labile species, decomposing at temperatures above 200 °C, as the cell efficiency is restored after an annealing at 200 °C.

In the attempt to find a theoretical support for these qualitative assumptions, B-O complexes have been the subject of a number of studies, whose starting point is that under irradiation with 1 MeV electrons the formation of B-O complexes is favoured by the presence of Si self-interstitials, which gives rise to a complex interaction pattern between O_i and B.

These models foresee the initial formation of a fast diffusing interstitial boron-silicon self-interstitial complex B-I, which has been clearly identified by ESR measurements, DLTS and optical absorption spectroscopy. The B-I defect, in turn, diffuses in the silicon matrix and reacts with interstitial O_i to form the complex (Bi-Oi), to which DLTS studies and theoretical computations associate a deep level at $E_c - 0.23$ eV [165, 166].

Under light exposure a different mechanism is supposed to occur, involving the formation of a fast diffusing oxygen dimer $(Oi)_2$, which reacts with B leading to the complex BO_{i2}, which has a calculated deep gap level at $E_c - 0.41$ eV [167]. This energy level fits reasonably well with the experimental value of the fundamental LID recombination centre, lying between $E_v + 0.35$ and $E_c - 0.45$ eV, giving another, indirect, evidence that LID is governed by recombination at B-O complexes.

It has been further suggested [168] that the formation of the electrically active $B-O_{i2}$ complex actually occurs by light activation of a poorly electrically active precursor $(B-O_{i2})^*$

$$(B-O_{i2})^* \xrightarrow{h\nu} B-O_{i2} \tag{2.100}$$

which pre-exists in the B-doped silicon wafers used for the fabrication of the solar cells.

Although the reaction sequence is still unknown, it is supposed that the formation of this precursor species is due to B-self-interstitials interactions during the cooling stage of the growth process of B-doped CZ silicon, which is associated with oxygen precipitation and self-interstitials emission.

Oxygen interaction is not limited to metals, it also interacts with point defects, as is the case of oxygen-vacancy complexes, as discussed in Section 2.3.3, and which has attracted theoretical interest recently [99, 169, 170].

A process involving the gettering of metallic impurities at oxygen precipitates will be discussed in Chapter 5, for its strategic role in microelectronic device fabrication processes.

2.3.4.4 Nitrogen

The hardening effect of nitrogen in silicon is well known, but its very low solubility ($4.5 \pm 1 \cdot 10^{15}$ at cm^{-3}), as compared to that of oxygen, and its very low segregation coefficient ($\sim 10^{-3}$) are severe, technical limits to direct nitrogen incorporation in the silicon lattice in melt growth processes using either nitrogen or ammonia as cover gases. Therefore, ion implantation followed by thermal annealing is very frequently used to dope silicon with nitrogen. Due to these physical limits the interest in nitrogen as a silicon and germanium dopant is relatively recent [171, 172], while its fundamental properties have been for years the subject of intense basic research, making them relatively well known.

It was deduced from the earliest IR measurements on both implanted and melt-doped samples that the predominant mode of nitrogen incorporation in the silicon lattice is the formation of N–N pairs [173], which give rise to two LVMs at 766 and 962 cm^{-1}. The substitutional species N_{Si}, which is also IR active with a LVM at 653 cm^{-1}, is a minority component. The preliminary conclusion of these studies was that pairing may occur either on sites adjacent to a silicon atom, or via the substitution of one silicon atom in a split interstitial configuration or a bridge bond configuration (see Figure 2.23a–c) [172].

Subsequent experimental and theoretical studies showed, however, that direct N-N bonding is unable to account for the IR high frequency mode at 962 cm^{-1}. The most probable nitrogen configuration is similar to that occurring in silicon nitride (Si_3N_4), with a trifold coordinated N atom in an N–Si pair presenting a split interstitial configuration, next to another N–Si split interstitial in antiparallel configuration [174, 175]. Bonding occurs via sp^2-like orbitals on the nitrogen atoms and sp^3-like orbitals on the silicon atoms.

Nitrogen-implanted silicon exhibits also a near band edge PL emission at 1.1223 eV [176] measured at 4.2 K, whose intensity increases with the implantation dose but does not depend on the electrically active nitrogen concentration, measured with spreading resistance measurements.

This PL emission wavelength corresponds to that also present in the PL emissions of intrinsically nitrogen-doped natural diamond, after e-beam or neutron irradiation and annealing at 800–1000 °C, or in nitrogen implanted and annealed diamond. Both types of sample exhibit strong PL emissions at 575, 610, 637 nm (2.156, 2.032 and 1.946 eV, respectively) [177] and 1046 nm (1.853 eV) [178], which are due to NV centres, whose structure was studied using optically detected magnetic resonance (ODMR), and whose formation is favoured by vacancies made available by particle or electron irradiation.

Nitrogen-implanted silicon is also shown to present a shallow level with donor properties at 0.017 ± 0.002 eV associated with a limited fraction (less than 1%) of the implanted dose [179], thus corresponding to the electrical activity of nitrogen in a substitutional position.

The formation of several electrically active N-O$_x$ complexes, differing in the number of oxygen atoms involved [180], is also observed in N-doped, high oxygen CZ silicon, whose signatures are absorption bands of electronic nature in the far-IR between 190

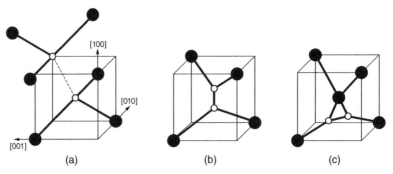

(a)   (b)   (c)

Figure 2.23   *(a)–(c) Proposed configurations of the N-N centre in silicon. Stein, 1986, [172]. Reproduced with permission from Cambridge University Press*

and 270 cm^{-1} (0.0236 and 0.0335 eV, respectively) [181] and whose rate of formation is enhanced in the 600–700 °C temperature interval.

Today, the potential of nitrogen as a n-type doping- and hardening-agent is fully recognized and current research work is addressed at a better understanding of its role in the transport and mechanical properties of silicon [178, 182–185] and in defect dynamics in silicon and diamond [186, 187].

This is the case of the N-V complexes, which are actually colour centres and are supposed to be of potential use as individual qbits in advanced quantum computing machines. The key features of the N-V centre in diamond are that it can be seen as an individual species by ODMR, it can behave as a light emitter and it can be precisely positioned with nanometre separation on a diamond surface by ion implantation [3].

As a further example, N behaves, like oxygen, as a hardening agent, but at concentrations a hundredfold lower than those required when using oxygen ([O] > 5 · 10^{17} at cm^{-3}) [173] via a dislocations locking process. It has been demonstrated, using stress–strain measurements on undoped and N-doped silicon, that the locking of dislocations by nitrogen persists in deformation tests carried out at temperatures in excess of 900 °C, with significant advantages over oxygen.

This result allows the evaluation of the interaction energy E_{int} of nitrogen with dislocations, using the standard theory for thermally activated dislocations unlocking

$$E_{int} = k_B T_c \ln \left(\frac{N v^*}{\Gamma} \right) \qquad (2.101)$$

where T_c is the unlocking temperature, k_B is the Boltzmann constant, N is the number of locking points along the dislocation line, v^* is the vibration frequency of the dislocation line and Γ is the experimental rate of dislocation release from the impurity locking atmosphere [188]. At a temperature of 1200 K, the value of E_{int} is between 3 and 4 eV, depending on the number of locking points.

2.4 Defects and Non-Stoichiometry in Compound Semiconductors

2.4.1 Structural and Thermodynamic Properties

Compound semiconductors are an extremely large group of materials, but three families, the III–V, II–VI and I–VII are particularly interesting, as they have the advantage over elemental semiconductors of the IV group of being direct band-gap semiconductors. Most of these materials crystallize with the (cubic) zinc blende structure, with the notable exception of some high melting compounds (AlN, GaN and InN) which crystallize with the (hexagonal) wurtzite structure.

As could be expected, an increasing polarity in their bonding features is observed passing from the III–V to the I–VII compounds, these last represented by the cuprous halides and α-AgI, which is the low temperature phase (up to 408 K) of AgI. The β-AgI phase is stable in a very restricted temperature interval (408–419 K) above which the γ-AgI phase is stable, which is a typical fast ion conductor [18].

The fact that the cuprous halides (Cu Br, CuCl and CuI) and α-AgI present a semiconductor character different from the other silver halides which crystallize with the rock salt structure and present pure ionic conductivity, has been known since the early works of Cardona [189]. Only recently [190], however, these semiconductors gained a renewed interest thanks to the high exciton binding energies (190 meV vs. 23 meV of GaN) and the exceptionally low lattice misfit with silicon (< 0.4%) as compared to GaN (17%). This last property allows, at least potentially, their heteroepitaxial, lattice matched, deposition on silicon substrates in the practical absence of misfit and threading dislocations, with a tremendous advantage over III–V compounds.

Vacancies and self-interstitials, as in Group IV semiconductors, are the predominant native defects in compound semiconductors, which, as in ionic compounds, are distributed both in the cationic and anionic sublattices. As an example, in GaAs it is possible to predict the existence of Ga and As vacancies (V_{Ga} and V_{As}) and of Ga and As interstitials, promoted by Schottky or Frenkel type of processes.

Different from ionic solids, neutral vacancies are not necessarily created in pairs, due to the predominant covalent character of the bonds, and the resulting imbalance gives rise to non-stoichiometry [191] (see Section 2.4.3).

Different from elemental semiconductors, self-interstitials may sit on tetrahedral (T) interstices of the cationic and anionic sublattice

$$Ga_{Ga} \rightleftharpoons Ga_{TGa} + V_{Ga} \tag{2.102}$$

$$Ga_{Ga} \rightleftharpoons Ga_{TAs} + V_{Ga} \tag{2.103}$$

at least in the case of the III–V semiconductors.

Due the predominant covalent character of bonds in compound semiconductors of the III–V and II–VI groups, one can also conceive and theoretically predict the presence of antistructural defects, briefly called antisites [192]. In the case of GaAs, as an example, it is expected to find atoms of Ga sitting in As positions and of atoms of As sitting in Ga positions

$$Ga_{Ga} + V_{As} \rightleftharpoons V_{Ga} + Ga_{As} \tag{2.104}$$

and

$$As_{As} + V_{Ga} \rightleftharpoons V_{As} + As_{Ga} \tag{2.105}$$

whose distribution and concentration is ruled by the energetics of the site exchange process through the law of mass action.

However, neither antisites nor interstitials were securely observed in II–VI semiconductors [193] until recently.

Defects (and impurities) can behave as donors, acceptors or deep levels and their presence is therefore associated with shallow and deep electronic levels in the gap. They can be neutral or charged species and may interact with other defects and with metallic and non-metallic impurities.

As in Group IV semiconductors, the experimental determination of the equilibrium (as well as the effective-) concentration of defects is often very difficult, due to the variety of defect species present and to their very low concentrations. The evaluation of the defect concentration as a function of the temperature is therefore preferably carried out by the use of empirical or *ab initio* calculations, complicated by the circumstance that in compound semiconductors the presence of d-electrons should also be accounted for.

A simple macroscopic model of the formation energy of virtual neutral defects in compound semiconductors was developed by van Vechten at the beginning of the 1970s [192–194], but has also been used recently with substantial [195] success.

The basic features of this model are the preliminary assumption that the formation energy of a neutral vacancy by a Schottky process in the cationic sublattice of an A(III)B(V) compound

$$A_A \rightleftharpoons V_A^x + A_{surf} \tag{2.106}$$

corresponds to the energy needed for the formation of a macroscopic cavity of atomic volume $\Omega = \frac{4}{3}\pi r_w^3$ (where r_w is the Wigner–Seitz radius of the missing species) in the crystal lattice. Thus, the formation energy is given by the surface energy of this cavity

$$H(V^x) = E_s(hkl)A_n \tag{2.107}$$

where $H(V^x)$ is the energy needed to transport an atom from a lattice position to the surface of the cavity, $E_s(hkl)$ is the surface energy of a specific crystal face and A_n is the surface area of the neutral point defect cavity inside the crystal, having the shape of a polyhedron, not a sphere. As already mentioned, we use an x label to indicate a neutral species.

Considering, further, that many properties of semiconductors can be understood by regarding a semiconductor as a metal of small energy gap centred on the Fermi level E_F [196], it is possible to assume that $E_s(hkl)$ is the sum of two terms

$$E_s(hkl) = E_s^m(hkl) + E_s^b(hkl) \tag{2.108}$$

of which $E_s^m(hkl)$ is the energy needed to remove an atom with a symmetrical charge density and $E_s^b(hkl)$ is the contribution for formation of dangling bonds inside the cavity.

The $E_s^m(hkl)$ term can be evaluated from the surface tension σ of the melt, when available,

$$E_s^m(hkl) = 4/3\pi r_w^3 \sigma \tag{2.109}$$

while the $E_s^b(hkl)$ term is calculated using the empirical equation

$$E^b = \frac{\Phi + E_G/2}{N} \tag{2.110}$$

where Φ is the work function, E_G is the band gap and N is an empirical factor related to the coordination of the missing atom ($N = 32$ for fourfold coordination).

In spite of the roughness of the approximations adopted, the results for Si and Ge, used as reference, are in fairly good agreement with the experimental results, as can be seen in Table 2.8 [194, 197–200]. There are good reasons, therefore, to be confident about the values of the formation energies of vacancies and vacancy pairs in compound semiconductors reported in Table 2.9 calculated using this empirical model [195].

The formation energies of individual antisites and of pairs of antisites has been also calculated by Van Vechten [192] and by Dobson and Wager [201] under the assumption

Table 2.8   *Calculated formation energies (in eV) of vacancies in Si and Ge*

	Σ (erg cm^{-2})	E^m	E^b	$H(V^x)$	H_V (exp.)	References
Si	734	1.63	0.74	2.38	2.44	[194]
Ge	607	1.47	0.43	1.90	1.9	[197–200]

Table 2.9 *Energy gap, melting enthalpies, vacancy and vacancy pairs formation energies in III–V and II–VI compounds*

Compound	E_G	ΔH_f	$H(V_A)$	$H(V_B)$	$H(V_C)$	$H(V_{A\text{-}B})$	$H(V_{A\text{-}C})$	$H(V_{B\text{-}C})$
GaN	3.39	—	4.22	1.69	—	5.91	—	—
GaP	2.27	—	2.96	2.43	—	5.39	—	—
GaAs	1.42	0.91	2.47	2.35	—	4.82	—	—
InP	1.34	0.65	3.17	2.10	—	5.27	—	—
ZnSe	2.70	—	2.75	2.5	—	5.25	—	—
CdS	2.58	—	3.06	2.01	—	5.07	—	—
CdTe	1.49	—	2.51	2.40	—	4.91	—	—
HgTe	—	0.45	0.47	0.43	—	0.9	—	—
CuInSe$_2$	1.02	0.86	1.04	3.28	2.37	—	4.32	5.65
CdZnTe	—	—	—	—	—	—	—	—
CuGaSe$_2$	2.55	—	1.01	2.37	2.02	—	4.74	4.75
CuInS$_2$	1.53	0.64	1.06	3.46	2.08	—	4.52	5.54

Fiechter, 2004, [195]. Reproduced with permission from Elsevier

Table 2.10 *Formation enthalpy (in eV) of individual antisites and antisites pairs in III–V semiconductors and electronic enthalpy terms for GaAs and InP*

Compound	$\Delta H^o_{A_B}$	$\Delta H^o_{B_A}$	$\Delta H^o_{A_B B_A}$	$\Delta H^e_{A_B}$	$^{30}_{14}Si$
GaP	0.68	0.38	1.06	—	—
GaAs	0.35	0.35	0.70	1.87	1.87
InP	0.89	0.42	1.11	2.31	1.84
InAs	0.57	0.33	0.90	—	—

Adapted from Van Vechten, 1975, [192] and Dobson and Wager, 1989, [201].

that the pair formation enthalpy $\Delta H_{A_B B_A}$ is given by

$$\Delta H_{A_B B_A} = \Delta H^o_{A_B B_A} + \Delta H^e_{A_B B_A} \tag{2.111}$$

where the first term is the expression of the disorder associated with the formation of two antisites and the second is the electronic contribution due to the defect ionization.

In the case of neutral pairs and individual electronic contributions independent of the nature of the defect species, one could suppose that the overall electronic contribution to the enthalpy is zeroed by electrostatic energy compensation and that the pair formation enthalpy is solely due to the disorder term.

The enthalpy of formation of individual antisites was then calculated and their sum used as a measure of the enthalpy of formation of the defect pair. Results for some wide gap semiconductors are reported in Table 2.10 [192, 201].

From the calculated values of the individual electronic enthalpy contribution of the twofold ionized defect species in InP and GaAs [201], it is possible to see that the electronic contribution is dependent on the nature of the individual defect species, its charge and its electronic energy levels. Therefore, at least in the case of InP, their contribution to the total pair formation enthalpy cannot be neglected.

Total energy calculations based on the initial definition of a set of independent defect reactions, which include vacancies, antistructural defects and self-interstitials [202] give a more reliable prediction of defect thermodynamics in a number of systems, including

GaAs, GaP, ZnSe, ZnTe, as a function of the non-stoichiometry and of the electronic properties of defects.

In fact, as a defect can behave either as a shallow donor, a shallow acceptor or as a deep level, its formation enthalpy and its charge depend obviously on the position of the corresponding electronic level within the gap. Therefore, as a charge exchange occurs between the redox couple of defect species and the electron reservoir of the matrix crystal

$$A_B^x + e \rightleftharpoons A_B^-$$

(2.112)

the occupation statistics depend on the electrochemical potential, that is on the Fermi level [202] and the electronic energy contributions are of the order of magnitude of the energy gap, that is of several eV for GaAs and InP, as can also be seen in Table 2.10.

Results concerning defects in p-type and n-type GaAs are reported in Figure 2.24 [202], which displays the dependence of the concentration of the different defects on the stoichiometry (S) and of their energies of formation ε_μ on the electrochemical potential μ.

To remove, at least in part, the influence of the temperature dependence, the defect concentrations $\overline{C}$ are given in terms of their relative value C/C_{max}, where $C_{max} = e^{-\frac{\varepsilon}{kT}}$.

It is particularly interesting to see that Ga_{As} and As_{Ga} antisites, not vacancies or self-interstitials, are the dominant defects in non-stoichiometric, p-type GaAs.

In non-stoichiometric, n-type GaAs, Ga vacancies and As antisites compete in As-rich GaAs, while Ga antisites are the predominant defects in Ga-rich GaAs [202]

Figure 2.25 displays the formation energies of defects in GaAs, which increase with the increase in n-type doping for the majority of the defects, with the exception of Ga_{As} and V_{Ga} defects.

Modelling the growth and the subsequent processing of a compound semiconductor implies, therefore, a deep concern about the thermodynamics of defects, which is much more challenging than for elemental semiconductors.

2.4.2   Defect Identification in Compound Semiconductors

Preparation conditions, non-stoichiometry, impurities and impurity–point defects complexes severely affect the nature and the concentration of native point defects present in compound semiconductors, making their experimental study more challenging than in Group IV semiconductors.

The main spectroscopic techniques used to study native defects in compound semiconductors are ESR, PL, DLTS, PAC and PAS. The first, preferably in the ODMR or photo-EPR configuration, is the most appropriate to determine the microscopic structures of native defects [193], but the use of DLTS is the preferred approach for the investigation on deep levels. As in the case of defects in elemental semiconductors, their effective concentration is so low that often, if not always, they are detectable only after irradiation with high energy electrons or fast neutrons.

2.4.3   Non-Stoichiometry in Compound Semiconductors

Non-stoichiometry in solids was the subject of a number of experimental and theoretical studies in the second half of the last century, addressed at understanding the correlations between structural constraints and defectivity in transition metal oxides, with a large impact on their subsequent technological applications. More recently, the synthesis and further

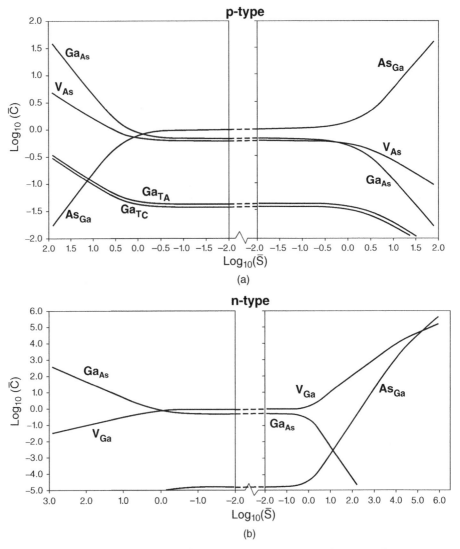

Figure 2.24 *Calculated concentration of vacancies, antisites and interstitials in GaAs as a function of the deviations from the stoichiometry, (a) p-type, (b) n-type. Jansen and Sankey, 1989, [202]. Reproduced with permission from American Physical Society*

processing of non-stoichiometric silicon oxide (SiO_{2-x}) opened a new route towards the fabrication of silicon quantum dots, with consequences relative to the development of full silicon-based optoelectronic devices [203]. Non-stoichiometric phases are particularly abundant in compound semiconductors, their thermodynamic and structural features being the main influence on their growth and functional behaviour.

The concept of stoichiometry of chemical compounds cannot be directly transferred to crystalline phases, where structural constraints predefine a constant ratio between the lattice sites belonging to their sublattices. Nevertheless, the relative lattice site occupation

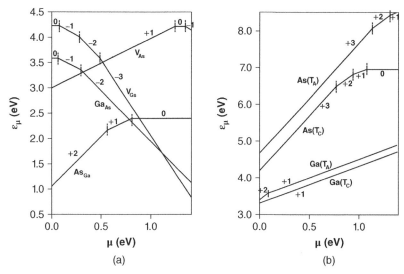

Figure 2.25 *(a,b) Calculated formation energies ε_μ as a function of the chemical potential of interstitial defects in GaAs (Ga_{TA} and Ga_{TC} for Ga interstitials in tetrahedral cationic and anionic interstices). The zero in the scale of the chemical potential μ is taken in correspondence to the valence band edge and the value of the band gap is the experimental value. Jansen and Sankey, 1989, [202]. Reproduced with permission from American Physical Society*

depends on many factors, including the presence of lattice defects and/or of substitutional and interstitial impurities and the oxidation states of metallic and non-metallic species populating the lattice sites.

As an example, in the fcc (face centered cubic) ionic phases of the alkali halides, the ratio of the cationic and anionic sites is 1, while it is 0.5 in the case of the cubic phase of CaF_2. The relative site occupation should satisfy electroneutrality conditions and the charge states of crystal defects (vacancies and interstitials), impurities and atomic components modulate the lattice sites occupation of both sublattices, as already shown in Section 2.1, satisfying full equilibrium conditions even in the case of non-stoichiometry.

In general, stoichiometry conditions may be fully respected only when the phase components are atomic species thermodynamically stable in a single oxidation state, or in environmental conditions such as to prevent a change in the oxidation state of the atomic components or the lattice sites occupancy ratio.

As an example, the impurity-free alkali halide phases are fully stoichiometric when prepared by room temperature crystallization of a saturated aqueous solution. Instead, when NaCl, taken as the prototype of alkali halides, is equilibrated at high temperatures in a Na vapour- rich atmosphere, it becomes a non-stoichiometric $Na_{1+x}Cl$ phase, presenting an excess of anionic vacancies

$$Na^g \underset{\longleftarrow}{\overset{NaCl}{\rightleftharpoons}} Na_{Na} + V_{Cl} \tag{2.113}$$

As a further, classical example of solid state chemistry, NiO could be prepared virtually stoichiometric at the thermodynamic oxygen equilibrium pressure $p^o_{O_2} = \exp-(\Delta G^o_{NiO}/2RT)$, where ΔG^o_{NiO} is the Gibbs free energy of NiO formation, but

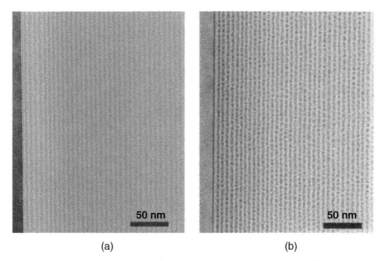

(a)   (b)

Figure 2.26   *(a) As grown SiO_{2-x}/SiO_2 layers and (b) ordered arrays of silicon nano dots after annealing at 1100 °C under N_2 atmosphere. Heitmann et al., 2002, [203]. Reproduced with permission from Elsevier*

it becomes non-stoichiometric at any pressure above $p_{O_2}^o$

$$\tfrac{1}{2}\,O_2(p_{O_2} > p_{O_2}^o) \xrightarrow{\text{NiO}} O_O^x + V_{Ni}^x \qquad (2.114)$$

(where O_O^x is an O^{2-} ion in a regular anionic position and V_{Ni}^x is a twofold positively charged vacancy in an empty cationic site) with the further oxidation of Ni^{2+} to Ni^{3+}, in view of the following defect reaction

$$V_{Ni}^x + 2Ni_{Ni}^x \rightarrow V_{Ni}^{2-} + 2Ni_{Ni}^+ \qquad (2.115)$$

where Ni_{Ni}^x is a Ni^{2+} ion in a regular Ni position, Ni_{Ni}^+ is a Ni^{3+} ion in a regular Ni position and V_{Ni}^{2-} is a neutral vacancy in an empty Ni position.[28]

In the case of non-stoichiometric silica, restoration of stoichiometry might be achieved with a suitable thermal annealing under the appropriate covering atmosphere, with the segregation of a secondary Si phase

$$Si_{1+x}O_2 \rightarrow xSi + SiO_2 \qquad (2.116)$$

This process, which will be discussed in Chapter 5, became of substantial technological interest in the case of metal-rich, non-stoichiometric phases of SiO_2, SiO_xN_y, Si_3N_4 and SiC. With these materials, the thermal annealing of non-stoichiometric layers of nanometric width induces the segregation of the silicon excess in the form of arrays of Si nanodots [203–208], electrically and optically active, embedded in a stoichiometric dielectric matrix (see Figure 2.26 [203]).

---

[28] This and the former reaction are written using the Kroeger–Vink formalism [22].

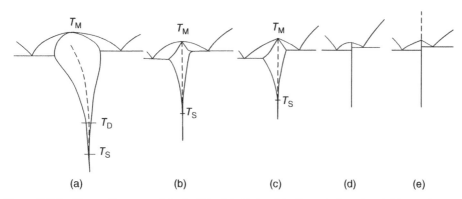

Figure 2.27 *Phase diagrams of Berhollide (a), Daltonide (b and c) and stoichiometric compounds (d and e). Fedorov, 2012, [209]. Reproduced with permission from P. Federov*

Coming back to basic issues concerning the thermodynamics of non-stoichiometric phases, the term Berthollide can be appropriately used to label a phase of variable composition [209], to distinguish it from the almost stoichiometric Daltonides phases. The first are phases whose mean composition (dotted line in Figure 2.27a) differs (either higher or lower) from that of the stoichiometric phase and varies strongly vwith the temperature, as is again seen in Figure 2.27a.

The range of stability of the Berthollide phase is between the melting temperature T_M and T_D, the temperature at which it converts to a Daltonide phase.

Daltonides are still non-stoichiometric phases whose mean composition deviates, however, very little from that of the stoichiometric phase, see Figure 2.27b,c, over its entire stability interval between T_M and T_S, where T_S is the temperature below which the truly stoichiometric compound is stable, see Figure 2.27a–c.

Figure 2.27d,e, finally, show examples of a system which behaves as a truly stoichiometric phase (or as a line phase) over the entire range of its existence, as $T_S > T_M$ and the non-stoichiometric phase would be virtually stable only at temperatures higher than the melting point of the stoichiometric phase.

These qualitative arguments might be implemented noting that the derivative dT/dx along the solidus (and liquidus) line does not show a break for $T = T_M$ for the cases (a) and (b), while it does show a break for $T = T_M$ for the case (c) and (e), this discontinuity indicating that T_M is the true melting point of a stoichiometric phase.

Accounting for the thermodynamically necessary presence of point defects in a solid phase for $T > 0$ K, Daltonide phases are non-stoichiometric phases whose behaviour is dominated exclusively by point defects, while Berthollide phases, spanning a very wide homogeneity range, may be considered as solid solutions of two virtual polymorphs of the individual component.

As an example, let us imagine that the phases stable in an A–B system are a non-stoichiometric $A_xB_{1-x}(\beta)$ phase and a stoichiometric A_3B_4 phase. The β-phase may be considered a solution of a virtual stoichiometric AB phase (not necessarily thermodynamically stable) and A_3B_4, provided that AB and A_3B_4 present close structural relationships. This situation finds many examples in metallic oxide systems, as is the case of the Fe–O systems.

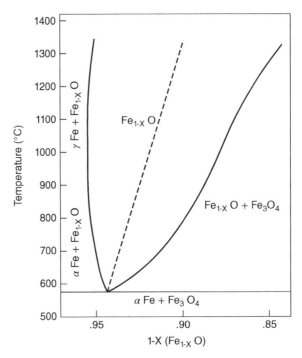

Figure 2.28   *Phase diagram of the Fe–O system in the range of stability of the wüstite phase, (---) line of average composition*

There, the non-stoichiometric wüstite $Fe_{1-x}O$ phase (with $0.03 < x < 0.17$) whose mean composition is always different from that of the stoichiometric FeO phase, see Figure 2.28 is a true Berthollide phase. It may be treated as a solution of a virtually stoichiometric FeO phase and Fe_3O_4, considering the similarities of the statistical assemblies that constitute these cubic crystals [210], where the oxygen sublattice is continuous and the composition shift with a decreasing concentration of Fe is accommodated by a cubic defect cluster, the so called Koch–Cohen cluster, see Figure 2.29.

This cluster has the size of the cubic cell of the virtually stoichiometric FeO, is the embryo of the Fe_3O_4 phase [4] and contains the proper number of iron vacancies and three-fold charged cation interstitials needed to fit the composition of the non-stoichiometric phase and the average charge state of iron.

Because there is not yet a comprehensive thermodynamic treatment of non-stoichiometric semiconductors, in spite of the challenges associated with the control of their non-stoichiometry [211] during their growth, we may rely for them on the approach used by Anderson already 40 years ago [212] to deal with non-stoichiometric oxides.

By analysing the thermodynamic conditions for the stability of stoichiometric and non-stoichiometric phases in terms of their Gibbs free energy surfaces in the (G, T, x) space, it can be easily demonstrated that the composition of a line (stoichiometric) phase is sensibly constant over a wide range of chemical potentials and the Gibbs energy surface is sharply curved at its minimum.

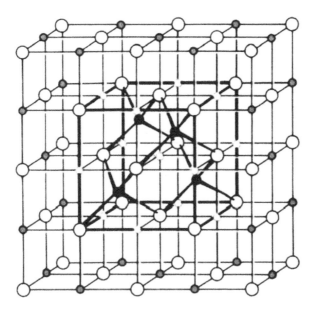

Figure 2.29   *The Koch–Cohen cluster, (●) Fe^{2+}_{oct}, (●) Fe^{3+}_{tetr}, (○) Fe vacancy, (○) oxygen*

By definition, a stoichiometric phase is highly ordered and its small deviations from order may be well understood by the use of the Kroeger and Vink theory of defects in ionic solids [22] already mentioned in Chapter 1. By contrast, the Gibbs energy surfaces of a disordered, non-stoichiometric compound should be rather flat at their minima.

It can also be deduced that the thermodynamic equilibrium of a homogeneous, non-stoichiometric phase is bivariant, while it is univariant $\mu_i = \mu(T)$ for a stoichiometric phase. Finally, there is no need that the absolute minimum of the Gibbs energy curve of an $A_x B_{1-x}$ compound, where the curvature takes its maximum, should coincide with the stoichiometric composition. The deviation of the absolute minimum composition from the stoichiometry depends on the difference in formation energy of point defects thermodynamically stable in the $A_x B_{1-x}$ compound [211].

Berthollide semiconductor phases can be treated, instead, with the approach used for non-stoichiometric oxides [210]. As an example, the almost flat Gibbs energy surface of a widely non-stoichiometric semiconductor (solid line in Figure 2.30) might be the convolution of a series of coexistence tangents at the Gibbs energy minima of different stoichiometric, ordered line phases, merging in an almost continuous envelope.

At the composition x_1, the difference $\tilde{\Delta}G$ between the Gibbs energy of a line phase and the corresponding energy of a pseudo-homogeneous phase may be so small that their distinction depends only on the accuracy of the measurements.

We have, therefore, three possible approaches to the equilibrium thermodynamics of non-stoichiometric phases.

The first relates to true Daltonide phases, where point defects play an exclusive role and for which the (shallow) width of the homogeneity region is entirely determined by the enthalpies of formation of point defects.

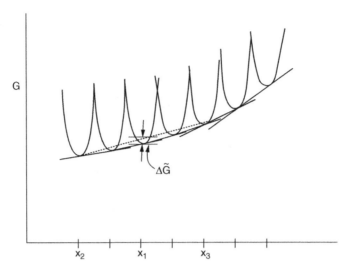

Figure 2.30   *Gibbs energy versus composition diagram of a wide non-stoichiometric compound as the convolution of the coexistence tangents at the free energy minima of different stoichiometric, ordered line phases*

The second, which accounts for Berhollide systems spanning a very large homogeneity range, relies on the stability of solid solutions of virtual isomorphs.

The third, which is a variant of the second one, is based on the hypothesis of the stability of a sequence of phases having very similar, almost indistinguishable structures, whose envelope thermodynamically features the behaviour of a homogeneous phase of variable composition.

These conclusions are based on topical examples brought from transition metal oxides, but these arguments are applicable also to compound semiconductors.In Section 1.7.10, we assumed, as a thermodynamically (and structurally) well acceptable condition for the stability of the non-stoichiometric, homogeneous, variable composition CdTe phase, the presence of point defects. The homogeneous phase may be considered a dilute solution of Cd_{Cd}, Te_{Te}, V_{Cd}, V_{Te}, Te_i and Te_{Cd} antisites in the CdTe lattice, their concentration being determined by the relative enthalpies of formation, which are relatively high ($> 4\,eV$), see Table 2.10 and 2.11 [195, 211, 213, 214], a condition that holds for a number of III–V and II–VI semiconductors. Although the width of the homogeneous region of these compounds is generally very small (see Table 2.11), a Berthollide type of behaviour seems to occur above a transition temperature T_s above 700 °C in the case of CdTe, as could be observed in Figure 2.31, with a marked deviation of the mean composition from the stoichiometric one. Berthollide features are also assigned by the shift of the melting point composition from the stoichiometry and by the continuous variation of the derivative (dT/dx) across the melting point.

Looking deeper into the defect thermodynamics of this system (see Figure 2.32) [215], one can, at first, note that at 900 °C, that is in the range of temperatures where the mean composition of the phase deviates from the stoichiometry, the Cd vacancies concentration is only slightly increasing with increase in the Te excess (i.e. with decrease in the Cd pressure in the diagram), while that of Te_{Cd} antisites increases by orders of magnitude.

Table 2.11 Maximum width $\Delta\delta_{max}$ of the homogeneity region for selected compound semiconductors and formation energies of vacancies and Schottky pairs. The defect concentration was calculated from the values of $\Delta\delta_{max}$

Material	GaAs	GaN	InP	CdTe	CdSe	CdS	PbTe
$\Delta\delta_{max}$ (mole fraction)	$\sim 2\cdot 10^{-4}$	—	$\sim 5\cdot 10^{-5}$	$\sim 1\cdot 10^{-4}$	$\sim 5\cdot 10^{-4}$	$\sim 1\cdot 10^{-4}$	$\sim 1\cdot 10^{-3}$
Side of $\Delta\delta$ max	As-rich	—	In-rich	Te-rich	Cd-rich	Cd-rich	Te-rich
Defect concentration (at cm^{-3})	$V_{Ga} \le 4.4\cdot 10^{18}$	—	$V_P \le 2\cdot 10^{17}$	$V_{Cd} < 1.4\cdot 10^{18}$	$V_{Se} \le 9.1\cdot 10^{18}$	$V_S \le 4.01\cdot 10^{18}$	$V_{Pb} \le 1.5\cdot 10^{19}$
Formation energy of vacancies (eV)	$H_{VGa} = 2.47$ $H_{VAs} = 2.35$	$H_{VGa} = 4.22$ $H_{VN} = 1.69$	$H_{VIn} = 3.17$ $H_{VP} = 2.10$	$H_{VCd} = 2.51$ $H_{VTe} = 2.4$	— —	$H_{VCd} = 3.06$ $H_{VS} = 2.01$	— —
Formation energy of a Schottky pair (eV)	4.82	5.91	5.27	4.91	4	5.07	N.d

Data from [195, 211, 213, 214]

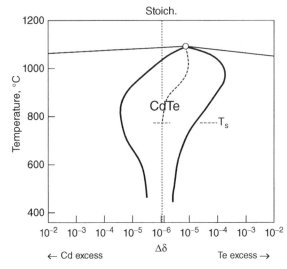

Figure 2.31 *Phase diagram of the CdTe system. At temperatures below T_S the system behaves as a true Daltonide phase – line of average phase composition. Rudolph, 2003, [211]. Reproduced with permission from John Wiley & Sons*

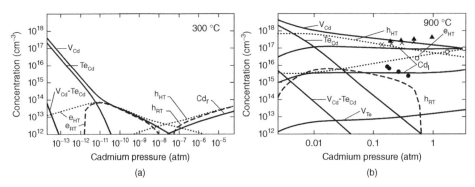

Figure 2.32 *Defect concentration within the existence region of the CdTe phase at (a) 300 and (b) 900 °C [215]. Berding, 1999, [214]. Reproduced with permission from American Physical Society. Reprinted with permission from Ref. [214], Copyright (1999) by the American Physical Society (licence #26770 May 5, 2014)*

Accommodation of Te excess occurs with antisites formation, more than with an increase in vacancies in the lattice.

In addition, although the solid Cd_2Te_3 and $Cd Te_2$ phases are unstable, the semiconducting melt in equilibrium with the solid CdTe phase is a 100% polyassociated liquid (see Section 1.7.10), and complexes having the composition Cd_2Te_3 and $Cd Te_2$ are calculated to be stable [215].

The non-stoichiometric, Te-rich CdTe phase might therefore be considered a solid solution of stoichiometric CdTe and of a virtual Cd_2Te_3 polymorph.

The Bi-Te alloys, which are materials used for thermoelectric heating and cooling applications [216–218], seem, instead, to follow the Anderson model of pseudo-homogeneous phases in the composition interval between 50 and 57 at% of Te, see Figure 2.33

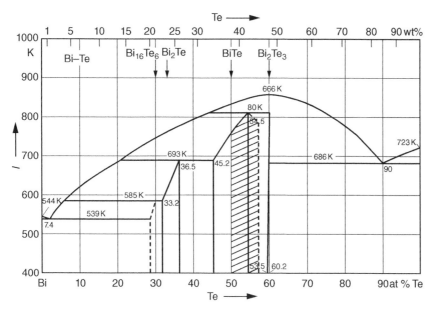

Figure 2.33 *The phase diagram of Bi–Te. The dotted region superimposed on the phase diagram represents the composition interval where the non-stoichiometric phase has been observed by Brebrick [219]*

Samples falling within this composition interval, which ranges between the stoichiometric phase BiTe and the stoichiometric phase Bi_4Te_5 (which decomposes at the peritectic temperature of 813 K) are actually a homogeneous solid solution of the two limiting phases which, after annealing at 450 °C, show X-ray evidence of a continuous shift of their hexagonal lattice parameters [219]. The continuous change in the lattice parameter with the composition is evidence of a peculiar property of Bi-Te alloys, whose structure is made up of close-packed metal atom layers and close-packed chalcogenide atom layers stacked perpendicularly to the hexagonal axis, in proportion dictated by the composition.

Although quite different from CS which accommodates continuous deviations from the stoichiometry of ReO_3-type structures by minute displacements of the ReO_6 octahedra, which represent their basic structural units [4], these alloys exhibit a different, but very effective, structural manifold which allows continuous composition changes by a stacking mechanism, without involving point defects, only extended defects.

References

1. Queisser, H.J. and Haller, E.E. (1998) Defects in semiconductors: some fatal, some vital. *Science*, **281**, 945–950.
2. Ourmazd, A., Ahlborn, K., Ibek, K. and Honda, T. (1985) Lattice and atomic structure imaging of semiconductors by high resolution transmission electron microscopy. *Appl. Phys. Lett.*, **47**, 685–687.
3. Wrachtrup, J., Jelexko, F., Grotz, B. and McGuiness, L. (2013) Nitrogen –vacancy centers close to surfaces. *MRS Bull.*, **38**, 149–154.

4. Rao, C.N.R. and Gopalakrishnan, J. (1997) *New Directions in Solid State Chemistry*, Cambridge University Press, Cambridge.
5. Estreicher, S.K., Gibbons, T.M., Kang, B. and Bebek, M.B. (2014) Phonons and defects in semiconductors: phonon trapping, phonon scattering, and heat flow at heterojunctions. *J. Appl. Phys.*, **115**, 011901.
6. Crawford, J.H. Jr., and Slifkin, L.M. (1972) *Point Defects in Solids*, vol. **1** and **2**, Plenum Press, New York.
7. Lannoo, M. and Burgoin, J. (1981) *Point Defects in Semiconductors*, vol. **1** and **2**, Springer-Verlag, Berlin.
8. Agullo-Lopez, F., Catlow, C.R., Towsend, P.D. (1988) *Point Defects in Materials*, Academic Press, London
9. Pizzini, S. (2002) *Defect Interaction and Clustering in Semiconductors*, Scitec Publications.
10. Pichler, P. (2004) *Intrinsic Point Defects, Impurities, and Their Diffusion in Silicon*, Springer.
11. Fistul, V.I. (2004) *Impurities in Semiconductors: Solubility, Migration and Interaction*, CRC Press, Boca Raton, FL.
12. Drabold, D.A. and Estreicher, S. (2007) *Theory of Defects in Semiconductors*, Springer.
13. McCluskey, D. and Haller, E.E. (2012) *Dopants and Defects in Semiconductors*, CRC Press, Boca Raton, FL.
14. Car, R. and Parrinello, M. (1985) Unified approach for molecular dynamics and density functional theory. *Phys. Rev. Lett.*, **55**, 2471.
15. Wang, C.Z., Chan, C.T. and Ho, K.M. (1991) Tight-binding molecular-dynamics study of defects in silicon. *Phys. Rev. Lett.*, **66**, 189–192.
16. Estreicher, S.K., Backlund, D. and Gibbons, T.M. (2010) Theory of defects in Si and Ge: past, present and recent developments. *Thin Solid Films*, **518**, 2413–2417.
17. Koizumi, K., Boero, M. and Oshiyama, Y.A. (2011) Self-diffusion in crystalline silicon: a Car-Parrinello molecular dynamics study. *Phys. Rev. B*, **84**, 205203.
18. Farrington, G. and Briant, J.L. (1979) Fast ion transport in solids. *Science*, **204**, 1371–1379.
19. Beamont, H. and Jacob, P.W.M. (1966) Energy and entropy parameters for vacancy formation and mobility in ionic crystals from conductance measurements. *J. Chem. Phys.*, **45**, 1496–1502.
20. Chadwick, A.V. (1991) Electrical conductivity measurements of ionic solids. *Philos. Mag. A*, **64**, 983–998.
21. Blöchl, P.E., Smargiassi, E., Car, R. *et al.* (1993) First-principles calculations of self-diffusion constants in silicon. *Phys. Rev. Lett.*, **70**, 2435–2438.
22. Kröger, F. A. and Vink, H.J. (1956) Relations between the concentration of imperfections in crystalline solids in *Solid State Physics*, Vol. **3**, pp. 307–435, Seitz, F., Turnbull, D. (Eds). Academic Press, New York
23. Tan, Y. and Gösele, U. (1985) Point defects, diffusion processes, and swirl defect formation in silicon. *Appl. Phys. A*, **3**, 1–17.
24. Dreyfus, R.W. and Nowick, A.S. (1962) Ionic conductivity of doped NaCl crystals. *Phys. Rev.*, **126**, 1367–1377.

25. Vineyard, G.H. (1957) Frequency factors and isotope effects in solid state rate processes. *J. Phys. Chem. Solids*, **3**, 121–127.
26. Hooton, E. and Jacobs, P.W.M. (1988) Ionic conductivity of pure and doped sodium chloride crystals. *Can. J. Chem.*, **66**, 830–835.
27. Weber, J.R., Janotti, A. and Van de Walle, C.G. (2013) Dangling bonds and vacancies in germanium. *Phys. Rev. B*, **87**, 0352031–0352039.
28. Istratov, A.A., Hieslmair, H. and Weber, E.R. (1999) Iron and its complexes in silicon. *Appl. Phys. A*, **69**, 13–44.
29. Gösele, U. and Tan, T.Y. (1983) The nature of point defects and their influence on diffusion processes in silicon at high temperatures. *MRS Symp. Proc.*, **14**, 45–59.
30. Seeger, A. (1998) Self-interstitials in silicon, in *Semiconductor Silicon 1998*, The Electrochemical Society Inc, pp. 215–219.
31. Song, E.G., Kim, E. and Lee, Y.H. (1993) Fully relaxed point defects in crystalline silicon. *Phys. Rev. B*, **48**, 1486–1489.
32. Antonelli, A., Kaxiras, E. and Chadi, D.J. (1998) Vacancy in silicon revisited: structure and pressure effects. *Phys. Rev. Lett.*, **81**, 2088–2091.
33. Jones, R., Carvalho, A., Palmer, D. *et al.* (2009) The selfinterstitial in silicon and germanium. *Mater. Sci. Eng., B*, **159–160**, 112–116.
34. Leung, W.K., Needs, R.J., Rajagopal, G. *et al.* (1999) Calculations of silicon self-interstitial defects. *Phys. Rev. Lett.*, **83**, 2351–2354.
35. Prawer, S., Rosenblum, I., Orwa, J.O. and Adler, J. (2004) Identification of the point defects in diamond as measured by Raman spectroscopy: comparison between experiment and computation. *Chem. Phys. Lett.*, **390**, 458–461.
36. Watkins, G.D. (1975) Electron paramagnetic resonance of point defects in solids with emphasis on semiconductors, in: *Point Defects in Solids*, Vol. **2**, J.H. Crawford Jr,,L.M. Slifkin, eds. pp. 333–392 Plenum Press, New York
37. Watkins, G.D. and Troxell, J.R. (1980) Negative-U properties for point defects in silicon. *Phys. Rev. Lett.*, **44**, 593–596.
38. Watkins, G.D. (1999) EPR of defects in semiconductors: past, present, future. *Phys. Solid State*, **41**, 746–750, ISSN no 1063-7834.
39. Collins, A.T. (1992) The characterization of point defects in diamond by luminescence spectroscopy. *Diamond Relat. Mater.*, **1**, 457–469.
40. Dannefaer, S., Mascher, P. and Kerr, D. (1986) Monovacancy formation enthalpy in Silicon. *Phys. Rev. Lett.*, **56**, 2195–2198.
41. Dannefaer, S., Avalos, V. and Andersen, O. (2007) Grown-in vacancy-type defects in poly- and single crystalline silicon investigated by positron annihilation. *Eur. Phys. J. Appl. Phys.*, **37**, 213–218.
42. Haesslein, H., Sielemann, R. and Zistl, C. (1998) Vacancies and self-interstitials in germanium observed by perturbed angular correlation spectroscopy. *Phys. Rev. Lett.*, **80**, 2626–2629.
43. Ehrhart, P. and Zillgen, H. (1999) Vacancies and interstitial atoms in e^--irradiated germanium. *J. Appl. Phys.*, **85**, 3503–3511.
44. Vanhellemont, J., Spiewak, P., Sueoka, K. *et al.* (2008) Intrinsic point defect properties and engineering in silicon and germanium Czochralski crystal growth. The 5th International Symposium on Advanced Science and Technology of Silicon Materials (JSPS Si Symposium), Kona, Hawaii, November 10–14, 2008.

45. Watkins, G.D. (1994) 35 years of defects in semiconductors. What next? *Mater. Sci. Forum*, **143–147**, 9–20.
46. Deák, P., Gali, A., Sólyom, A. *et al.* (2005) Electronic structure of boron-interstitial clusters in silicon. *J. Phys. Condens. Matter*, **17**, S2141–S2153.
47. Kelton, K.F. (2003) Diffusion influenced nucleation: a case study of oxygen precipitation in silicon. *Philos. Trans. R. Soc. London, Ser. A*, **361**, 429–446.
48. Bronner, G.B. and Plummer, J.D. (1987) Gettering of gold in silicon: a tool for understanding the role of silicon interstitials. *J. Appl. Phys.*, **61**, 5286–5298.
49. Seeger, A. (1971) Investigation of point defects in silicon and germanium by non-irradiation techniques. *Radiat. Eff.*, **9**, 15–24.
50. Bracht, H., Stolwijk, N.A. and Mehrer, H. (1995) Properties of intrinsic point defects in silicon determined by zinc diffusion experiments under nonequilibrium conditions. *Phys. Rev. B*, **52**, 16542–16560.
51. Ural, A., Griffin, P.B. and Plummer, J.D. (2001) Silicon self-diffusion under extrinsic conditions. *Appl. Phys. Lett.*, **79**, 4328–4330.
52. Mayburg, S. (1954) Vacancies and interstitials in heat treated germanium. *Phys. Rev.*, **95**, 38–43.
53. Frank, W., Gösele, U., Mehrer, H., Seeger, A. (1984) in *Diffusion in Crystalline Solids*, p.63 G. E. Murch and A. S. Nowick Edt. Academic Press, New York.
54. Fuchs, H.D., Walukiewicz, W., Haller, E.E. *et al.* (1995) Germanium Ge^{70}/Ge^{74} isotope heterostructures: an approach to self-diffusion studies. *Phys. Rev. B*, **51**, 16817–16820.
55. Wang, L., Wolk, J.A., Hsu, L. *et al.* (1997) Gallium self-diffusion in gallium phosphide. *Appl. Phys. Lett.*, **70**, 1831–1833.
56. Bracht, H., Haller, E.E. and Clark-Phelps, R. (1998) Silicon self-diffusion in isotope heterostructures. *Phys. Rev. Lett.*, **81**, 393–396.
57. Ural, A., Griffin, P.B. and Plummer, J.D. (1999) Self-diffusion in silicon: similarity between the properties of native point defects. *Phys. Rev. Lett.*, **83**, 3454–3457.
58. Huger, E., Tietze, U., Lott, D. *et al.* (2008) Self-diffusion in germanium isotope multilayers at low temperatures. *Appl. Phys. Lett.*, **93**, 162104.
59. Bracht, H., Schneider, S. and Kube, R. (2011) Diffusion and doping issues in germanium. *Microelectron. Eng.*, **88**, 452–457.
60. Frank, F.C. and Turnbull, D. (1956) Mechanism of diffusion of copper in germanium. *Phys. Rev.*, **10**, 617–618.
61. Jacob, M., Pichler, P., Ryssel, H. and Falster, R. (1997) Determination of the vacancy concentration in the bulk of silicon wafers by platinum diffusion experiments. *J. Appl. Phys.*, **82**, 182–191.
62. Zimmermann, H. and Ryssel, H. (1992) The modeling of Pt diffusion in silicon under non-equilibrium conditions. *J. Electrochem. Soc.*, **139**, 256–262.
63. Gösele, U., Frank, W. and Seeger, A. (1980) Mechanism and kinetics of the diffusion of gold in silicon. *Appl. Phys.*, **23**, 361–368.
64. Lerner, L. and Stolwijk, N.A. (2005) Vacancy concentrations in silicon determined by the indiffusion of iridium. *Appl. Phys. Lett.*, **86**, 011901.
65. Zimmermann, H. and Ryssel, H. (1992) Gold and platinum diffusion: the key for the understanding of intrinsic point defects in silicon. *Appl. Phys. A*, **55**, 121–134.

66. Tang, M., Colombo, L., Zhu, J. and Diaz de la Rubia, T. (1997) Intrinsic point defects in crystalline silicon: tight-binding molecular dynamics studies of self-diffusion, interstitial-vacancy recombination, and formation volumes. *Phys. Rev. B*, **55**, 14279–14289.
67. Gösele, U., Schroer, E., Werner, P.,Tan, T. Y. (1996) Low-temperature diffusion and agglomeration of oxygen in silicon in: *Early Stages of Oxygen Precipitation in Silicon*, R. Jones (ed), pp. 243–261. Kluwer Academic Publishers, Dordrecht.
68. Zimmermann, H. (1998) Vacancy distribution in silicon and methods for their accurate determination. *Defects Diffus. Forum*, **153-155**, 111–136.
69. Bracht, H., Stolwijk, N.A., Mehrer, H. and Yonenaga, I. (1991) Short-time diffusion of zinc in silicon for the study of intrinsic point defects. *Appl. Phys. Lett.*, **59**, 3559–3561.
70. Habu, R., Iwasaki, T., Harada, H. and Tomiura, A. (1994) Diffusion behavior of point defects in Si crystal during melt growth IV: numerical analysis. *Jpn. J. Appl. Phys.*, **33**, 1234–1242.
71. Morehead, F.F. (1987) The diffusivity of selfinterstitials in silicon. *MRS Symp. Proc.*, **104**, 99–104.
72. Justo, J., Bazant, M.Z., Kaxiras, E. *et al.* (1998) Interatomic potential for silicon defects and disordered phases. *Phys. Rev. B*, **58**, 2539–2550.
73. Probert, M.J. and Payne, M.C. (2003) Improving the convergence of defect calculation in supercells: an ab initio study of the neutral silicon vacancy. *Phys. Rev. B*, **67**, 0752041-11.
74. Car, R., Bloch, P. and Smargiassi, E. (1992) Ab initio molecular dynamics of semiconductor defects. *Mater. Sci. Forum*, **83-87**, 433–446.
75. Koizumi, K., Boero, M., Shigeta, Y. and Oshiyama, A. (2011) Self-diffusion in crystalline silicon: a Car-Parrinello molecular dynamics study. *Phys. Rev. B.*, **84**, 205203-1–205203-10.
76. Falster, R., Voronkov, V.V. and Quast, F. (2000) On the properties of the intrinsic point defects in silicon: a perspective from crystal growth and wafer processing. *Phys. Status Solidi B*, **222**, 219–244.
77. Pesola, M., von Boehm, J., Mattila, T. and Nieminen, R.M. (1999) Computational study of interstitial oxygen and vacancy-oxygen complexes in silicon. *Phys. Rev. B*, **60**, 11449–11463.
78. Bracht, H., Fage Pedersen, J., Zangenberg, N. *et al.* (2003) Radiation enhanced silicon self-diffusion and the silicon vacancy at high temperatures. *Phys. Rev. Lett.*, **91**, 245502.
79. Obeidi, S. and Stolwijk, N.A. (2001) Diffusion of iridium in silicon: changeover from a foreign-atom-limited to a native-defect-controlled transport mode. *Phys. Rev. B*, **64**, 113201–113204.
80. Giese, A., Bracht, H., Stolwijk, N.A. and Baither, D. (2000) Microscopic defects in silicon induced by zinc out-diffusion. *Mater. Sci. Eng., B*, **71**, 160–165.
81. Sinno, T., Brown, R.A., von Ammon, W. and Dornberger, E. (1998) Point defect dynamics and the oxidation-induced stacking-fault ring in czochralski-grown silicon crystals. *J. Electrochem. Soc.*, **145**, 302–318.
82. De Souza, M.M. and Sankara Narayanan, E.M. (1998) Self-diffusion in silicon. *Defects Diffus. Forum*, **153–155**, 69–80.

83. Tan, T.Y. and Gösele, U. (2000) in *Handbook of Semiconductor Technology*, vol. **1**, p. 261, K.A Jackson, W. Schröter (Edt), Wiley-VCH Verlag GmbH, Weinheim.

84. Puska, M.J. and Nieminen, R.M. (1983) Defect spectroscopy with positrons: a general calculational method. *J. Phys. F: Met. Phys.*, **13**, 333–346.

85. Brandt, W. and Dupasquier, A. (1983) *Positron Solid State Physics*, North-Holland, Amsterdam.

86. Saarinen, K., Corbelt, C., Hautojarvit, P. *et al.* (1990) Vacancies and voids in deformed GaAs studied by positron lifetime spectroscopy. *J. Phys. Condens. Matter*, **2**, 2453–2459.

87. Gösele, U., Conrad, D., Werner, P. *et al.* (1997) Point defects, diffusion and gettering in silicon. *MRS Symp. Proc.*, **469**, 13–24.

88. Werner, P., Grossmann, H.J., Jacobson, D.C. and Gösele, U. (1998) Carbon diffusion in silicon. *Appl. Phys. Lett.*, **73**, 2465–2467.

89. Pizzini, S. (2002) Chemistry and physics of defect interaction in semiconductors. *Solid State Phenom.*, **85–86**, 1–66. doi: 10.4028/www.scientific.net/SSP.85–86.1

90. Reiss, H., Fuller, C.S. and Morin, F.J. (1956) Chemical interactions among defects in Ge and Si. *Bell Syst. Tech. J.*, **35**, 535–636.

91. Istratov, A.A., Flink, C., Hieslmair, H. *et al.* (1998) Intrinsic diffusion coefficient of interstitial copper in silicon. *Phys. Rev. Lett.*, **81**, 1243–1246.

92. Hall, R.H. and Racette, H. (1964) Diffusion and solubility of copper in extrinsic and intrinsic germanium, silicon, and gallium arsenide. *J. Appl. Phys.*, **35**, 379–397.

93. Estreicher, S.K. (1990) Copper, lithium, and hydrogen passivation of boron in *c*-Si. *Phys. Rev. B*, **41**, 5447–5449.

94. Kalejs, J.P. and Rajendran, S. (1989) Model for diffusion and trapping of hydrogen in crystalline silicon. *Appl. Phys. Lett.*, **55**, 2763.

95. Herrero, C.P., Stutzmann, M., Breitschwert, A. and Santos, P.V. (1990) Trap-limited hydrogen diffusion in doped silicon. *Phys. Rev. B*, **41**, 1054–1058.

96. Borenstein, J., Corbett, J. and Pearton, S.J. (1993) Kinetic model for hydrogen reactions in boron-doped silicon. *J. Appl. Phys.*, **73**, 2751–2754.

97. Wu, R.Q., Yang, M., Feng, Y.P. and Ouyang, Y.F. (2008) Effect of atomic hydrogen on boron-doped germanium: an ab initio study. *Appl. Phys. Lett.*, **93**, 0821007.

98. Gorka, B. (2010) Hydrogen passivation of polycrystalline thin silicon solar cells. Berichte Helmoltz Centre, Berlin, ISSN-1868-5781.

99. Coutinho, J., Jones, R., Briddon, P.R. and Öberg, S. (2000) Oxygen and dioxygen centers in Si and Ge: density-functional calculations. *Phys. Rev. B*, **62**, 10824–10840.

100. Casali, R.A., Ruecker, H. and Methfessel, M. (2001) Interaction of vacancies with interstitial oxygen in silicon. *Appl. Phys. Lett.*, **78**, 913.

101. Furuhashi, M. and Taniguchi, K. (2005) Diffusion and dissociation mechanisms of vacancy-oxygen complex in silicon. *Appl. Phys. Lett.*, **86**, 142107.

102. Backlund, D.J. and Estreicher, S.K. (2010) Ti, Fe, and Ni in Si and their interactions with the vacancy and the *A* center: a theoretical study. *Phys. Rev. B*, **81**, 235213-1–235213-8.

103. Xiang-Ti, M. (1989) Two Si-H IR bands caused by vacancy-oxygen-hydrogen complexes in neutron transmutation doped FZ-Si grown in hydrogen atmosphere. *Semicond. Sci. Technol.*, **4**, 892–894.

104. Wright, A.F. and Wixon, R.R. (2008) Density functional-theory calculations for silicon vacancy migration. *J. Appl. Phys.*, **103**, 083517-1–083517-5.

105. Henry, M.O., Campion, J.D., McGuigant, K.G. *et al.* (1994) A photoluminescence study of Zn-0 complexes in silicon. *Semicond. Sci. Technol.*, **9**, 1375–1381.

106. Voronkov, V.V. and Falster, R. (2002) Intrinsic point defects and impurities in silicon crystal growth. *J. Electrochem. Soc.*, **149**, G167–G174.

107. Bracht, H. (2015) Self and dopant diffusion in silicon, germanium and its alloys, in *Silicon, Germanium, and Their Alloys: Growth, Defects, Impurities, and Nanocrystals* (eds G. Kissinger and S. Pizzini), Taylor & Francis pp.159–215.

108. Haller, E.E. (1978) Hydrogen in germanium. Proceedings of the 10th International Conference on Defects and Radiation Effects in Semiconductors, Nice, France, September 11–14, 1978.

109. Cerofolini, G.F. and Ottaviani, G. (1989) Hydrogen in silicon: state, reactivity and evolution after ion implantation. *Mater. Sci. Eng., B*, **4**, 19–24.

110. Pearton, S.J. (1987) Hydrogen in crystalline silicon, in *Oxygen, Carbon, Hydrogen and Nitrogen in Crystalline Silicon*, MRS Symposia Proceedings, vol. **59**, Cambridge University Press, Cambridge, pp. 457–468.

111. Van de Walle, C.G. and Neugebauer, J. (2003) Universal alignment of hydrogen levels in semiconductors, insulators and solutions. *Nature*, **423**, 626–628.

112. Weber, J., Hiller, M. and Lavrov, E.V. (2006) Hydrogen in germanium. *Mater. Sci. Semicond. Process.*, **9**, 564–570.

113. Stavola, M. (2008) Hydrogen in silicon and germanium. Proceedings of the 5th International Symposium on Advanced Science and Technology of Silicon Materials (JSPS Si Symposium), Kona, Hawaii, November 10–14, 2008, pp. 337–343.

114. Endrös, A., Krüler, W. and Grabmaier, J. (1989) An hydrogen-carbon related deep donor in crystalline n-type Si. *Mater. Sci. Eng., B*, **4**, 35–39.

115. Seager, C.H., Anderson, R.A. and Panitz, J.K.G. (1987) The diffusion of hydrogen in silicon and mechanism for "unintentional" hydrogenation during ion beam processes. *J. Mater. Res.*, **2**, 96–106.

116. Dürr, M. and Höfer, U. (2013) Hydrogen diffusion on silicon surfaces. *Prog. Surf. Sci.*, **88**, 61–101.

117. Ballutaud, D., Jornard, F., LeDuigou, J. *et al.* (2000) Diffusion and thermal stability of hydrogen in homoepitaxial CVD diamond films. *Diamond Relat. Mater.*, **9**, 1171–1174.

118. Pearton, S.J., Corbett, J.W. and Stavola, M. (1992) *Hydrogen in Crystalline Semiconductors*, Springer-Verlag, Berlin.

119. Estreicher, S.K., Docaj, A., Bebek, M.B. *et al.* (2012) Hydrogen, H in C -rich Si, and the diffusion of vacancy complexes. *Phys. Status Solidi A*, **209**, 1872–1879.

120. Neegard Walternburg, H. and Yates, J.T. (1995) Surface chemistry of silicon. *Chem. Rev.*, **95**, 1589–1673.

121. Kern, W. (1993) Wafer cleaning technology, in *Handbook of Semiconductors*, Noyes Publishing, Westwood, NJ.

122. Hartung, J. and Weber, J. (1989) Defects created by ion implantation into silicon. *Mater. Sci. Eng., B*, **4**, 47–50.

123. Meda, L., Cerofolini, G.F., Bresolin, C., Dierckx, R., Donelli, D., Orlandini, M., Anderle, M., Canteri, R., Ottaviani, G., Tonini, R., Claeys, C., Vanhellemont, J., Pizzini, S., Farina, S. (1990) State and evolution of hydrogen implanted silicon in: *Semiconductor Silicon 90*. H.R. Huff, K.G. Barreclought and J.-i. Chikawa, Edt, The Electrochemical Society, Pennington, NJ, pp. 456–471.

124. Hartung, J. and Weber, J. (1993) Shallow hydrogen related donors in silicon. *Phys. Rev. B*, **48**, 14161–14166.

125. Pantelides, S., Ramamoorthy, M., Reboredo, F. (1999) Nucleation and growth of extended defects in silicon, in *Defects in Silicon III* Electrochemical Society Proceedings, **99**(1), 3–19. The Electrochemical Society, Pennington, NJ.

126. Fukata, A., Kasuya, A. and Suezawa, M. (2001) Formation energy of vacancy in silicon determined by a new quenching method. *Physica B*, **308**, 1125–1128.

127. Peng, C., Zhang, H., Stavola, M. *et al.* (2011) Microscopic structure of a VH_4 center trapped by C in Si. *Phys. Rev. B*, **84**, 195205.

128. Sato, K., Castaldini, A., Fukata, N. and Cavallini, A. (2012) Electronic level scheme in boron-and phosphorous–doped silicon nanowires. *Nano Lett.*, **12**, 3012–3017.

129. Ceccaroli, B. and Pizzini, S. (2012) Processes in *Advanced Silicon Materials for Photovoltaic Applications*, pp. 21–78, S. Pizzini (Edt) John Wiley & Sons, Ltd.

130. Boyle, R. (2008) FT-IR Measurement of Interstitial Oxygen and Substitutional Carbon in Silicon Wafers. Application Note 50640, Thermo Fisher Scientific, Madison, WI.

131. Abe, T. (1998) History and future of silicon crystal growth, in *Semiconductor Silicon 1988*, The Electrochemical Society, pp. 157–178.

132. Taylor, W.J., Tan, T.Y. and Gösele, U. (1993) Carbon precipitation in silicon: why is so difficult? *Appl. Phys. Lett.*, **62**, 3336–3339.

133. Endros, A. (1989) Charge state dependent hydrogen carbon related deep donor in crystalline silicon. *Phys. Rev. Lett.*, **63**, 70–73.

134. Kamiura, Y., Maeda, T., Yamashita, Y. and Nakamura, M. (1998) Formation of carbon related defects during the carbon enhanced annhihilation of thermal donors in silicon. *Jpn. J. Appl. Phys.*, **37**, L101–L104.

135. Scholz, R., Gösele, U., Hu, J.Y. and Tan, T.Y. (1998) Carbon induced undersaturation of silicon self-interstitials. *Appl. Phys. Lett.*, **72**, 200–202.

136. Newman, R.C. (1998) Oxygen aggregation and interaction with carbon and hydrogen, in *Semiconductor Silicon 1988*, The Electrochemical Society, pp. 257–271.

137. Tan, Y. and Gösele, U. (2000) Point defects, diffusion and precipitation in *Handbook of Semiconductor Technology*, Vol. **1**, K.A Jackson, W. Schröter (Edt.), Wiley-VCH Verlag GmbH, Weinheim, pp. 260–290.

138. Narushima, T., Yamashita, A., Ouchi, C. and Iguchi, Y. (2002) Solubilities and equilibrium distribution coefficients of oxygen and carbon in silicon. *Mater. Trans.*, **43**, 2120–2124.

139. Bean, A.R. and Newman, R.C. (1971) The solubility of carbon in pulled silicon. *J. Phys. Chem. Solids*, **32**, 1211–1219.

140. Nozaki, T., Yatsurugi, Y. and Akiyama, N. (1970) Concentration and behavior of carbon in semiconductor silicon. *J. Electrochem. Soc.*, **117**, 1566–1568.

141. Endo, Y., Yatsurugi, Y. and Akiyama, N. (1972) Infrared spectrophotometry for carbon in silicon as calibrated by charged particle activation. *Anal. Chem.*, **44**, 2258–2262.

142. Gösele, U. (1986) The role of carbon and point defects in silicon, in *Oxygen, Carbon, Hydrogen and Nitrogen in Crystalline Silicon*, MRS Symposia Proceedings, vol. **59**, Cambridge University Press, Cambridge, pp. 419–431.

143. Zhu, X.-Y., Lee, S.-M., Kim, J.-Y. *et al.* (2001) Structural and vibrational properties of impurities in crystalline silicon. *Semicond. Sci. Technol.*, **16**, 41–49.

144. Hu, S.M. (1986) Oxygen precipitation in silicon, in *Oxygen, Carbon, Hydrogen and Nitrogen in Crystalline Silicon*, MRS Symposia Proceedings, vol. **59**, Cambridge University Press, Cambridge, pp. 249–267.

145. Mikkelsen, J.C. Jr. (1986) An Overview of oxygen in silicon, in *Oxygen, Carbon, Hydrogen and Nitrogen in Crystalline Silicon*, MRS Proceedings, vol. **59**, Cambridge University Press, Cambridge, pp. 3–5.

146. Mikkelsen, J.C. Jr. (1986) Diffusivity and solubility of oxygen in silicon, *Oxygen, Carbon, Hydrogen and Nitrogen in Crystalline Silicon*, MRS Proceedings, Cambridge University Press, Cambridge vol. **59**, pp. 19–30.

147. Borghesi, A., Pivac, B., Sassella, A. and Stella, A. (1995) Oxygen precipitation in silicon. *J. Appl. Phys.*, **77**, 4169–4244.

148. Hieslmair, H., McHugo, S.A., Istratov, A.A., Weber, E.R., Gettering of transition metals in crystalline silicon in: EMIS Datareviews Series, Chapter 15, vol. **20**, *Properties of Crystalline Silicon*, R. Hull (Edt) (Inspec, Exeter, 1999), pp. 775–808. ISBN: 0 85296 933 3.

149. Newman, R.C. (2000) Oxygen diffusion and precipitation in Czochralski silicon. *J. Phys. Condens. Matter*, **12**, R335–R365.

150. Kaiser, W. and Thurmond, S.D. (1961) Solubility of oxygen in germanium. *J. Appl. Phys.*, **31**, 115–118.

151. Hrostowsi, H.J. and Kaiser, R.H. (1959) The solubility of oxygen in silicon. *J. Phys. Chem. Solids*, **9**, 214–216.

152. Gass, J., Müller, H.H., Stussi, H. and Schweitzer, S. (1980) Oxygen diffusion in silicon and influence of different dopants. *J. Appl. Phys.*, **51**, 2030–2037.

153. Mikkelsen, J.C. Jr., (1982) Excess solubility of oxygen in silicon during steam oxidation. *Appl. Phys. Lett.*, **41**, 871.

154. Itoh, Y. and Nozaki, T. (1985) Solubility and diffusion coefficient of oxygen in silicon. *Jpn. J. Appl. Phys.*, **24**, 279–284.

155. Huang, X., Terashima, K., Sasaki, H. *et al.* (1993) Oxygen solubility in silicon melts: the influence of Sb additions. *Jpn. J. Appl. Phys.*, **32**, 3671–3674.

156. Artacho, E. and Yndurain, F. (1995) Theory of interstitial oxygen in silicon and germanium. *Mater. Sci. Forum*, **196–201**, 103–107.

157. Kimerling, C. (1986) Structure and properties of the oxygen donors, in *Oxygen, Carbon, Hydrogen and Nitrogen in Crystalline Silicon*, MRS Symposia Proceedings, vol. **59**, Cambridge University Press, Cambridge, pp. 83–94.

158. Stavola, M. and Lee, K.M. (1986) Electronic structure and atomic symmetry of the oxygen donors in silicon, in *Oxygen, Carbon, Hydrogen and Nitrogen in Crystalline Silicon*, MRS Symposia Proceedings, vol. **59**, Cambridge University Press, Cambridge, pp. 96–110.

159. Tan, T.Y. (1986) Exigent accommodation volume of precipitation and formation of oxygen precipitates in silicon, in *Oxygen, Carbon, Hydrogen and Nitrogen in Crystalline Silicon*, MRS Symposia Proceedings, vol. **59**, Cambridge University Press, Cambridge, pp. 269–279.

160. Pizzini, S., Guzzi, M., Grilli, E. and Borionetti, G. (2000) The photoluminescence emission in the 0.7-0.9 range from oxygen precipitates, thermal donors and dislocations in silicon. *J. Phys.: Condens. Matter*, **12**, 10131–10143.
161. Schmidt, J. and Cuevas, A. (1999) Electronic properties of light-induced recombination centers in boron-doped Czochralski silicon. *J. Appl. Phys.*, **86**, 3175–3180.
162. Glunz, S.W., Rein, S., Lee, J.Y. and Warta, W. (2001) Minority carriers degradation in boron-doped Czochralski silicon. *J. Appl. Phys.*, **90**, 2397–2404.
163. Schmidt, J. and Bothe, K. (2004) Structure and transformation of the metastable boron- and oxygen-related defect center in crystalline silicon. *Phys. Rev. B*, **69**, 024107.
164. Schmidt, J., Bothe, K., MaDonald, D. *et al.* (2006) Electronically stimulated degradation of silicon solar cells. *J. Mater. Res.*, **21**, 5–12.
165. Adley, J., Jones, R., Palmer, D.W. *et al.* (2004) Degradation of Boron-doped Czochralski grown silicon solar cells. *Phys. Rev. Lett.*, **93**, 0555041.
166. Carvalho, A., Jones, R., Sanati, M. *et al.* (2006) First principle investigations of a bistable boron-oxygen interstitial pair in Si. *Phys. Rev. B*, **73**, 245210-1–245210-7.
167. Sanati, M. and Estreicher, S.K. (2006) Boron–oxygen complexes in Si. *Physica B*, **376–377**, 133–136.
168. Voronkov, V. and Falster, R. (2010) Latent complexes of interstitial boron and oxygen dimers as a reason for degradation of silicon-based solar cells. *J. Appl. Phys.*, **107**, 053509.
169. Van Kemp, R., Sieverts, E.G. and Amerlaan, C.A.J. (1986) The electronic structure of the oxygen-vacancy complex in silicon. *Mater. Sci. Forum*, **10–12**, 875–880.
170. Ewels, C.P., Jones, R. and Öberg, S. (1995) A first principle investigation of vacancy oxygen defects in silicon. *Mater. Sci. Forum*, **196–201**, 1297–1301.
171. Pavlov, P.V., Zorin, E.I., Tetelbaum, D.I. and Khokhlov, A.F. (1976) Nitrogen as dopant in silicon and germanium. *Phys. Status Solidi A*, **35**, 11–36.
172. Stein, H.J. (1986) Nitrogen in crystalline silicon, in *Oxygen, Carbon, Hydrogen and Nitrogen in Crystalline Silicon*, MRS Symposia Proceedings, vol. **59**, Cambridge University Press, Cambridge, pp. 523–535.
173. Itoh, T. and Abe, T. (1988) Diffusion coefficient of a pair of nitrogen atoms in float zone silicon. *Appl. Phys. Lett.*, **53**, 39.
174. Jones, R., Öberg, S., Berg Rasmussen, F., Bech Nielsen, B. (1994) Identification of the dominant nitrogen defect in silicon, *Phys. Rev. Lett.*, **72**, 1882.
175. Berg Rasmussen, F. and Bech Nielsen, B. (1994) Microstructure of the nitrogen pair in crystalline silicon studied by ion channeling. *Phys. Rev. B*, **49**, 16353–16360.
176. Alt, H.C. and Tapfer, L. (1984) Photoluminescence study of nitrogen implanted silicon. *Appl. Phys. Lett.*, **45**, 426.
177. Lim, H., Parj, S., Cheong, H. *et al.* (2006) Photoluminescence of natural diamonds. *J. Korean Phys. Soc.*, **48**, 1556–1559.
178. Rogers, L.J., Amstrong, S., Sellers, M.J. and Manson, N.B. (2008) Infrared emission of the NV centre in diamond: Zeeman and uniaxial stress studies. *New J. Phys.*, **10**, 103024-1–103024-15.
179. Mitchell, B., Shewchun, J., Thompson, D.A. and Davies, J.A. (1975) Nitrogen–implanted silicon. II. Electrical properties. *J. Appl. Phys.*, **46**, 335–343.

180. Voronkov, V.V., Porrini, M., Collareta, P. *et al.* (2001) Shallow thermal donors in nitrogen-doped silicon. *J. Appl. Phys.*, **89**, 4289–4293.
181. Alt, H.C., Gomeniuk, Y.V., Bittersberger, F. *et al.* (2006) Analysis of electrically active N-O complexes in nitrogen-doped CZ silicon crystals with FTIR spectroscopy. *Mater. Sci. Semicond. Process.*, **9**, 114–116.
182. Yang, D., Lu, J., Shen, Y. *et al.* (2001) Investigation of as-grown nitrogen-doped Czochralski silicon. *Proc. SPIE*, **4412**, 116–119.
183. Murphy, J.D., Alpass, C.R., Giannattasio, A. *et al.* (2006) Nitrogen in Silicon: transport and mechanical properties. *Nucl. Instrum. Methods Phys. Res., Sect. B*, **253**, 113–117.
184. Alpass, C.R., Murphy, J.D., Falster, R.J. and Wilshaw, P.R. (2009) Nitrogen diffusion and interaction with dislocations in single-crystal silicon. *J. Appl. Phys.*, **105**, 013519.
185. Alpass, C.R., Jain, A., Murphy, J.D. and Wilshaw, P.R. (2009) Measurements of dislocation locking by near-surface ion-implanted nitrogen in Czochralski silicon. *J. Electrochem. Soc.*, **156**, H669.
186. Rasmussen, F.B., Öberg, S., Jones, R. *et al.* (1996) The nitrogen-pair oxygen defect in silicon. *Mater. Sci. Eng., B*, **36**, 91–95.
187. Kulkarni, M.S. (2008) Defect dynamics in the presence of nitrogen in growing Czochralski silicon crystals. *J. Cryst. Growth*, **310**, 324–335.
188. Sumino, K., Yonenaga, I., Imai, M. and Abe, T. (1983) Effects of nitrogen on dislocation behavior and mechanical strength in silicon crystals. *J. Appl. Phys.*, **54**, 5016–5020.
189. Cardona, M. (1963) Optical properties of the silver and cuprous halides. *Phys. Rev.*, **129**, 68–78.
190. Ahn, D. and Chuang, S.L. (2013) High optical gain of I-VII semiconductor quantum wells for efficient light emitting devices. *Appl. Phys. Lett.*, **102**, 121114.
191. Nishizawa, J.-I. (2003) Stoichiometry control and point defects in compound semiconductors. *Mater. Sci. Semicond. Process.*, **6**, 249–252.
192. Van Vechten, J.A. (1975) Simple theoretical estimates of the enthalpy of antistructure pair formation and virtual-enthalpies of isolated antisite defects in zinc-blende and wurtzite type semiconductors. *J. Electrochem. Soc.*, **122**, 423–429.
193. Newmark, F. (1997) Defects in wide band gap II-VI crystals. *Mater. Sci. Eng.,R*, **21**, 1–46.
194. Phillips, J.C. and Van Vechten, J.A. (1973) Macroscopic model of formation of vacancies in semiconductors. *Phys. Rev. Lett.*, **30**, 220–223.
195. Fiechter, S. (2004) Defect formation energies and homogeneity ranges of rock salt-, pyrite-,chalcopyrite- and molybdenite–type of compound semiconductors. *Sol. Energy Mater. Sol. Cells*, **83**, 459–477.
196. Van Vechten, J.A. (1972) Scaling theory of melting temperature of covalent crystals. *Phys. Rev. Lett.*, **29**, 769.
197. Pinto, H.M., Coutinho, J., Torres, V.J.B. *et al.* (2006) Formation energy and migration barrier of a Ge vacancy from ab initio studies. *Mater. Sci. Semicond. Process.*, **9**, 498–502.
198. Śpiewak, P., Vanhellemont, J., Sueoka, K. *et al.* (2008) First principles calculations of the formation energy and deep levels associated with the neutral and charged vacancy in germanium. *J. Appl. Phys.*, **103**, 086103.

199. Kamiyama, E., Sueoka, K. and Vanhellemont, J. (2013) Formation energy of intrinsic point defects in Si and Ge and implications for Ge crystal growth. *ECS J. Solid State Sci. Technol.*, **2**, P104–P109.

200. Hiraki, A. (1966) Experimental determination of diffusion and formation energies of thermal vacancies in germanium. *J. Phys. Soc. Jpn.*, **21**, 34–41.

201. Dobson, T.W. and Wager, J.F. (1989) Enthalpy of formation of antisite defects and antistructure pairs in IIIV compound semiconductors. *J. Appl. Phys.*, **66**, 1997–2001.

202. Jansen, R.W. and Sankey, O.F. (1989) Theory of relative native- and impurity-defect abundances in compound semiconductors and the factors that influence them. *Phys. Rev. B*, **39**, 3192–3206.

203. Heitmann, J., Scholz, R., Schmidt, M. and Zacharias, M. (2002) Size controlled nc-Si synthesis by SiO/SiO_2 superlattices. *J. Non-Cryst. Solids*, **299–302**, 1075–1078.

204. Pavesi, L., Dal Negro, L., Mazzoleni, C. *et al.* (2000) Optical gain in silicon nanocrystals. *Nature*, **408**, 440–444.

205. Hartel, A.M., Hiller, D., Gutsch, S. *et al.* (2011) Formation of size-controlled silicon nanocrystals in plasma enhanced chemical vapor deposition grown SiO_xN_y/SiO_2 superlattices. *Thin Solid Films*, **520**, 121–125.

206. Sarikov, A. and Zacharias, M. (2012) Gibbs free energy and equilibrium states in the Si/Si oxide systems. *J. Phys. Condens. Matter*, **24**, 385403–385410.

207. Luna Lopez, A., Lopez, J.C., Vazquez Valerdi, D.E. *et al.* (2012) Morphological, compositional, structural and optical properties of Si-nc embedded in SiOx films. *Nanoscale Res. Lett.*, **7**, 604.

208. Löper, P., Canino, M., Qazzazie, D. *et al.* (2013) Silicon nanocrystals embedded in silicon carbide: investigation of charge carrier transport and recombination. *Appl. Phys. Lett.*, **102**, 033507.

209. Fedorov, P.P. (2012) Berthollide formation conditions. *Russ. J. Inorg. Chem.*, **57**, 959–969.

210. Roth, W.L. (1970) Extended defects in magnetic structures in: *The Chemistry of Extended Defects in Non Metallic Solids* L. Eyring, M. O'Keeffe (Edt) North Holland, Amsterdam, pp. 445–455.

211. Rudolph, R. (2003) Non-stoichiometry related defects at the melt growth of semiconductor compound crystals-a review. *Cryst. Res. Technol.*, **38**, 542–554.

212. Anderson, J.S. (1970) Non stoichiometry and ordered phases, thermodynamic considerations in *The Chemistry of Extended Defects in Non Metallic Solids*, Eyring, L., O'Keeffe, M. (Edt) North Holland, Amsterdam pp. 1–14.

213. Berding, M.A., Sher, A. and Cher, A.B. (1990) Vacancy formation and extraction energies in semiconductor compounds and alloys. *J. Appl. Phys.*, **68**, 5064–5076.

214. Berding, A. (1999) Native defects in CdTe. *Phys. Rev. B*, **60**, 8943–8950.

215. Moskvin, P.P., Raskovets'ki, L.V., Kavertsev, S.V. *et al.* (2003) Polyassociative thermodynamic model of A^2B^6 semiconductor melt and P-T-X equilibria in Cd-Hg-Te system: 2 phase equilibria in initial two-component system. CdTe system. *Semicond. Phys., Quantum Electron. Optoelectron.*, **6**, 23–27.

216. Wright, D.A. (1958) Thermoelectric properties of bismuth telluride and its alloys. *Nature*, **181**, 834.

217. Goldsmid, J. (1961) Recent studies of bismuth telluride and its alloys. *J. Appl. Phys.*, **32**, 2198–2202.
218. Poudel, B., Hao, Q., Ma, Y. *et al.* (2008) High-thermoelectric performance of nanostructured bismuth antimony telluride bulk alloys. *Science*, **320**, 634–638.
219. Brebrik, R.F. (1970) Evidence for a continous sequence of structures in the Bi-Te system in: *The Chemistry of Extended Defects in Non Metallic Solids*, Eyring, L., O'Keeffe, M. (Edt) North Holland, Amsterdam pp. 183–197.

3

Extended Defects in Semiconductors and Their Interactions with Point Defects and Impurities

3.1 Introduction

Extended defects (EDs) may be classified as 1D (line) defects, 2D (surface) defects and 3D (volume) defects. Dislocations are typically 1D defects, grain boundaries (GBs), twins and stacking faults (SFs) are typically 2D defects and precipitates, voids and inclusions are typically 3D defects. Single crystalline semiconductors for microelectronic and optoelectronic applications are seldom totally free of EDs or, at least, dislocation-free. Czochralsky (CZ) or float zone (FZ) silicon and germanium ingots are almost unique exceptions among single crystal semiconductors with their very low density of dislocations ($N_D \approx 10^2$ cm^{-2}), while single crystal GaAs, GaN, SiC and InP, as typical examples, are strongly dislocated, with dislocation densities up to 10^8 cm^{-2} or more.

Different from point defects, which are equilibrium species (although not necessarily present in equilibrium concentration, see Chapter 2), dislocations, SFs, twins, GBs and internal surfaces are non-equilibrium defects, and their density, therefore, does not depend on the law of mass action.

The formation of dislocations occurs to relieve local strain conditions caused by thermal or mechanical stresses during the crystal growth (see Chapter 4) or during subsequent thermal or mechanical treatments (see Chapter 5). GB formation typically occurs when a crystalline semiconductor is grown in the absence of seeding. Precipitate formation occurs, eventually, in supersaturated solutions of single or multiple impurities, after a proper nucleation process.

The presence of EDs adds to the internal energy of the material an excess energy contribution ΔE_{exc} that depends on the nature of the particular defect and on its absolute density. As EDs are not equilibrium defects, it should be theoretically possible to prepare extended defect-free materials, although it is often impossible to avoid the conditions which

Physical Chemistry of Semiconductor Materials and Processes, First Edition. Sergio Pizzini.
© 2015 John Wiley & Sons, Ltd. Published 2015 by John Wiley & Sons, Ltd.

favour their homogeneous or heterogeneous nucleation, such as non-uniform temperature conditions during bulk crystal growth, lattice misfit in heteroepitaxial growth, impurity supersaturation during thermal cycling and mechanical and thermal stresses in device manufacturing processes.

While the structural and physical properties of EDs in semiconductors have been treated in a number of books [1–8] and in many thousands of papers over the last 50 years, the physics and the chemistry of the interaction processes involving EDs, atomic defects and impurities still deserves basic and technological attention.

The challenge was, and still is, the design and deliberate use of interaction processes involving impurities and EDs to improve or modify the materials properties, for a specific technological application [9–15].

Among EDs, dislocations have caused increasing concern in semiconductors technology because they are at the origin of severe degradation effects of elemental and compound semiconductor devices.

While, with the improvement in crystal growth techniques, the density of growth-induced dislocations in elemental and compound semiconductors has been drastically reduced, there is still concern [16–19] about the generation of slip and, consequently, of dislocations, in the active area of microelectronic devices manufactured with large diameter silicon wafers (from 300 up to 450 mm). In these massive and thick substrates the thermal and gravitational stresses increase so dramatically such that they induce slip and dislocation generation during high temperature treatments, which are common in ultra large scale integration (ULSI) engineering.

In areas other than silicon microelectronic devices deliberately induced dislocations could be beneficial, since dislocations behave as deep sinks for impurities and could be, and are, used in impurity gettering processes (see Chapter 5).

Dislocations interact irreversibly with atomic defects and impurities if the interaction process, which can be described with a chemical reaction

$$ED + X \rightleftharpoons ED - X \tag{3.1}$$

brings the system to a state of minimum energy, with a Gibbs free energy of reaction $\Delta G_{3.1} < 0$.

As EDs–point defect/impurity interactions are shown to be thermodynamically favoured processes, there is good reason to believe that, in the majority of cases, EDs are thermodynamically stable when they are '*reacted*'.

It follows, also, that the experimental study of '*clean*' EDs is difficult, if not impossible.

Considering the physico-chemical relevance of interaction processes involving EDs, this chapter will only deal briefly with the fundamentals of ED physics, while more emphasis will be given to the reactivity of EDs towards point defects and impurities. Gettering and hydrogenation processes,will be discussed in Chapter 5.

3.2   Dislocations in Semiconductors with the Diamond Structure

3.2.1   Geometrical Properties

Dislocations are the most frequent, although unwanted, EDs in semiconductors and can have an intrinsic or extrinsic character. Intrinsic dislocations are generated as a

consequence of plastic deformation of a sample, extrinsic ones by coalescence of supersaturated self-interstitials. Their structural and physical properties have been systematically
studied on plastically deformed samples, where the presence of dislocations is accompanied by point defects and point defect clusters, which share with dislocations the properties
of being paramagnetic centres. As an example, in a silicon sample plastically deformed
at 650 °C to get a dislocation density of 10^9 cm^{-2}, the measured paramagnetic centres
density is around 10^{16} cm^{-3}, of which between 65 and 80% are point defect clusters.

As is known, well before their direct observation by transmission electron microscopy
(TEM), the presence of dislocations in synthetic crystals was theoretically suggested to
explain the three- or even more orders of difference between the calculated and the experimental values of the maximum resolved shear-stress needed to induce plastic flow in
metals [2, 20].

In addition to TEM (see Figure 3.1a) and high resolution transmission electron
microscopy (HRTEM), dislocations may be easily imagined with SEM by anisotropic
chemical etching procedures (see Figure 3.1b) or, thanks to their electrical and optical
activity, by electron beam induced current (EBIC), light beam induced current (LBIC),
photoluminescence (PL) and cathodoluminescence (CL) measurements.

Electron spin resonance (ESR), PL and deep level transient spectroscopy (DLTS) are,
the main spectroscopic tools used to assess their electronic properties. The special feature of ESR is that it makes possible the distinction between dislocation-related and point
defect-related ESR signals, thanks to the anisotropy of the low temperature dislocation
spectrum, although it yields information only on silicon. In fact, the high nuclear spin of
the isotope Ge^{73} and strong spin–orbit interactions are impeding factors for ESR measurements in Ge and similar problems occur with III–V compounds [4].

Using a geometrical description, two fundamental types of dislocations (edge and screw)
can be formally generated by halving a cubic crystal along an $x - y$ lattice plane and shear-
ing the crystal along the x or y direction (see Figure 3.2) [21]. As a consequence of the
shear, a local deformation of the crystal occurs, associated with the nucleation of a dislocation, formally consisting of a line of unsaturated lattice atoms. In the case of elemental

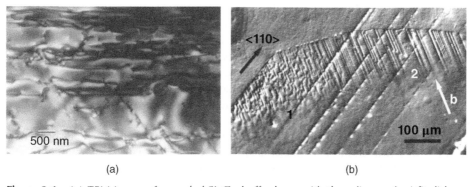

(a)   (b)

Figure 3.1 (a) TEM image of a graded Si–Ge buffer layer with threading and misfit dislocations: the silicon/germanium interface is the source of dislocations. (b) A SEM microphoto-
graph of arrays of dislocations introduced by standard four-point bending of FZ samples at
950 K. Selective etching was carried out using the basic Sirtl etchant. Courtesy of E. Yakimov,
Institute of Microelectronics Technology RAS

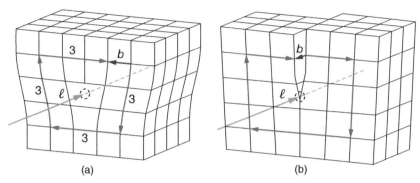

Figure 3.2   *Burgers loops in a faulted crystal containing an edge dislocation (a) and a screw dislocation (b). Blumenau, 2002, [21]. Reproduced with permission from A. T. Blumenau*

semiconductors of the IV group, the lattice atoms belonging to the dislocation line are only threefold coordinated and, consequently, characterized by the presence of broken (dangling) bonds and unpaired electrons.

The identification of the structural properties of a dislocation in a crystalline solid can be obtained using the classical Burgers loop procedure, which consists in drawing a closed loop of *n* lattice distances *a* in the faulted region of the crystal containing a glide or a screw dislocation (Figure 3.2). It is immediately apparent that to close the loop in the dislocated crystal $(n + 1)$ integer steps are necessary.

The difference gives the size and the direction of a vector *b*, called the Burger's vector, which is a measure of the dislocation width and characteristic of the dislocation line ℓ in one particular plane of the crystal.

Because a dislocation line connects a region where slip occurred to a still perfect region of the crystal, it cannot end in a perfect region of the crystal, but either forms a closed loop or terminates at an internal or external surface or at another dislocation.

A vector ξ, tangent to the dislocation line, is used to follow the spatial evolution in character and in direction of the dislocation. For an edge dislocation *b* is orthogonal to ξ, while it is parallel to ξ for a screw dislocation.

When the angle ϑ between *b* and ξ is neither zero nor an integral multiple of $\pi/2$, the dislocation has a mixed character (neither edge nor screw).

Thus, the couple of *b* and ξ vectors permits one to describe entirely the geometrical properties of the dislocations at any point in the crystal lattice.

Limiting our concern to dislocations in diamond type structures [23], it can be demonstrated that they take the most stable configuration when the line vector ξ is directed along the <110> direction.

As the diamond lattice results from the co-penetration of two sublattices, displaced by (1/4, 1/4, 1/4), two sets of {111} planes are present, called, respectively, the shuffle-set and glide-set on which a perfect dislocation (see Figure 3.3) [22] can be nucleated and expand, see Figure 3.4 [21].

Dislocations on the glide-set are either screw or 60° edge dislocations and the latter is so called because its Burgers vector makes an angle of 60° with its line vector. These are the most frequently found in crystals with the diamond structure after plastic deformation at high temperatures.

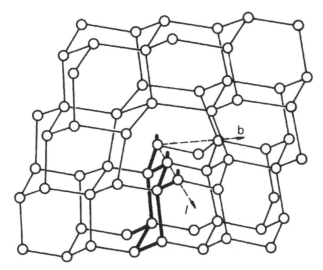

Figure 3.3 *Three-dimensional view of a perfect 60° dislocation. Hornstra, 1958. [22]. Reproduced with permission from Elsevier*

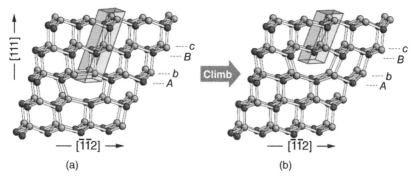

Figure 3.4 *The glide (a) and the shuffle (b) structure of a perfect 60° dislocation in a diamond-type of crystal. The extra half-plane is enclosed in a box. Depending where the half-plane terminates, the dislocation expands in a glide-set or in a shuffle-set. Blumenau, 2002, [21]. Reproduced with permission from A. T. Blumenau*

60° perfect dislocations spontaneously dissociate into a pair of partial dislocations [23, 24] (see Figure 3.5) with the topological consequence of the formation of a faulted region in between, which is called an intrinsic SF that binds the 30° and 90° partials. This defect takes this name because the stacking sequence in the faulted region (box in the centre of the image) connecting the partial dislocations differs from the stacking sequence of the {111} planes in the undisturbed region of the bulk (left and right). The average distance between partials is of the order of 50 Å (the width of the SF ribbon), so as to allow a simplifying assumption that partials are energetically virtually decoupled [24]. The energy of the intrinsic SF in silicon was measured by Ray and Cockayne [25] and shown to amount to 51 ± 5 mJ m^{-2}. The corresponding experimental value for diamond is 290 ± 40 mJ m^{-2} [26].

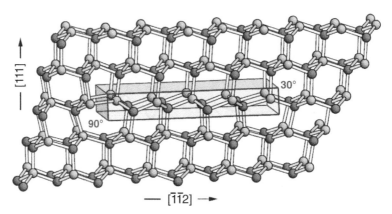

Figure 3.5 *Three-dimensional view of a dissociated 60° dislocation. The intrinsic stacking fault binding the 30° and 90° partial dislocations is included in the box. Blumenau, 2002, [21]. Reproduced with permission from A. T. Blumenau*

Dislocations lying in the shuffle set cannot dissociate into partials, as in this case the process would involve the formation of a thermodynamically unstable high-energy SF. Instead, under applied stress, shuffle dislocations may transform into glide dislocations.

The study of the mechanism of shuffle-glide transformation demonstrates that the transition occurring under the application of stress in the gigapascals range is thermally activated and involves the deformation of the core of a perfect shuffle dislocation with the formation of a 30° partial which is stable only in the glide set [27].

The dislocation dissociation is ruled by the Frank energy criterion [28], that is based on the argument that the energy of a dislocation is proportional to the square of the Burgers vector b^2. Although the most stable dislocations are those with the shortest Burgers vectors, the Frank criterion in its original version states that a dislocation with a Burgers vector b_1 dissociates when

$$b_1^2 > b_2^2 + b_3^2 \tag{3.2}$$

where b_2 and b_3 are the Burger's vectors of the partials, under the condition that for topological reasons dissociation occurs with the conservation of the overall Burgers vector.

$$b_1 = b_2 + b_3 \tag{3.3}$$

In its generalized version, the Frank criterion states that a perfect dislocation with a Burgers vector b will dissociate into two perfect dislocations A and B if

$$k(\beta)b^2 > k(\beta_A)b_A^2 + k(\beta_B)b_B^2 \tag{3.4}$$

where $k(\beta)$, as will be seen in the next section, is a function of the mechanical properties of the material.

A consequence of the Frank criterion is also that the average of the sum of the angles of the partials between their Burgers vectors and the dislocation line corresponds to the angle of the undissociated, perfect dislocation. This is the reason why the dissociation of a perfect 60° dislocation causes the set-up of a couple of 30° and 90° partials.

According to the Frank criterion, moreover, a dissociated dislocation is unstable if the interaction force between partials is attractive, while it is stable when the interaction is repulsive, that is dislocations with the same Burgers vector tend to repel each other in order to reduce the total excess of elastic energy, dislocations with opposite Burgers vectors, instead, attract each other.

Dislocations in Si and Ge could be electrically charged due to the thermal ionization of gap dislocation levels [29], as will be discussed in Section 3.2.5. Repulsion between equally charged isolated dislocations may therefore also occur.

3.2.2   Energy of Regular Straight Dislocations

The evaluation of the energy of dislocations has been a matter of systematic concern since the Peierls–Nabarro (PN) first modelling [30, 31] and is still a matter of debate or reconsideration [32, 33], in view of the availability of modern atomistic modelling tools of the dislocation core [34, 35].

Figure 3.6 reproduces the original Peierls scheme [30], which corresponds to that adopted in the present modelling works. It represents a section of the crystals and shows that the slip plane P, lying between the lattice planes A and B, divides the crystal into an upper part, confined by the plane A and a lower part, confined by the plane B. Plane A contains one row of atoms more than plane B, due to the formal insertion of a half-plane S oriented along the z-axis and running orthogonal to the plane of the figure. The intersection of the half-plane S with the slip plane P corresponds to the centre of the core of the dislocation line, which also runs orthogonally to the plane of the figure.

The energy of a regular straight dislocation is the sum of the of its purely elastic strain energy E_s due to lattice deformation, of the core energy E_c and of the electronic energy of dangling bonds E_{DB}.

$$E_{tot} = E_s + E_c + E_{DB} \tag{3.5}$$

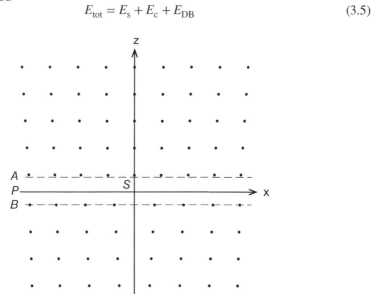

Figure 3.6   *Peierls model of a perfect dislocation in the glide set*

In order to calculate the elastic strain energy term E_s, one has to observe first that the atoms lying on the plane A are subjected to two forces:

1. that originating from the interaction with the other atoms of the top half crystals, which tends to distribute the compression uniformly over the plane A and to extend the dislocation width
2. that due to the interaction with the atoms of the bottom half crystal and in particular with the atoms of the plane B. This force brings as many atoms of A into alignment with B, that is to shorten the dislocation width. At the thermodynamic equilibrium these two forces balance and the actual dislocation width and energy are the result of the balance.

The elastic energy contribution can then be evaluated using linear elasticity theory, which shows that the elastic energy per unit length of a dislocation contained in a cylinder of radius R and length L is given by [22]

$$\frac{E_s(R)}{L} = \frac{k(\beta)b^2}{4\pi} \ln \frac{R}{R_c} \quad R \geq R_c \tag{3.6}$$

where β is the angle between the Burger's vector b and the line direction ℓ of the dislocation.

The energy factor $k(\beta)$ in Eqs. (3.6) and (3.4) is a function of the mechanical properties of the crystal, considered an isotropic medium

$$k(\beta) = \left(\mu(\cos^2\beta + \frac{\sin^2\beta}{1-\nu} \right) \tag{3.7}$$

where μ is the shear modulus and ν the Poisson ratio. Once $k(\beta)$ is known or can be calculated, the elastic contribution can be evaluated.

The other two terms in Eq. 3.5 should be evaluated by quantum mechanical calculations, as the continuum elasticity theory is unable to describe correctly the system at the atomic scale.

The PN model has, however, been used for the calculation of the elastic misfit energy and of the core energy for perfect and partial dislocations in silicon [36] under the approximation that dangling bonds are entirely saturated[1] and that the mean core energy could be indirectly estimated.

The results are given in terms of the elastic misfit energy $W(u)$ as a function of the displacement u, which varies between 0 and a (where a is a fraction of the lattice constant) and lead to values of the core energy E_c, approximately equal to $W(a/2)$ and of the Peierls misfit energy to $W_p \approx E_{tot}$, reported in Table 3.1 [36].

It can be seen that the misfit energy of perfect (undissociated) dislocations in the glide set is much higher than that of the corresponding partials, thus achieving the thermodynamics of the spontaneous dislocation dissociation as discussed in the previous section.

It can be seen that the PN model, as expected, gives only the correct order of magnitude of the core energies which, actually, are largely underestimated with respect to the corresponding values calculated using atomistic methods.

As an example, the core energies of screw dislocations in silicon evaluated using semi-empirical potentials and *ab initio* methods [37] are around $0.5\,\text{eV}\,\text{Å}^{-1}$ for the shuffle set and $0.7\,\text{eV}\,\text{Å}^{-1}$ for the glide set.

---

[1] The approximation concerning dangling bonds is tenable, as dislocation cores may be reconstructed (see Section 3.2.4).

Table 3.1 *Calculated total misfit energy W$_p$ and core energy E$_c$ for perfect and partial dislocations in silicon*

Dislocation type		W_p(eV/Å^{-1})	E_c(eV/Å^{-1})
Glide	60°	3.77	0.169
Glide	Screw	3.44	0.110
Shuffle	60°	0.116	0.538
Shuffle	Screw	0.148	0.408
Glide	Partial 30°	0.343	0.062
Glide	Partial 90°	0.093	0.093

Data from Joós, *et al.*, 1994, [36]

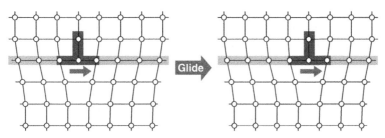

Figure 3.7 *Motion of a dislocation in the glide set. Blumenau, 2002, [21]. Reproduced with permission from A. T. Blumenau*

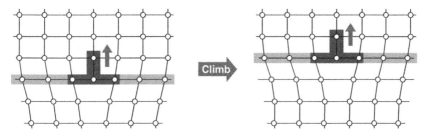

Figure 3.8 *Schematic representation of a dislocation climbing process: it is evident that climbing requires the formation of an interstitial or the presence of a vacancy. Blumenau, 2002, [21]. Reproduced with permission from A. T. Blumenau*

The corresponding calculated core energies (eV Å^{-1}) of dislocations in diamond [21] amount to 3.32 for shuffle screw, 2.13 for 60° glide and 2.60 for 60° shuffle.

3.2.3 Dislocation Motion

Under intentional or unintentional applied stress, dislocations glide or climb, as shown schematically in Figures 3.7 and 3.8 [21], and in crystals with the diamond structure dislocation gliding or climbing may occur either on the close packed planes corresponding to the shuffle-set (type I) or in the glide-set (type II) (see Figure 3.4).

Dislocations motion was first modelled theoretically by Peierls [30] and a few years later by Nabarro [31], under the assumption that when dislocations have to glide on their slip plane, the bonds of one row of atoms close to the dislocation line should be broken and subsequently rebonded as the dislocation glides one atomic distance. This is equivalent to assuming that dislocations have to surmount an energy barrier (a Peierls hill) when passing from the initial equilibrium position (a Peierls valley oriented along the <110> direction) to the neighbouring equilibrium position.

Using simplifying assumptions (no displacement normal to the glide plane and the static interaction energy localized at the dislocation core), the values of the Peierls stress σ_p, which is the stress required to move the dislocation across the Peierls hill, is given by

$$\sigma_p = \frac{1}{b^2}\left[\frac{\partial W}{\partial \alpha}\right] = \frac{2\mu}{1-v}\exp-\frac{4\pi\xi}{b} \tag{3.8}$$

where μ is the shear modulus of the material, v is the Poisson ratio, $\xi = \frac{a}{2(1-v)}$ and a is the interplanar spacing. Therefore, the model predicts that the Peierls stress should decrease with increasing b^2, that is with the square of the dislocation width.

The PN model allowed a significant advance in the understanding of the phenomenology of the dislocations motion, even with unrealistically high values of the Peierls stress [38, 39], and only recently has the physics of the process found a more substantial, although probably not yet definitive, assessment [40–45] using both analytical and numerical methods.

A critical assessment of all these issues [34] shows that the most serious shortcomings of the available models are:

1. most calculations are isotropic or pseudo-isotropic
2. most calculations are 1D, while the displacement $u(x)$ across the glide plane deviates from the direction of b
3. in fcc crystals dislocations are often dissociated and the Burgers vectors of the partials are not parallel, which makes a 2D treatment mandatory
4. the atomic interaction energy is not localized at the position of the atomic nuclei but in a more extended range
5. the artificial subdivision of the misfit energy in elastic and atomic terms is adequate for the description of equilibrium conditions but untenable under dislocation motion
6. covalent bonding in silicon and in other semiconductors gives rise to a high barrier to dislocation motion.

Another issue of relevant interest concerns the controversy regarding the dislocation nucleation and motion in the glide and in the shuffle set.

It is known that at low temperatures and under gigapascals stresses, perfect dislocations nucleate in the shuffle set, while at high temperatures and under lower stresses dislocations nucleate in the glide set [46].

It is also known that while theoretical arguments predict that slip in silicon should occur on the *shuffle* planes, experimental observations indicate that slip does occur on the glide planes and involves dislocation partials originated by perfect dislocation dissociation [23].

The physical grounds behind these features are that edge dislocations [47] spontaneously dissociate in the glide set in 30° and 90° partials, with the set-up of a low energy SF, as was already shown.

In addition, at least in silicon, the dislocations do not move by rigid translation but rather via a mechanism called dislocation drag associated with the formation of a kink (step a) and its propagation (step b and c) (see Figure 3.9). Within the elastic theory, kinks are dislocation segments which bring a part of a dislocation lying on a low energy direction to an adjacent low energy site, dragging the entire dislocation through a unit translation.

Kinks are shown to be equilibrium defects and their density is, therefore, an equilibrium property of the system. As the activation energy for the kinks mobility is lower in the glide set [47], thermodynamics and kinetics favour the dislocation motion in the glide set, at least at temperatures higher than the brittle/ductile transition temperature which ranges between 900 and 1075 K in undoped silicon [48].

Eventually, as the formation energy of the SFs in the shuffle set is very high, dislocation dissociation is prevented but at temperatures below the brittle/ductile transition under gigapascals confining pressures undissociated perfect dislocations glide in the shuffle set [37, 46, 49].

Gliding is not the only movement that dislocations make under an applied stress. When a dislocation line moves upwards or downwards, crossing one or more sets of planes, starting from a reference glide plane, it climbs. Also climbing from one plane to the neighbouring gliding plane occurs thanks to the formation of kinks.

3.2.4   Dislocation Reconstruction

Dislocations not only interrupt the translational symmetry of the lattice, and consequently change the topology of the whole system, but their presence should be associated with the formation of a continuous array of dangling bonds and of three-coordinated atoms in their core. In the case of silicon, one could expect a volumetric density n_{db} of dangling bonds close to a value of $5 \cdot 10^{14}$ cm^{-3}, for a dislocation density $N_D = 10^7$ cm^{-2}.

Thanks to the anisotropy of the low-temperature ESR dislocation spectra, which allows distinction between dislocation-related and point defect-related ESR signals, the results of ESR measurements on dislocated silicon show that a continuous periodic array of dangling bonds in the core of dislocations is a very poor representation of the real core configuration. Dislocations present, at most, only a fraction of 1% of the theoretically predicted unpaired electrons density [50]. In addition, ESR measurements show that all or most of the active ESR centres which can be ascribed to dislocations belong to vacancies introduced in their

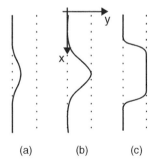

(a)   (b)   (c)

Figure 3.9   *Kink formation and dislocation drag*

cores [51, 52]. Theoretical studies [53–58] show, further, that both 30° and 90° partial dislocations behave as if their core were partially or totally reconstructed, reconstruction occurring by suitable pairing of dangling bonds with associated partial elastic strain relief.

Dislocation reconstruction, considered from the thermodynamics view point, is the process followed by a faulted crystal to reach thermodynamic equilibrium conditions, and the evaluation of the structural features and the energetics of this process is of relevant basic and technological interest.

As can be observed in Figure 3.10 [21] for a 30° partial dislocation, reconstruction is suggested to occur by formation of dimeric units crossing the dislocation line that lead to fourfold coordination of each core atom. During the reconstruction process defects like anti-phase defects (APDs) or solitons (Figure 3.11) [59], and soliton–kinks complexes can also be formed.

APDs are topological defects occurring as a result of breaking the mirror symmetry of the reconstructed configuration normal to or along the dislocation line, as is seen in Figure 3.11 for the 90° partial, where the APD appears as a fault in the dimerization sequence. Each ADP is associated with a dangling bond at the atom labelled 12 and with reconstructed (5–10, 7–14), stretched (5–6, 6–7) and contracted (10–11, 12–13) bonds.

The reconstruction energies (i.e. the energy gain per unit bond length to shift from the unreconstructed to the reconstructed configuration) and the APD formation energies (i.e. the energy gain associated with the formation of the defect) have been calculated for 90° and 30° partials [60, 61]. The results are reported in Table 3.2, which shows that the core reconstruction of 90° and 30° partials and the formation of APD are thermodynamically spontaneous processes.

It should be noted, however, that these results are relative to single, isolated dislocations in a model cell, while the actual features of a dislocated sample are those of dislocation arrays embedded in the matrix crystal (see Figure 3.12), where dislocations interact within each other, with point defect and with point defect clusters.

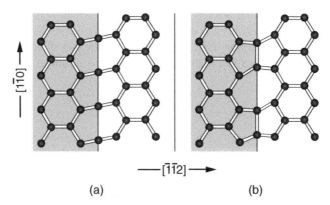

(a)   (b)

Figure 3.10   *Core reconstruction of a 30° partial projected into the (111) glide plane (a) unreconstructed partial (b) reconstructed partial. The shaded areas embed the stacking fault binding the 30° partial with the 90° one (not seen in the figure) The reconstructed bonds are 18% stretched in comparison with the bulk silicon structure. Blumenau, 2002, [21]. Reproduced with permission from A. T. Blumenau*

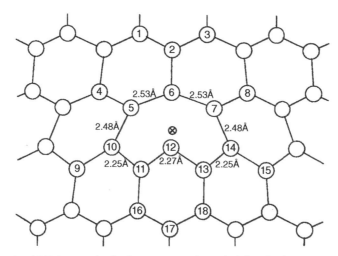

Figure 3.11   *An ADP (atmospheric downstream plasma) defect in the reconstructed core of a 90° partial dislocation in silicon. The bond lengths between atoms around the defect core are calculated equilibrium values. Usmerski & Jones, 1993, [59]. Reproduced with permission from Taylor & Francis*

Table 3.2   *Theoretical estimates of the dislocation reconstruction energies (E_{rec} in eV/bond), ADP formation energies and dangling bond E_{DB} energies in eV*

Formation energies	EDIP	Ab initio
E_{rec} 30° partial (eV/b)	0.33	0.43
		0.53 ± 0.015 [61]
E_{APD} 30° partial (eV)	0.34	0.43
E_{rec} 90° partial (eV/b)	0.84	0.87
90° E_{APD} (eV)	0.41	—

Adapted from Justo *et al.*, 1998, [60] and Engerness and Arias, 1997, [61].

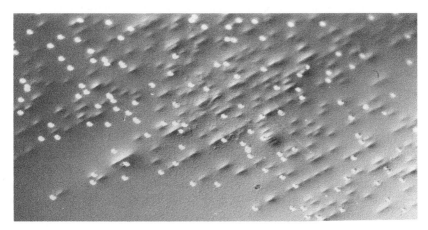

Figure 3.12   *Overlapping of defects and dislocations on a Si (111) sample dislocated at 920 K and anisotropically etched to reveal defects and dislocations. Courtesy of V. Eremenko and E. Yakimov, Institute of Microelectronics Technology RAS*

The extension of results of theoretical modelling to the real word is therefore not at all straightforward, but it could be of substantial support for the analysis of experiments carried out on dislocated samples, where contributions from intrinsic dislocations should be subtracted from a background of extrinsic signals, arising from bulk properties, impurity contamination and dislocation–impurity interactions.

3.2.5   Electronic Structure of Dislocations in Si and Ge, Theoretical Studies and Experimental Evidences

It has been shown in the previous section that most of the geometrically possible broken bonds in the core of dislocations are reconstructed or are lacking of unpaired electrons, leaving less than 1% of the core sites occupied by unpaired electrons.

There is, however, theoretical [24, 51, 52, 62–67] and experimental evidence that dislocations in Group IV semiconductors (Si, Ge and Si-Ge alloys) are electrically and optically active. One of the most important results of these studies is the theoretical prediction [51, 68, 69], which has been experimentally confirmed by DLTS measurements [70], of the presence of shallow 1D-band states due to the strain field of the SF bonding the partials of dissociated 60° dislocations (see later in this section).

In addition to shallow states, the presence of impurity-related deep gap states and *native* deep gap states associated with dislocations, but not intrinsic to their cores, such as reconstruction defects and vacancies or self-interstitials incorporated in their cores, has also been revealed, but not always explicitly assigned, by a number of experimental techniques, such as the Hall effect, photoconductivity and DLTS, which are, however, not sensible to individual dislocations.

Concerning deep gap states associated with impurities, it should be mentioned again that dislocations are normally dirty or 'reacted', due to their ability to work as deep sinks for impurities which are ubiquitously present, at least in trace amounts, in any kind of matrix in which dislocations form. To get 'clean' dislocations it is necessary to extract impurities and drive them to deeper impurity sinks. It has been shown that a good practice is a dual P-gettering/hydrogenation process,[2] where phosphorus gettering is used to extract the metallic impurities from the dislocation core and hydrogenation to passivate the residual dangling bonds at dislocations and point defects introduced by plastic deformation [71, 72], see Chapter 5.

The study of the properties of individual dislocations may by approached with a number of different methodologies. One is the study of a few individual dislocations aligned in the gate of a micrometric MOSFET device [73], which allowed, by tuning the channel width, analysis of the effect of a different number of dislocations and evidenced the occurrence of electrical conduction along single dislocation lines, caused by a 2D carrier confinement, in good agreement with theoretical predictions [63].

The more conventional way is to study the electrical and optical activity of individual dislocations using EBIC [74] (see Figure 3.13), LBIC, X-ray beam induced current (XBIC) and PL measurements (see Figure 3.14) [75].

The basic principles of EBIC measurements can be found in the literature [76, 77], where it is shown that the EBIC (but also LBIC and XBIC) contrast is the figure of merit of

---

[2] Passivation and gettering techniques will be discussed in full detail in Chapter 5.

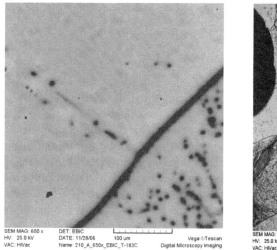
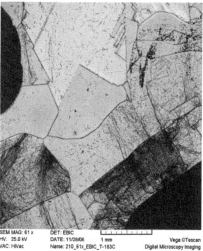

Figure 3.13 *EBIC contrast maps of recombining dislocations in the same region of a multicrystalline silicon wafer at different magnifications (T = 90 K). Courtesy of M.Acciarri, Department of Material Science, University of Milano Bicocca*

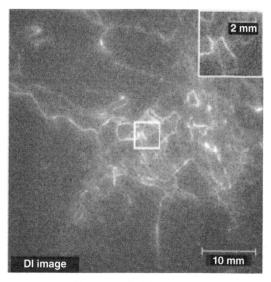

Figure 3.14 *Room temperature PL image of dislocations in a multicrystalline silicon wafer. Mankovics et al., 2012, [75]. Reproduced with permission from John Wiley & Sons*

these measurements

$$C = \frac{I - I^o}{I^o} \tag{3.9}$$

where I is the EBIC (but also LBIC and XBIC) current measured at the defect and I^o the corresponding current measured in a defect-free region. As the measured contrast of impurity contaminated dislocations is inversely proportional to the defect/impurity lifetime τ_{disl} at dislocations

$$C \propto \frac{1}{\tau_{disl}} \tag{3.10}$$

the defect lifetime could, therefore, be experimentally determined.

The lifetime of defects/impurities incorporated or trapped at dislocations is different from that of isolated defects or impurities in the bulk of the silicon matrix, due to their bonding features with the dislocation core or interaction with the dislocation strain field.

Evaluation of the minority carriers recombination at dislocations, carried out in the framework of the Shockley–Read–Hall (SRH) recombination theory [78, 79] shows that if the contrast increases with decrease in the temperature, see Figure 3.15, [80] then recombination is due to shallow levels intrinsic to the core states of dislocations (0.075 eV in the figure), if it decreases, recombination is due to mid-gap levels (0.55 eV in the figure) of intrinsic or extrinsic origin.

There is also quantitative evidence that the EBIC contrast of 'clean' dislocations is below the instrumental EBIC detection limit, while it is very strong in the case of metal-decorated dislocations [74].

EBIC contrast studies, therefore, are a key to quantitatively revealing the presence of shallow and deep states responsible for the recombination activity of dislocations, but they are unable to provide a correlation between the electronic structure and the chemical nature of the species responsible for the recombination.

Therefore, and quite obviously, only spectroscopic measurements may be employed to find the physical or chemical attributes of a deep level, although it may be difficult to unequivocally attribute its origin to a single source. As said before, residual dangling bonds at the dislocation core, unpaired mobile electrons, overlapping point defects and point defect clusters and (metallic) impurities segregated at dislocations [74] may simultaneously contribute with a measurable spectroscopic signal.

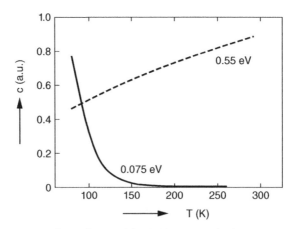

Figure 3.15   *Temperature dependence of the EBIC contrast for low contamination levels, with the indication of the energy of the level responsible of the recombination. Kittler et al., 1995, [80]. Reproduced with permission from AIP Publishing*

A proper selection of clean and impurity contaminated samples, good preparation techniques of individual dislocation arrays, excellent structural, physical and analytical characterization methods and theoretical modelling expertise are the tools needed to solve the problem concerning the electronic structure of dislocations, a solution which is, however, not yet at hand, as the following examples seem to show.

ESR measurements on plastically deformed silicon demonstrated, in fact, as shown in the last section, that dislocations cores are almost fully reconstructed, leaving to dangling bonds-related gap states a limited contribution to the electronic structure of dislocations. Secondary defects, such as vacancies and self-interstitials in the core of dislocations and reconstruction defects (APD) are considered to be responsible for the total ESR signal of clean dislocations [50] and, therefore, possible sources of native deep levels.

Further, comparison of the DLTS spectra of as-grown, dislocated and hydrogen pas- sivated p-type Si samples [70, 79–83] shows that among the three main peaks, labelled T1, T3 and T4, present in the DLTS spectrum of a dislocated CZ sample (see Figure 3.16 and Table 3.3) [70] only T1 at $E_T - E_V = 0.28\,\mathrm{eV}$, is the fingerprint of defects or impu- rities (presumably oxygen) trapped in the strain field of dislocations [81, 83]. A broad signal in the temperature range of the T3 and T4 peaks is present also in the undislo- cated and hydrogen-passivated sample and belongs to impurities and defect complexes. The additional presence of two shallow acceptor and donor levels (STh and STe) at $E_V + 0.07\,\mathrm{eV}$ and $E_C - 0.03\,\mathrm{eV}$) is also detected on thermally annealed dislocated samples (see Figure 3.16b), in good agreement with theoretical predictions.

Once depurated from extrinsic contributions, the results of DLTS measurements seem to show that the electronic structure of *clean* dislocations is relatively simple, and may be

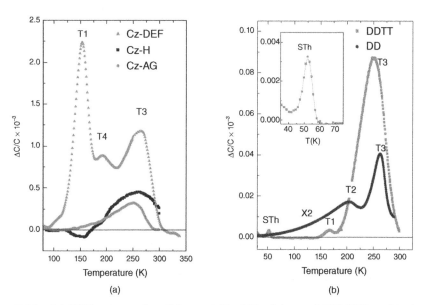

Figure 3.16 (a) DLTS spectra of undislocated (Cz-AG), dislocated (CZ-DEF) and hydrogen annealed (CZ –H) p-type CZ silicon (b) DLTS spectra of dislocated and thermally annealed (DDTT) samples. Castaldini et al., 2005, [81]. Reproduced with permission from John Wiley & Sons

Table 3.3 *DLTS signatures of deep and shallow states in undislocated and dislocated CZ silicon (the dislocation density in both DD and DDTT samples is 10^7 cm^{-2}) AG = as grown CZ samples: DD = dislocated by stretching and bending at 670 °C, DDTT further annealed at 800 °C in argon*

Sample	$E_T - E_V$ (eV)	CZ-AG N_T (cm^{-3})	CZ-DD N_T (cm^{-3})	CZ-DDTT N_T (cm^{-3})
T3a	0.44	10^{12}	—	—
T3b	0.36	—	$2 \cdot 10^{13}$	$1 \cdot 10^{13}$
T2	0.45	—	$2 \cdot 10^{12}$	—
T1	0.28	—	—	$7 \cdot 10^{12}$
T0	0.09	$3 \cdot 10^{11}$	—	—
STh	0.07	—	—	X
	$E_C - E_T$ (eV)	—	—	—
STe	0.06	—	—	—

Adapted from Castaldini *et al.*, 2005, [70] and Cavalcoli and Cavallini, 2007, [82].

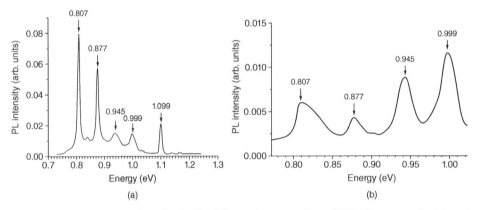

Figure 3.17 *PL spectra of a plastically deformed (111) oriented FZ silicon sample (a) and of a plastically deformed (111) oriented CZ silicon sample (measurement temperature 12 K). The emission at 1.099 eV is the signature of the band to band emission of the silicon matrix. Pizzini* et al., *2004, [84]. Reproduced with permission from Trans Tech Publications*

entirely accounted for by two shallow levels, while the deep level associated with the trap T1 at $E_T - E_V = 0.28$ eV is evidence of an impurity–dislocation interaction.

Low temperature PL measurements on dislocated silicon add important and perhaps decisive information on the electronic structure of dislocations in silicon and on the effect of oxygen contamination on their optical activity [84–86].

The PL spectrum of 60° dislocations at 12 K in FZ and CZ silicon (see Figure 3.17) [84–86] shows a quartet of PL lines at 0.807 eV (D1), 0.877 eV (D2), 0.945 eV (D3) and 0.999 eV (D4), of which D3 is the phonon replica of D4, in addition to the band to band emission at 1.099 eV, whose value fits well with the literature value of the energy gap of Si. These bands are systematically present in (metallic) impurity-free dislocated samples, but their width is shown to be different in FZ and CZ samples, with a consistent broadening of all the emissions of the dislocated CZ samples.

Considering that the temperature range of dislocation generation by plastic deformation (from 650 to 800 °C), designed in order to have straight, non-intersecting 60° dislocations without threading arms, is compatible with the precipitation of oxygen from supersaturated oxygen solutions of CZ crystals (see Section 3.5.3) it would be reasonable to suppose that broadening of the PL lines is associated with oxygen precipitates heterogeneously nucleated at dislocation.

From Figure 3.18 [84] it is possible to observe that the deconvolution of the broad D1 line of a dislocated CZ sample could be well fitted by three emissions at 0.807, 0.816 and 0.820 eV, which are also the independently measured emissions of oxygen microprecipitates segregated from oxygen-supersaturated (undislocated) CZ silicon samples [85].

The TEM image of one of these microprecipitates is displayed in Figure 3.19 [86] and shows the presence of an expanding dislocation loop, arising from the condensation of self-interstitials ejected from the SiO_2 particle during the precipitation process (see Section 3.5.3). As the intensity of the PL emission at 0.807 eV from oxygen microprecipitates was detectable only in the simultaneous presence of dislocation loops, it seems possible to conclude that the D1 line at 0.807 eV is the PL emission from native dislocation states.

Although the nature of the electronic levels involved in the D1 emission is still under debate [83, 87], a good fit between the DLTS and PL data is obtained if the D1 emission is assigned to the radiative recombination of an exciton trapped at the shallow levels of the dislocation core through a deep level at $E_V + 0.28 + 0.04$ eV, the level of the DLTS trap T1.

Concerning the remaining emissions, the D4 emission can be assigned to the direct recombination of excitons trapped at shallow levels, which leads to an emission at $hv \approx 0.99 \pm 0.030$ eV, while the D2 one remains unassigned.

It seems, therefore, confirmed that the intrinsic electronic properties of dislocations are ruled by two native shallow states and a gap state, but that the actual optical and electronic behaviour of dislocated silicon is dominated by oxygen and other impurities.

The experimental determination of the electronic structure of dislocations in high purity (HP) germanium is less challenging than for silicon, thanks to the circumstance that in H_2-grown germanium oxygen is present in very low concentration and its lowest growth

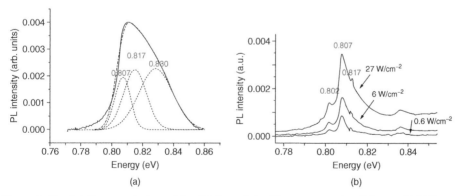

Figure 3.18   *(a) Deconvolution of the D1 line of the PL spectrum of a dislocated CZ-Si sample (measurement temperature 12 K). (b) PL emission of submicrometric oxygen precipitates. Pizzini et al., 2004, [84]. Reproduced with permission from Trans Tech Publications*

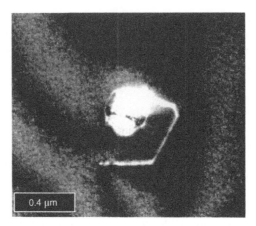

Figure 3.19   *TEM micrograph of a sub-micrometric oxide (0.4 µm) precipitate in CZ silicon with a punched-off dislocation loop. Binetti et al., 2002, [86]. Reproduced with permission from AIP Publications*

temperature should preserve, at least in part, its unintentional contamination during the growth process arising from the crucible and the growth chamber (see Chapter 4).

Because of the impossibility of using ESR measurements, due to the already mentioned impeding factors associated with the high nuclear spin of the Ge^{73} isotope and a strong spin–orbit interaction, DLTS and PL are the main spectroscopic tools used to study the electronic structure of dislocations in Ge.

Three weakly overlapped DLTS broad bands of traps located at $E_V + 0.025$ eV, $E_V + 0.1$ eV and $E_C - 0.09$ eV are found in HP p-type and n-type Ge samples [88] with a typical dislocation density $N_D \approx 10^4$. A DLTS band at $E_V + 0.33 \pm 0.02$ eV, attributed to copper impurities, was also found, in addition to shallow states at $E_C - 0.09$ and $E_V + 0.10$ eV [89] of which that at $E_c - 0.09$ eV is attributed to the strain field of dislocations, similar to the case of dislocations in silicon.

The assignment to Cu contamination, which behaves as a donor impurity, of the band at $E_V + 0.33$ eV was confirmed by a recent study on Cu-doped dislocated Ge [90] that reports the presence of two additional Cu-related traps at $E_v + 0.04$ eV and $E_c - 0.26$ eV, assigned to substitutional copper in three charge states, that is $Cu_s^{0/-}$ for the trap level at $E_v + 0.04$ eV, $Cu_s^{-/2-}$ for the trap level at $E_v + 0.33$ eV and $Cu_s^{2-/3-}$ for the trap level at $E_c - 0.26$ eV.

These results indicate that copper is an ubiquitous contaminant of germanium, its source being probably the metallic copper feedthroughs in the crystal growth chamber.

Additional information on Ge dislocation energetics comes from PL measurements (Figure 3.20) which show the presence of a broad PL band lying between 0.45 and 0.6 eV, which is the convolution of at least four PL lines, with a main emission (the d8 band) centred at 0.5 eV at 80 K (0.512 eV at 4 K), and of another broad band centred around 0.72 eV [91].

As the emission at ≈ 0.7 eV is the convolution of the indirect band to band emission of Ge and of a direct band to band emission of Ge induced by the strain field of dislocations [91], clearly visible as a shoulder of the main band, only the d8 band at 0.512 eV at 4 K

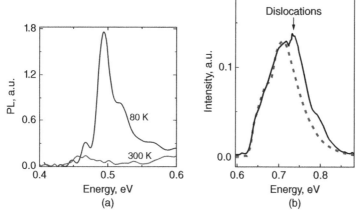

Figure 3.20   *Photoluminescence spectra of 60° dislocations in germanium. In (a) is reported the d8 band. In (b) the shoulder is attributed to dislocation emission. Arguirov et al., 2011, [91]. Licenced under the Creative Commons CC-BY 3.0 Licence and with permission from T. Arguirov*

should be directly associated with dislocations in germanium [92, 93]. In analogy with the D4 band of dislocated silicon, it could be assigned to a direct radiative transition within the shallow levels at $E_C - 0.09$ and $E_V + 0.10$ eV.

There is, therefore, experimental evidence that the fingerprint of the intrinsic electronic properties of dislocated, *impurity free,* silicon and germanium are their shallow states, in good agreement with the theoretical predictions.

3.3   Dislocations in Compound Semiconductors

The high density of SFs, misfit and threading dislocations in bulk substrates and epitaxial layers (see Figure 3.21 [94]) has been for years a significant problem in compound semiconductors and a hint for the development of optimized crystal growth methods and for a better understanding of the nature of dislocations in compound semiconductors [95–102].

Microelectronic and optoelectronic devices are particularly sensitive to threading dislocations because when they cross the active device region they cause shorts and local recombination processes.

As the result of a massive experimental work carried out over the last two decades, 4″ diameter GaAs and InP single crystals grown with the vertical gradient freeze (VGF) method (see Chapter 4) with a dislocation density of 10^2 and 10^3 cm^{-2} are today commercially available. Furthermore, the density of misfit and threading dislocations in nitride semiconductors heteroepitaxially deposited on foreign single crystalline substrates may be reduced in the active region of the device to acceptably low values by the use of intermediate buffer layers, as will also be discussed in Chapter 4. As the electrical activity of dislocations in compound semiconductors is, in general terms, relatively modest, dislocation reconstruction is supposed to be a relevant phenomenon in these materials.

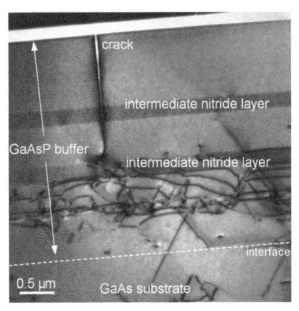

Figure 3.21   *TEM bright field image of misfit dislocations in a buffer nitride layer of GaAsP on a GaAs substrate. Schöne et al., 2008, [94]. Reproduced with permission from AIP*

3.3.1   Electronic Structure of Dislocations in Compound Semiconductors

The core structures of edge and screw dislocations in compound semiconductors are more complex than those of Group IV elemental compounds and are strongly correlated with the crystallographic structure of the host phase, as the following examples show.

The apparently simplest cases are given by GaAs, InP and by all semiconductors crystallizing with the zinc blende (ZB) cubic structure, which present the same type of dislocations typical of the diamond structure [22, 103].

Different is the case of nitride semiconductors and of 2H- or 6H-SiC crystallizing with the wurtzite (hexagonal) structure, which will be discussed in some detail because of the technological relevance of this family of compounds.

An interesting example is given by GaN heterostructures grown along non-polar planes, which present high densities of threading dislocations and partial 60° dislocations loops embedding basal stacking faults (BSFs) [104] which represent the major defects in these structures. As not all threading dislocations in these structures are electrically active, it could be deduced that a reconstruction process works also in nitride hosts.

Most of the information on the core properties of perfect and dissociated dislocations in GaN arises from theoretical studies, which indicate that SFs give rise to shallow quantum well-like regions, while the strain field of undissociated threading 60° dislocations induces the presence of deep levels.

These studies indicate also that dislocation cores are non-stoichiometric and that core configuration and stoichiometry depend on doping (and therefore on the Fermi level) and on the growth stoichiometry [105], that is on the nitrogen-rich or gallium-rich growing environment. This last could be managed either by control of the nitrogen pressure during the high pressure growth of bulk GaN crystals [106], or the nitrogen flux during the plasma assisted molecular beam epitaxy (PAMBE) [107] or molecular beam epitaxy (MBE) (see details in Chapter 4).

As the scope of this section is to outline the basic questions concerning the structures and energetics of dislocations in compound semiconductors, the attempt made here is to address attention to issues concerning core non-stoichiometry and energetics of dislocations in wurtzite type structures, to which GaN belongs, as a relevant example of problems common to all compound semiconductors.

Several types of core structures can be foreseen to be stable, of which the full core (stoichiometric) structure is schematically illustrated in Figure 3.22a [105], showing the presence of an asymmetric central row of alternating, under-coordinated Ga and N atoms. Open core structures and non-stoichiometric Ga-vacancy or N-vacancy structures may be generated by formally removing Ga and N atoms from the central row.

The formation energy of these structures, calculated with the density functional approach [105] is obtained using the following equation

$$E^{f}(q, E_F) = E_{tot}(q) - n_{Ga}\mu_{Ga} - n_{N}\mu_{N} + qE_V + qE_F \tag{3.11}$$

(see also Ref. [104]) where q is the charge, E_F the Fermi level, $E_{tot}(q)$ the total energy of the structure (the defect cell in the figure), E_V the valence band maximum in GaN to which the Fermi level is referred. The other terms have their conventional thermodynamic meaning.

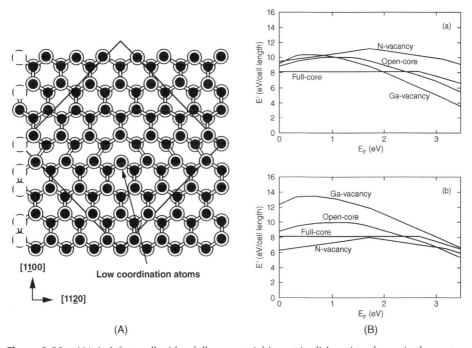

(A)   (B)

Figure 3.22   *(A) A defect cell with a full core, stoichiometric dislocation shown in the centre. (B) Dependence of the formation energy of non-stoichiometric dislocations on the Fermi level (a) nitrogen-rich growth (b) Ga-rich growth. Wright and Grossner, 1998, [105]. Reproduced with permission from AIP*

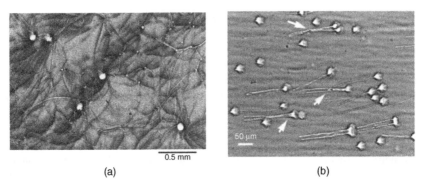

0.5 mm

(a)   (b)

Figure 3.23   *(a) X-ray topographic image of micropipes in SiC. Vetter and Dudley, 2006, [108]. Reproduced with permission from Taylor & Francis. (b) Synchrotron radiation radiographic image of interacting micropipes in a SiC crystal. Gudkin et al., 2009, [109]. Reproduced with permission from John Wiley & Sons*

The results of the calculations are displayed in Figure 3.22b where one can see that full core dislocations are stable over a wide range of p-doping for N-rich growth conditions, while N-vacancy cores are more stable for Ga-rich growth conditions.

The calculated density of states (DOS) function shows that all core configurations of partial dislocations lead to deep states and are therefore electrically active [104].

The second example refers to SiC (see Figure 3.23) [108, 109] and also GaN [110], both presenting a particular type of screw dislocation with a core width several times larger than the unit cell. Due to the extremely high excess energy stored in such wide cores, during the crystal growth preferential sublimation occurs from the core region, forming hollow core dislocations called micropipes or nanopipes, these last being a kind of nanotube which are supposed to behave as sinks for normal screw dislocations [110].

The interaction between micropipes and dislocations can be discussed as a case of the interaction of a dislocation with a surface. In this case the attractive interaction force F^{sd} is given by

$$F^{sd} = -\frac{\mu c^2 r^2}{2\pi d(d+r)(d+2r)} \qquad (3.12)$$

and the repulsive force

$$F^{rep} = \frac{n\mu c^2}{2\pi(d+r)}$$

where μ is the shear modulus, r the radius of the micropipe, d the distance of the screw dislocation from the surface of the micropipe, c the length of the unit cell along the c-axis of the hexagonal SiC crystal ($\approx$1.5 nm for a 6H-SiC crystal) and nc is the diameter φ of the micropipe (Figure 3.23).

It can be seen from Figure 3.24 that the net interaction force increases with the diameter of the micropipe and that the interaction is positive for d/r ratios larger than $\approx$0.1. Of course, the efficiency of the screw dislocation capture depends on the mobility of dislocations, and thus on the temperature. This example is a good introduction to the next section, entirely devoted to the discussion of the interaction of point defects and impurities with dislocations.

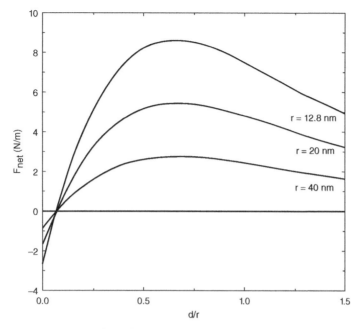

Figure 3.24   *Net interaction force between a micropipe (φ = 3c) and a screw dislocation. Pirouz, 1998, [110] Reproduced with permission of Taylor and Francis*

3.4   Interaction of Defects and Impurities with Extended Defects

3.4.1   Introduction

The interaction of impurities with dislocations is a well known effect in metals and semi-conductors as it affects both the local distribution of impurities, with the set-up of the so-called Cottrell atmosphere [112] and the electrical and mechanical properties of the material.

It has already been mentioned that thermodynamics favours the stability of impurity-decorated dislocations which, therefore, work as sinks for impurities. It is also known that impurity-decorated dislocations may work, directly or indirectly, as optically radiative centres. Dislocations could, therefore, be suitably *engineered* for impurity gettering processes and optoelectronic applications [84, 85, 113–133].

Impurity interaction occurs not only at dislocations, but also at GBs of polycrystalline semiconductors and at the surface of 3D defects, such as precipitates and voids.

For this reason, the knowledge of the energetics and dynamics of the defect diffusion, migration and impurity-bonding processes at EDs, often occurring in physically non-homogeneous environments, as is the case for systems subjected to thermal and mechanical stresses or to photon, particle or electron irradiation [113, 114], has been and still is the obvious prerequisite for the understanding and forecasting of the impact of these interactions [134, 135].

These kinds of concerns, which hold not only for elemental semiconductors, but also for compound semiconductors like SiC and GaN, where it was found almost impossible to avoid the presence of high densities of dislocations, triggered in the last decades a renewed interest in the properties of EDs, which added value to the previous knowledge, but opened new questions about the reactivity and the optical properties of EDs.

As an example, while it is already well recognized that shallow- and deep-impurities could be removed from bulk silicon by interaction with other defects or with dislocations, the physics and chemistry of processes involving gettering at GBs and strained interfaces is not yet completely elucidated.

A detailed description of these processes requires a deep understanding of the reaction coordinates of the process and a precise knowledge of the nature of the intermediate metastable species and the structure and bonding characteristics of the reaction product(s).

This is a problem of tremendous complexity already in the case of processes involving atomic species (point defects and impurities) and is even more complicated when EDs are involved, as in this case mechanical, chemical and electrical (for charged defects) forces are simultaneously involved.

Other examples of key technological importance are the aggregation of interstitials and vacancies into micro defects, like A- and B-swirls and D-defects (vacancy voids), in CZ grown silicon crystals, which are known to induce dramatic consequences for the lifetime and the dielectric rigidity of the oxide grown on these substrates [111, 136, 137], an issue which will be discussed in more detail in Section 3.6.4.

Finally, one of the best examples of compatibility between EDs and good device performance is given by the use of multi-crystalline silicon in low-cost solar cells, where efficiencies close to those obtained with single crystalline silicon solar cells are today obtained, in spite of the presence of GBs and dislocations, which can be properly cured or passivated [138].

The situation is even more demanding when nanofabrication processes are involved. There, the external surface versus volume ratio plays a crucial role in doping processes and impurity redistribution.

It seems, therefore, very clear that every future development in semiconductor technology will continue to rely on a suitable share of experimental and theoretical efforts of both chemical and physical character, addressed at a deep understanding of the role of EDs in complex chemical environments.

3.4.2   Thermodynamics of Defect Interactions with Extended Defects

The analysis of the interaction processes between atomic defects, impurities and EDs may be treated using equilibrium thermodynamics, considering, however, that EDs are not equilibrium species and accounting for their local configuration (topology), charge state and the presence of normally anisotropic stress- and electrostatic fields (for charged EDs).

As a first approximation, an interaction process might be treated as:

1. a true chemical reaction between quasi-chemical species (atomic defects) and/or impurities with unsaturated bonds at the interface of 2D and 3D defects or in the core region of 1D defects, with the formation of chemical bonds or
2. a repartition of atomic defects and/or impurities between two phases, of which one is the perfect crystal matrix and the other is the extended defect, considered as a low dimensionality Gibbs type of phase, whose properties depend also on the type and extent of the charge or/and stress-induced fields.

The result of a point defect–ED interaction is the formation of a reacted core or a reacted interface or an impurity- or atomic defect-cloud around the extended defect, in the presence of a space charge region when the interaction is between charged species.

In most practical cases the features of the interaction process are even more compli-cated, if we consider that EDs (surfaces, interfaces, dislocations) undergo reconstruction processes to reach minimum energy conditions (see Section 3.2.4). Interaction with recon-structed defects implies, therefore, that any bonding interaction should be coupled to an unbonding process of reconstructed bonds. The unreconstruction process itself may be a highly energetic process and occurs only if a more stable final bonding state with an impu-rity is established.

An additional concern is the different equilibrium conditions established during the repartition of a chemical species between bulk phases and phases of lower dimensionality, as for the latter ones the assumption of chemical and physical homogeneity is generally invalid. This means that the chemical potential $\mu_i = \left(\frac{\delta G}{\delta n_i} \right)_{T, n_1, \ldots n_\partial}$ of the ith defect or impurity species and the internal pressure $P = \left(\frac{F}{V} \right)_T$ of the phase (with G and F the Gibbs and Helmoltz free energies, respectively) should be used, having taken into proper account the chemical and physical configuration of the local environment of a specific ED.

It should be noted that we have neglected systematically in Chapters 1 and 2 the influence of the internal pressure of the phase on the thermodynamics of solutions and, therefore, on defect interaction processes, under the tenable assumption that the composition of a macroscopic 3D phase where interaction occurs is homogeneous and the internal pressure of the phase is also constant.

Instead, the values of both the standard chemical potentials μ_i^o and P in the ED 'phase' are different from those pertaining to the same species in the homogeneous bulk phase, as a consequence of different local coordination, different bonding features and the presence of a mechanical stress field associated with the ED.

3.4.3   Thermodynamics of Interaction of Neutral Defects and Impurities with EDs

When defects and impurities behave as neutral species, they interact also with charged EDs as if they were neutral. This is a case of practical interest, as impurities like oxygen, carbon, sulfur and even hydrogen in silicon and germanium behave in this way.

In order to deal with these interaction processes, without considering the details of their microscopic interaction which must be investigated using quantum mechanical tools, we may define the ED environment as a chemically homogeneous phase of reduced dimen-sionality, whose size or width depends on the extent of the associated charge and stress fields. We shall, further, describe the process at the thermodynamic equilibrium as a repar-tition of an individual species i (an impurity or an atomic defect) between the bulk and the ED environment.

This approach could be used with full physical meaning when the low-dimensionality ED phase has some physical and chemical attributes of a 3D phase, as is the case for surfaces in equilibrium with impurity-diluted bulk phases, where the interface is micro-scopically ordered and behaves like a few nanometres thick, chemically homogeneous Gibbs-type of phase [139]. In these conditions, the chemical and hydrostatic equilibrium conditions imply that the chemical potential of the species i and the internal pressure are the same in both phases, which we label for convenience α and β, where α is the label of the bulk phase. Therefore

$$\mu_i^\alpha = \mu_i^\beta \tag{3.13}$$

and

$$P_\alpha = P_\beta \tag{3.14}$$

where $\mu_i = \mu_i^\circ + RT \ln a_i$ is the chemical potential of a generic dissolved species i, μ_i° is its standard chemical potential, a_i the activity of the species i and P_α and P_β are the internal pressures of the two phases. If we use as the concentration of the impurity i in solution its atomic fraction $x_i = \frac{n_i}{N}$, where n_i is the volumetric density of the species i (at cm^{-3}) and $N = \sum_i n_i$ i s the volumetric density of the solvent, the activity should be taken as the product $\gamma_i x_i$, where $\gamma_i = f(x_i)$ is the activity coefficient. This last takes the value of 1 in ideal solutions, is a constant in the case of ideally diluted solutions and is a function of the concentration in non-ideal solutions, as shown in Chapter 1.

Therefore, at constant temperature, and disregarding at first the effect of local strain fields, which will be considered later in this section, the equilibrium between a solute i in α and β could be described by

$$\mu_i^{0,\alpha} + RT \ln \gamma_i^\alpha(x) x_i^\alpha = \mu_i^{0,\beta} + RT \ln \gamma_i^\beta(x) x_i^\beta \tag{3.15}$$

from which a distribution coefficient can be deduced

$$K_{seg} = \frac{x_i^\beta}{x_i^\alpha} = \Gamma(x_i^\beta, x_i^\alpha) \exp -\frac{\mu_i^{0,\beta} - \mu_i^{0,\alpha}}{kT} \tag{3.16}$$

where x_i^β is the atomic fraction of the impurity i in the low dimensionality phase, x_i^α is its atomic fraction in the bulk phase and $\Gamma(x_i^\beta, x_i^\alpha)$ is the ratio of the activity coefficients of the species i in the two phases, which accounts for all the interactions occurring between the impurity atoms and the atoms of the matrix in both the α and β phases.

In Eq. (3.16) the $\mu_i^{0,\beta} - \mu_i^{0,\alpha}$ term represents the standard excess of Gibbs energy of mixing ΔG_{mix}° of the species i when dissolved in the low-dimensional phase β, taking as reference the bulk phase α, whose thermodynamic properties are supposed to be known, and is a measure of the interaction energy of the species i with the ED.

Equation (3.16) could have, also, a direct practical application for the determination of the thermodynamic properties of the β phase, when the experimental values of the $\frac{x_i^\beta}{x_i^\alpha}$ ratios (determined by a proper analytical technique) as a function of the temperature are available and can then be fitted using the algorithms developed for regular or non-regular solutions. To the author's knowledge, however, experimental results which could be used for this kind of analysis are not available in the literature.

In most practical cases, however, the situation looks quite different, as the presence of an ED is associated with the set-up of a local stress field.[3] In this case the system could

---

[3] Among the few exceptions to this rule, precipitate (3D) phases which are isomorphous with the matrix phase should not present interfacial stresses of mechanical origin.

reach equilibrium conditions not only via the redistribution of impurities within the core and the strain field region but also via the emission or absorption of point defects, in view of an at least partial recovery of the stress.

The distribution of impurities within the stress field region, in the form of a 3D impurity cloud, should follow the spatial evolution of the elastic interaction field, but it is not so trivial to distinguish the states associated with impurities or defects trapped at the dislocation core and those surrounding the dislocations [51].

To extend a thermodynamic treatment to impurities distributed in elastically strained phases one could use the method developed more than 40 years ago by Hirth and Lothe [140].

In this context, the repartition equilibrium of an impurity between a solution and the ED could be discussed as the repartition of the impurity within the ED and a solution of chemically non-interacting particles in the presence of a position (ξ, θ) -dependent elastic field, where ξ and θ are, respectively, the distance and the angle, in cylindrical polar coordinates, between the ED and the impurity. In these conditions, any impurity i embedded in an elastically strained phase is subjected to an internal pressure $P(\xi, \theta)$.

Provided the system behaves as an elastic continuum, the interaction between the impurity and the ED could be treated as the interaction between the stress field of the ED and the local elastic (stress) field of the impurity arising from a size effect (a difference between the radius r or the volume Ω of the impurity and that of the atoms of the host crystal), such that the interaction energy is zero when the radius/volume of the impurity i is the same as the atoms of the host crystal. Additionally, the condition is included that each site could be occupied by only one (solute) impurity atom or left empty by one lattice atom.

In these conditions, the atomic fraction x_i of i segregated in the β (dislocation) phase could be calculated to be

$$x_i^\beta = \frac{1}{[1 + \exp[P(\xi, \theta)\Delta\Omega - G^\circ]/kT]} \tag{3.17}$$

where $\Delta\Omega$ is the difference between the atomic volumes of the solute and solvent, $G^\circ = kT \ln \frac{n_o}{N - n_o}$, N is the atomic density of lattice sites per unit volume, n_o is the solute concentration (i.e the atomic density of the solute atoms per unit volume) in the α (bulk) phase and ξ varies from r_0 (the radius of the dislocation core) to R (the inter-dislocation half-distance).

By assuming that $P(\xi, \theta)$ varies only slightly between r_0 and R, that $n_o \ll N$ (diluted solution) and that isotropic conditions hold (P is only a function of ξ), Eq. (3.17) could be simplified:

$$x_i^\beta \approx \frac{1}{1 + (1/x_i^o)\exp(E_i/kT)} \tag{3.18}$$

where $x_o = (n_o/N)$ is the atomic fraction of the impurities in the bulk of the crystal and E_i is the average elastic interaction energy of the impurity with the ED, which could be calculated using the Bullough and Newman method [10] (Table 3.3).

For $E_i \gg kT$, which is, as an example, the case for the elastic interaction energy of dislocations with oxygen (see Table 3.1), where the off-core interaction energy is 0.3 eV, Eq. (3.18) simplifies to

$$x_i^\beta \approx x_i^o \exp -\frac{E_i}{kT} \tag{3.19}$$

where x_i^β is the mean value of the atomic fraction of i in the dislocation impurity cloud of thickness $(R - r_o)$.

A similar equation

$$x_i^c \simeq x_i^o \exp{-\frac{E_b}{kT}} \tag{3.20}$$

(where x_i^c is the atomic fraction of i in the dislocation core and x_i^o is the atomic fraction of i in the bulk and E_b is the binding energy) can be derived for the segregation of an impurity i in the core of the dislocation. Here the core region can be simulated as a cylinder of radius r_o, where the interaction energy is constant and corresponds to the binding energy E_b. Some numerical estimates of interaction energies of 90° partial dislocations with typical native and dopants impurities in silicon are reported in Table 3.4 [141] which shows that these impurities have a strong tendency to form solitonic defects in reconstructed dislocations.

Equation (3.20) was used [12, 14] to get a qualitative estimate of the probability p of finding an impurity i in the impurity cloud of a dislocation, for different values of the interaction energy.

Using for the off-core average elastic interaction strength between an impurity and a dislocation a value of 0.5 eV, which is close to that of oxygen in silicon, see Table 3.4, it can be deduced that an impurity present in the bulk at a concentration $x_o = 10^{-6}$ (or $5 \cdot 10^{16}$ at cm^{-3} in silicon) would not be found segregated at dislocations at temperatures higher than 500 °C (see Figure 3.25) [14].

Instead, it has been calculated [12] that for interaction energies of 1.5 eV, typical of in-core or solitonic interactions, a temperature close to the melting point of silicon, would be necessary to clean a dislocation embedded in a silicon matrix with an impurity contamination $x_o = 10^{-7}$ or $5 \cdot 10^{15}$ at cm^{-3}.

It is to be stressed, however, that the straightforward application of Eqs. (3.19) and (3.20) is a gross approximation, as the concentration of any solute interacting with the elastic field of the dislocation depends on the position within the impurity cloud and makes the use of the chemical potential improper.

A way to formally overcome this constraint would be the introduction of a new potential $\eta_1(x)$ having the structure of the electrochemical potential [142]

$$\eta_i(x) = \mu_i + e\varphi(x) \tag{3.21}$$

Table 3.4 *Values of interaction energies of some typical impurities with 90° partial dislocations in silicon*

Impurity	Bond strength (eV) off-core	Bond strength (eV) in-core	Bond strength (eV) in a solitonic site
O	0.3	—	1.2
O$_2$	—	2.5	—
2O → O$_2$	—	4.5	—
B	—	0.4	2.5
P	—	0.4	1.5
N	—	0.6	3.4

Lehto and Heggie, 1999, [141]. Reproduced with permission from The Institution of Engineering and Technology.

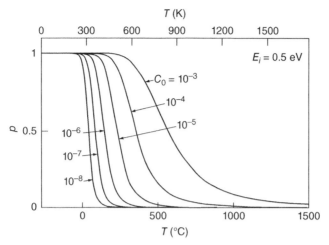

Figure 3.25 *Segregation probability p as a function of the of the impurity concentration. Sumino, 1999, [14]. Reproduced with permission from WILEY-VCH Verlag Berlin GmbH, Fed. Rep. of Germany*

by substituting the electrical energy term $e\varphi(x)$ with an elastic interaction energy term $P(\xi,\theta)\Delta\Omega$

$$\eta_i = \mu_i + P(\xi,\theta)\Delta\Omega = \mu_i^o + kT \ln x(\xi,\theta) + P(\xi,\theta)\Delta\Omega \qquad (3.22)$$

where, as before, μ_i^o is the standard chemical potential of i, θ the angle and ξ the distance of the impurity from the dislocation core.

In analogy with the equilibrium between electrified phases, the equilibrium condition between a bulk α phase and an impurity cloud β 'phase'

$$\eta_i^\alpha = \eta_i^\beta \qquad (3.23)$$

holds for the entire volume of the impurity cloud, and thus, for any value of $x(\xi,\theta)$. Therefore, assuming that $\mu_i^{o,\alpha} = \mu_i^{o,\beta}$, one obtains

$$\frac{x(\xi,\theta)}{x} = \exp - \frac{P(\xi,\theta)\Delta\Omega}{kT} \qquad (3.24)$$

where x is the impurity concentration in a region of zero internal pressure.

The assumption that $\mu_i^{o,\alpha} = \mu_i^{o,\beta}$ is tenable, as the solution of i in the dislocation cloud is thermodynamically equivalent to the solution of i in the bulk phase and we neglect the effect of the bonding at the dislocation core if $\xi > r_o$

For an edge dislocation the $P(\xi,\theta)\Delta\Omega$ term can be expressed as a position-dependent isotropic elastic interaction energy $E_i(\xi,\theta) = \frac{\varepsilon\,(\sin\theta)}{\xi}$, where

$$\varepsilon = \frac{\mu b}{3\pi}\frac{1+\nu}{1-\nu}\Delta\Omega = {}^1\!/_2\,\mu b\frac{1+\nu}{1-\nu}\delta r^3 \qquad (3.25)$$

where b is the modulus of the Burger's vector, μ the shear modulus, ν the Poisson's ratio and δr^3 the difference of the atomic volumes of the impurity and solvent atoms, which determines the strength of the isotropic strain field of the impurity [140].

In turn, the elastic interaction energy is given by the expression

$$E_i(\xi, \theta) = \tfrac{1}{2}\,\mu b \frac{1+v}{1-v}\,\delta r^3\,\frac{\sin\theta}{\xi} \tag{3.26}$$

and the distribution concentration equilibrium is given by the equation

$$\frac{x(\xi,\,\theta)}{x} = \exp-\frac{\varepsilon\sin\theta}{\xi kT} = \exp-\frac{\varepsilon\sin\theta}{bkT} \tag{3.27}$$

where ξ takes its minimum value for $\xi = r_o = b$ and its maximum value ξ_{max} at the inter-dislocations half-distance.

Some calculated radial elastic interaction energy profiles are reported in Figure 3.26a, for C, O and H impurities in silicon, using for μ and v the values of $6.81 \cdot 10^{10}$ Pa and

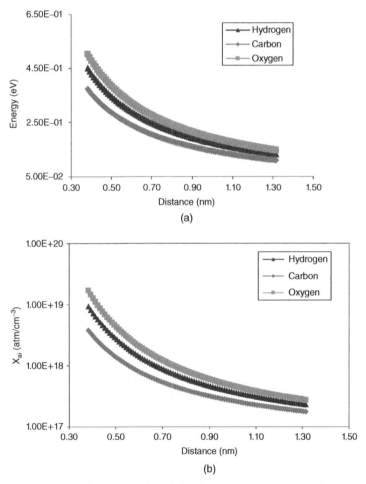

(a)

(b)

Figure 3.26 *(a) Calculated radial profiles of the elastic interaction energy for O, C and H, calculated using Eq. (3.26) (the core region diameter is taken equal to = 0.4 nm). (b) Calculated radial concentration profiles of C, O and H impurities*

0.210, respectively, for the length of the Burgers vector a value $b = 3.84 \cdot 10^{-10}$ m [143] and for the covalent radii of silicon, carbon and oxygen and hydrogen the values of 111, 77, 73 and 37 pm, respectively.

It appears that at a distance larger than 1.3 nm the interaction energy has been decreased by a factor of 10, which allows one to conclude that the radius of the impurity cloud is of the order of a few nanometres.

Further, for $\theta = \pm\pi/2$ and $\xi = b$

$$E_{i,\max} = \frac{1}{2}\mu\frac{1+\nu}{1-\nu}\delta r^3 \tag{3.28}$$

Estimates of the maximum value of the elastic interaction energy with an edge dislocation in silicon and of the cut-off radii, defined as the distance for which the interaction energy $E_{i,\max} \sim kT$, are reported in Table 3.5 [9].

From Eqs. (3.25)–(3.27) the equilibrium conditions for an isotropical section of the cloud ($\sin\theta = 1$) the radial concentration profile is given by

$$kT\ln\frac{x(\xi,\theta)}{x} = \frac{1+\nu}{1-\nu}\delta r^3\frac{\mu b}{2R} = \frac{\varepsilon}{R} \tag{3.29}$$

where $\xi = R$ the mean distance between the impurity and the dislocation core, and

$$\frac{x(\xi,\theta)}{x} = \exp-\frac{\varepsilon}{RkT} \tag{3.30}$$

which corresponds to the Hirsch and Lothe solution [140].

Some calculated radial concentration profiles relative to C, O and H impurities are reported in Figure 3.26b. It can be noted that the impurity concentration steadily increases when approaching the dislocation core, up to values higher by about 2 orders of magnitude with respect to that of the bulk for a distance r° from the dislocation core. This concentration increase can lead to supersaturation conditions, with the consequent segregation of a second phase, a phenomenon often experimentally observed, that will be discussed in Section 3.6, but of which one example is reported in Figure 3.27, which shows the segregation of $NiSi_2$ at a crossing point of two dislocations [144].

Table 3.5 *Estimates of the elastic (size effect) interaction energies of various impurities to an edge dislocation in silicon*

Impurity	$E_{\max}(T = 1473$ K) (eV)	Cut-off radius (nm)
Boron	0.75	2.4
Aluminium	0.23	2.5
Phosphorous	0.18	0.5
Arsenic	0.03	0.8
Carbon	1.0	3.1
Copper	0.28–0.47	0.9–1.4
Gold	0.85	2.6
Oxygen	0.57	1.7
Iron	0.23	0.7

Data from Bullough and Newman, 1963, [9].

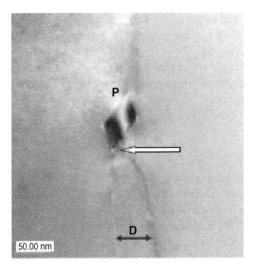

Figure 3.27   *A NiSi$_2$ precipitate at a dislocation dipole. Seibt et al., 2005, [144]. Reproduced with permission from John Wiley & Sons*

It should be noted that Eqs. (3.29) and (3.30) could be used to calculate the behaviour of substitutional impurities and interstitial impurities sitting in tetrahedral sites, where a δr^3 term well accounts for an isotropic strain field. For interstitial impurities sitting in bond centred (BC) positions some problems could arise.

We have equally to note that in some cases this kind of procedure loses its physical meaning, as for example the interaction of vacancies and self-interstitials with external or internal surfaces or a dislocation. In fact, in a 3D phase, the properties of a vacancy as well as its reactivity can be discussed under the assumption that the chemical identity of a vacancy is associated with the presence of four dangling bonds. Apart from some distortion, the vacancy maintains its chemical identity as soon as it approaches a surface or a dislocation, but the reacted vacancy is an integral part of the extended defect, which can be identified as a local change in the average coordination of the atoms where the vacancy has been trapped.

3.4.4   Kinetics of Interaction of Point Defects, Impurities and Extended Defects: General Concepts

A chemical reaction between an atomic or ionic species (also an atomic defect) and an ED in a solid phase can be formally described by the conventional approach used for chemical reactions, which are thought to occur in the presence of a quantum mechanical reaction barrier, whose shape, width and height depend on the nature, length and strength of bonds which should be broken and reconstructed.

Therefore, the kinetics of interaction reactions between point defects and EDs depends very closely on the local chemical and topological configuration of the ED centres involved. In addition, since the solubility, diffusivity and reactivity of the atomic defects and impurities[4] range within orders of magnitude, we expect a variety of interaction

---

[4] The solubility itself is different for unreacted and reacted impurities.

behaviours, the features of which should depend strongly on the defect nature and on the process temperature [10, 145].

Frequently, under light or particle irradiation or electron injection, reaction can involve excited *charged* states of the reactants, with possible differences in the intermediate or final states of the reaction. Different from liquid solutions, where reactants could easily rearrange their reciprocal positions and configuration to find that presenting the least energetic path for the completion of the process, in the case of solids, chemical reactions follow the principle that the elementary paths which are favoured are those that involve the least change in the final and initial atomic positions and electronic configuration. These reactions feature a topotactic character, as reproducible orientations and configurations must exist (and may exist in practice) between the parent and product structure, often at the expense of atomic defect formation or absorption.

These concepts, in the case of molecular solids, are coupled to the definition of a reaction cavity, in order to accomplish the condition that the reactants and the reaction products occupy a proper space in the matrix crystal [146]. As we will see in Sections 3.5.3 and 3.7.2, this concept has been intuitively used for the precipitation of SiO_2 and SiC in silicon with the Exigent Volume Model of Tan [147] and Hu [148], where it is supposed that the reaction excess- or defect-volume $\Delta\Omega$ is delivered by intrinsic atomic defect ejection or absorption. Another, major difference with respect to reactions occurring in liquids is that in solids the atomic match at the interface between the reaction product and the matrix could range from perfect (coherent) interfaces to random or fully incoherent interfaces, with rather different contributions to the reaction free energy contribution coming from the interfacial energy.

Having this background in mind and considering the ED as the active or inactive *substrate* on/at which the reaction occurs, processes involving EDs will be discussed using the conventional expressions for a chemical reaction.

An example of a reaction involving an active *substrate* is the *passivation* of GBs or external surfaces of a semiconductor sample with hydrogen (or with any other suitable chemical species). Such a reaction could be formally written, for the case of hydrogen passivation of an unreconstructed dislocation in silicon

$$H + ED_{Si}^{\cdot} \rightleftharpoons ED_{Si} - H \tag{3.31}$$

where a point is used to label a lone (unpaired) electron at a dangling bond of the ED and the effect of hydrogen is that of saturating dangling bonds with the formation of a Si-H bond.

Reaction (3.31) could also occur when the dangling bonds are totally or partially reconstructed.

In this case, if ΔG_{un} is the Gibbs free energy for the un-reconstruction reaction, the reaction occurs when the condition $\Delta G_{3.31} - \Delta G_{un} < 0$ is fulfilled. We would like to mention that this condition should hold for every passivation reaction, independently of the nature of the ED and of the passivating species, which could consist of hydrogen, oxygen, nitrogen, halogens and metallic atoms. Also the self-annihilation of vacancies and self-interstitials at external surfaces and the nucleation and growth of a precipitate could be treated in this way as new lattice positions are formed or destroyed as a consequence of the reaction.

Instead, in the case of the segregation of an impurity at an ED from a saturated or supersaturated solution with the formation of a second phase, the ED works as a heterogeneous nucleation centre, providing easier reaction paths to the process and maintaining its integrity and identity during the process. A typical example of this behaviour is illustrated by the nucleation of Pt at emerging dislocation sites in a silicon substrate, where Pt works as the catalyst for the dissociation of molecular hydrogen [149], see Section 3.5.2.

As in solid hosts the diffusion of the reactants to the reaction site could be rate determining, the reaction kinetics of chemical bond formation between the substrate and any reacting species can be studied only at temperatures high enough to suppress the diffusion contribution.

3.4.5   Kinetics of Interaction Reactions: Reaction Limited Processes

As emphasized in the last section, a true reaction-limited process occurring at the boundary of a 1D defect or at the interface of a 2D defect requires that the process of delivery of neutral or charged reactants at the reaction interface (which could also be charged) is fast and that the rate determining step is the formation of a chemical bond (ionic or covalent or mixed) between the reactants.

In all these cases the reaction rate is quantitatively described by an absolute rate constant $k(s^{-1}) = \omega \, \exp-(E_R/kT)$ where ω is a frequency factor and E_R is the height of the reaction barrier or the activation energy for the reaction.

In order to discuss reaction limited processes it should, however, be kept in mind that a chemical reaction event might occur provided the distance between reactants corresponds to their interaction length, that is the capture ratio. In turn, the interaction could be long-or short-range.

Long-range interaction might be of electrostatic (in the case of charged defects) or elastic nature, this last occurring when the strain energy around a defect can be reduced by impurity/defect bonding [150].

If the elastic interaction can be assimilated to a reaction between sphere-shaped defects, the interaction energy varies as r^{-3}, where r is the inter-defect distance. For electrostatic interactions, the interaction energy $E = k\frac{q_1 q_2}{r}$ varies with r^{-1}, k being the dielectric constant. As the capture radius r_c is defined as the value of r for which the interaction energy becomes larger than kT, for the particular case of silicon at room temperature, see Figure 3.28 [150], the elastic capture radius is of the order of magnitude of a few atomic distances, while the electrostatic capture radius is about 10 nm.

In the case of covalent bond formation (short-range interaction), the covalent capture radius is of the order of one atomic distance and impurities and/or defects should therefore necessarily be driven to the reaction site by diffusion or by elastic strain fields.

Thus, in most experimental circumstances the true extent of the interaction region can be roughly assimilated to the diffusion length $L_D = (D_i t)^{1/2}$, where D_i is the diffusion coefficient and t is the time, and the reaction kinetics is diffusion controlled (see next section) unless the bond formation event is a strongly activated process leading to slow reaction rates.

3.4.6   Kinetics of Interaction Reactions: Diffusion-Limited Reactions

If the diffusion process of reacting species towards the reaction site is slow with respect to the rate of the bond formation process, the overall rate of the interaction process is ruled by diffusion.

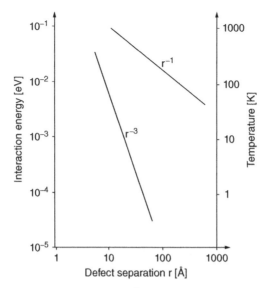

Figure 3.28 *Dependence of the elastic (r^{-3}) and coulombic (r^{-1}) interaction energies between defects as a function of their separation distance in silicon. Lannoo & Bourgoin, 1981, [150]. Reproduced with permission from Springer Science and Business Media*

In this case, a theory of diffusion-limited bimolecular chemical reactions, based on the determination of the joint distribution densities of A–B pairs in an unstructured medium [151, 152] can be applied to reactions between defects or impurity species in a semiconductor matrix.

The application of this theory is, however, restricted to those cases for which the initial distribution of the reacting species A and B is uniform, the actual distribution depends only on the A–B distance and the interaction is due to short-range chemical forces. Thus the potential energy of unreacted species is independent of the position and the rate of the diffusion-limited reaction depends on the number of pairs having reached the appropriate A–B separation r_0 for reaction. The r_0 value is within the order of magnitude of the bond length, that is it is the sum $r_0 = r_A + r_B$ of the covalent radii of A and B. When the separation distance is larger than r_0 no interaction can occur, so that the force between A and B is a step function.

The solution for these boundary conditions (uniform distribution of species, two isolated point defects or impurities A and B of concentration C_A and C_B, respectively) and slow reaction rates takes the form

$$\frac{dC_A}{dt} = \frac{dC_B}{dt} = -4\pi r_0^2 \left[1 + \frac{r_0}{(Dt)^{1/2}} \right] C_A C_B \tag{3.32}$$

where $D = D_A + D_B$, D_A and D_B are the effective diffusion coefficients of A and B and $(Dt)^{1/2}$ is the diffusion length

In the case of the interaction of a point defect or impurity (A) with an extended defect, on which exists a uniform distribution of active reaction sites, we might foresee two alternative conditions. The first is that the number of active sites at the ED is finite, the second is that each active site behaves as a virtually unsaturable trap at which a multiple trapping process occurs. In the first case the solution could be given by using Eq. (3.32) with $D = D_A$ and $C_B = N_{db}$ (the number of dangling bonds as available reaction sites). The theory of trap-limited processes was discussed in Section 2.3.3, when dealing with point defect–point defect and point defect–impurity interactions, and the results can be extended to point defect/impurity interactions with EDs. It will be seen, as an example, in Section 3.5.2, that dislocations behave as unsaturable traps with respect to their interaction with hydrogen.

Finally, as diffusion processes of impurities occur via hopping paths and involve interstitials and vacancies, the local concentration of atomic defects does influence the diffusion processes.

3.5   Interaction of Atomic Defects with Extended Defects: Theoretical and Experimental Evidence

3.5.1   Interaction of Point Defects with Extended Defects

While direct experimental observations concerning the features of the interactions of impurities with EDs are available, thanks to the use of micro- and nano-analytical techniques, as will be shown in the following, access to the same types of information for vacancies and interstitials is limited to indirect EPR evidence and to the outcomes of theoretical modelling studies, making the study of these processes quite challenging.

In addition, while impurities maintain their individuality when bonded, albeit changing their original coordination, vacancies and self-interstitials behave quite differently, as they can induce a change of configuration of the core and become an integral part of the ED itself.

We have already seen that no more than a few percent of the dislocation-related EPR signals can be ascribed to the presence of dangling bonds (unpaired electrons) in the core of dislocations, rather they can be ascribed to secondary defects, such as point defects, reconstruction defects and kinks, generated during plastic deformation or to vacancies trapped in the core partial dislocations (see Section 3.1). Thus, a problem arises concerning the topology and the energetics of the interaction of vacancies with dislocations.

Relative to the first issue, calculations carried out taking into account both elastic and quantum mechanical effects [153] show that a vacancy approaching a 90° or a 30° partial causes some strain relief and bonds at the dislocation core with the full saturation of its dangling bonds.

Concerning the energetics of the interaction, the study of the interaction of vacancies with the core of 30° partials, carried out using a combination of *ab initio* total energy calculations with finite temperature free energy calculations [154], shows that the formation energies of defects consisting of an isolated vacancy or of a vacancy in a di-vacancy in the dislocation core (2.36 and 1.87 eV, respectively) are definitely lower than the calculated formation energy of vacancies in the bulk (3.64 eV see Table 2.4). The thermodynamic meaning of these results is that the nucleation of vacancies in the dislocation core is more favourable than in the bulk, in good agreement with previous assumptions (see Section 2.3.2) that dislocations are sinks and sources of vacancies.

A similar approach has been used for the study of the interaction of vacancies and split interstitials with SFs using *ab initio* total energy calculations [155]. It has been shown that the formation energies of both defects are lower in the intrinsic SF, as compared with the bulk ($\Delta E_{sf} - \Delta E = -0.23$ eV for the vacancy and -0.15 eV for the self-interstitial), which is a good evidence that SFs are also sinks and sources of vacancies and self-interstitials [156].

3.5.2 Hydrogen-Dislocation Interaction in Silicon

Being a fast diffuser, hydrogen can react with both surface and bulk defects, even at temperatures close to the ambient. Unsaturated (dangling) bonds present at surfaces or interfaces are easily passivated by hydrogen, with the formation of Si-H bonds, using a diluted hydrogen plasma or catalytically dissociated molecular hydrogen [157–159]. The electrical deactivation of hydrogenated silicon surfaces has been explained in terms of a reduced surface recombination rate of minority carriers [160]. The recombination rate increases, however, after exposure of the passivated surfaces to air, accompanied by an increase in the surface barrier potential [161] reaching stationary conditions after 100–1000 h, depending on previous hydrogenation conditions.

It is also known that hydrogen interacts with defects (D) generated by hydrogen implantation [162] giving rise to a number of D–H complexes which behave as shallow donors and that hydrogen passivates the recombining activity of GBs and dislocations in Si single crystals and in poly/multicrystalline silicon [163–170].

For this reason hydrogenation is used as a standard procedure to improve the photovoltaic performance of multicrystalline solar cells [171–173] but it is extremely difficult to establish, case to case, whether hydrogen passivation relies on a cleaning action of native deep centres of dislocations or of deep levels associated with impurities and/or point defects gettered or bonded to dislocations.

Most of the information available about the effect of hydrogen passivation comes from lifetime, PL, CL or EBIC measurements, and is very often contradictory, as is particularly evident from results of PL or CL measurements in the emission range of radiative centres of dislocations.

As already seen in Section 3.2.5, the PL and CL spectra of dislocated silicon consist of four main bands (D1–D4) at 0.807, 0.870, 0.935 and 0.99 eV, respectively, whose intensity and width depends on sample preparation conditions [121, 126, 174], dislocation density, dislocation dissociation [175, 176] and impurity contamination [122, 125], while their position is substantially preparation and contamination independent.

When hydrogen plasma treatments at 300 °C are carried out on dislocated silicon, the effect depends strongly on the sample preparation conditions. As an example, when dislocations are generated by damaging the silicon surface with diamond tools, PL appears only after hydrogenation. This effect is apparently due to the fact that mechanical damage is responsible for not only the dislocation nucleation, but also the generation of additional point defects, which work as strong non-radiative recombination centres [119] and inhibit the dislocation luminescence emission.

In contrast, on plastically deformed samples[5] the D1 line intensity generally remains unaffected or increases after hydrogenation, while that of D2 increases [121].

---

[5] Plastic deformation is a standard method for the generation of dislocation arrays in semiconductor samples of known structure and impurity content.

The contrary happens with samples dislocated by laser melting and recrystallization, where the D1 line intensity is quenched by hydrogen passivation while that of the D2 band increases [177]. On Si–Ge epitaxial samples CL measurements generally indicate that the PL of misfit dislocations is quenched by hydrogen passivation [178], quenching being apparently less pronounced in the case of samples intentionally contaminated with impurities [178].

It may, therefore, be supposed that in the case of impurity contaminated samples the dislocation luminescence is suppressed by deep recombination centres associated with metal impurities, which can be passivated by hydrogen, with a consequent relief of the PL [179]. In the case of metallic impurity-clean reconstructed dislocations hydrogen is, instead, ineffective. What is not easily understandable is why in several cases hydrogen may quench or even suppress the dislocation luminescence, unless one assumes that Si-H bonds at the dislocation core remove from the gap the levels responsible for the D1 luminescence.

More insight about the very nature of the interaction process of hydrogen with dislocations might be inferred from the analysis of the depth profiles associated with deuterium diffusion measurements carried out on dislocated silicon samples of different origin, presenting a dislocation density ranging from 10^9 to 10^9 cm^{-2} [149, 158].

The deuterium diffusion experiments were associated with diffusion length measurements, which showed that the passivation yield decreases with the dislocation density [149], for dislocation densities between 10^3 and 10^8 cm^{-2}.

Independently of the type of sample investigated, the depth profiles of deuterium (see Figure 3.29) do not fit a standard erf profile, which would be expected in the case of a simple diffusion process, but follow a law of the type

$$[D] = [D_o] \exp(-\alpha x) \tag{3.33}$$

where $[D_o]$ is the surface deuterium concentration, α is a fitting constant and x is the depth.

This equation is formally equivalent to that used to fit the hydrogen diffusion toward an assembly of non-saturable traps [180].

$$[H_T](t) = 4\ \pi(Dt)R_o[T_o][H_o] \exp(-\alpha x) \tag{3.34}$$

where R_o is the capture radius, $[T_o]$ is the total trap concentration, $[H_o]$ is the constant hydrogen concentration at the surface and $\alpha = (4\ \pi\ R_o[T_o])^{1/2}$. As the values of α range from 1.49 to 0.82 m^{-1}, almost independently of the nature of the samples, the resulting average concentration of traps $[T_o]$ is close to $6 \cdot 10^{13}$ cm^{-3}, assuming a capture radius of 10 nm. Therefore, as the average measured concentration of deuterium is around $5 \cdot 10^{17}$ cm^{-3}, (see Figure 3.29) each single trap provides conditions for trapping around 10^4 atoms of deuterium, forming a small deuterium bubble. Apparently, dislocations behave as non-saturable traps, favouring the formation of deuterium bubbles, which behave as secondary hydrogenation sources.

The fact that the passivation yield of hydrogen decreases with the increase in the dislocation density indicates that with increase in the dislocation density the relative number of sites that can be passivated, which could be solitonic sites if the dislocations are reconstructed, decreases in favour of the centres responsible for heterogeneous nucleation of hydrogen/deuterium bubbles, which may be dislocation intersections, kinks and jogs.

This conclusion is supported by theoretical calculations [181] which show that on reconstructed partial dislocations the H–soliton complex is 0.19 eV less stable than the

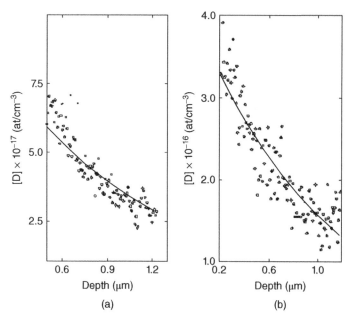

Figure 3.29   *Deuterium diffusion profiles in dislocated samples (a) $N_d = 10^8$ cm^{-2} and (b) $N_d = 10^6$cm^{-2}*

H–soliton–kink complex. This complex is very mobile along the dislocation line, it readily forms H–H pairs and then H_2 molecules, thus explaining the role of dislocations as dimerization centres for H atoms in Si.

3.5.3   Interaction of Oxygen with Dislocations

It is known that oxygen interacts with dislocations, catalyzing the formation of oxide precipitates [125, 182] or oxygen complexes [121] and influencing the electrical and/or optical properties of CZ and FZ silicon [71, 85, 121, 125, 174, 177] although often the experimental results are controversial.

As an example, it has been demonstrated that the D-band feature of the dislocation luminescence of a sample plastically deformed at 700 °C disappears in the case of long deformation times, a fact which is interpreted as due to the interaction of oxygen with the dislocation [183].

On the other hand, oxygen precipitation is often associated with dislocation generation [125], as can be seen in the TEM image of Figure 3.19.

This close connection between oxygen precipitation and dislocation nucleation and growth is not unexpected, as it could be forecast by the exigent volume law, which will be discussed in Section 3.7.2. Furthermore, the formation of Si-O bonds at a solitonic site in the dislocation core is also a spontaneous process, as the formation energy of a Si-O bond (1.2 eV) (see Table 3.4) is larger than the reconstruction energy (0.3–0.84 eV) of both 30° and 90° partials, see Table 3.2.

The energetics of the oxidation of the dislocation core suggest, therefore, that even in the case of FZ silicon the dislocation core is reacted with oxygen, dissociation requiring high temperature treatments. This conclusion is supported by theoretical calculations on the formation of an (ADP) soliton–oxygen complex in the reconstructed core of a 90° partial, which occurs with a total energy gain of about 1 eV [59]. As in the case of hydrogen, the most stable configuration of a solitonic site results from its interaction with a light impurity.

3.6   Segregation of Impurities at Surfaces and Interfaces

3.6.1   Introduction

Surfaces and interfaces of solid phases may be considered 2D systems, with specific thermodynamic, chemical, structural and physical properties. Their properties affect only slightly the average property of the bulk phase, as long as the volume/surface ratio is sufficiently large. When surface predominates over volume, as in the case of nanodots and nanowires, the effect may be so important as to make homogeneous doping an almost impossible task.

Another problem is the chemical reactivity of interfaces, which could be easily oxidized, with the formation of a surface layer of amorphous oxide, as is the case for the spontaneous oxidation of a surface of single crystal silicon, leaving a planar array of dangling bonds at the interface.

The properties of the internal surfaces, that is of GBs in multicrystalline, microcrystalline and nanocrystalline semiconductors deeply affect the mass and charge transport phenomena, which need to be analysed and thereafter formalized in term of general concepts to permit the design and then the fabrication of microelectronic and optoelectronic devices.

The segregation in the volume of single crystalline or polycrystalline semiconductors of gas bubbles and micro- or nano-precipitates from supersaturated solutions of impurities deserves significant interest for advanced technological applications, such as the smart cut technique or fabrication of quantum dots.

The main emphasis of the next sections will be on the structural and physical properties of GBs and precipitates and the physico-chemical aspects of their interactions with defects and impurities.

3.6.2   Grain Boundaries in Polycrystalline Semiconductors

Polycrystalline semiconductor ingots or thin films can be grown at higher growth rates and using a seed-less process, generally with growth furnaces less sophisticated than those used for the growth of single crystalline ingots, as will be seen in Chapter 4. This makes polycrystalline semiconductors less expensive than single-crystalline materials and interesting for their use in large surface applications, for instance photovoltaic cells.

In polycrystalline materials, however, the structural and chemical homogeneity is confined within grains of limited size whose shape depends on the specific environmental (for natural crystals) and experimental conditions of their growth. Each grain in a polycrystalline matrix is separated from its neighbour grains by internal surfaces or GBs, see Figure 3.30 which displays a picture of a multicrystalline silicon wafer, cut from an ingot grown with the directional solidification technique (see Chapter 4).

The main concern with the application of polycrystalline semiconductors in electronic devices (thin film transistors, solar cells, chemical sensors) is the influence of GBs

Figure 3.30   *Grain boundaries in a multicrystalline silicon wafer*

on majority carrier transport and minority carrier recombination processes and on the segregation of impurities, on which systematic and almost exhaustive work has been carried out in the last three decades [184–192].

As in the case of dislocations, the properties of GBs may be discussed in terms of their geometrical and electronic structure, which determine their measurable physical and chemical properties. Like dislocations, GBs are electrically active, behaving as deep recombination centres and could be imaged by EBIC and LBIC measurements.

In the EBIC contrast maps of Figure 3.31 [193], which displays the recombination activity of the same GBs in three different wafers cut at different heights of the same multicrystalline ingot, it is possible to see that the recombination activity varies from one GB to the neighbouring one and also along the same GB. Moreover, it is also evident that the contrast of the same GB observed at three different heights of the same ingot is not constant and takes a minimum value at the centre of the ingot.

As impurity contamination during a growth process is a thermodynamically spontaneous (irreversible) process, since it causes a decrease in the free energy of the system by an increase in its configurational entropy

$$\Delta S = R \sum_1^n x_i \ln x_i \tag{3.35}$$

it is reasonable to suppose that the variable contrast would depend not only on the structural and physical attributes of single GBs, but on different amounts of impurities segregated at GBs during the growth process.

In turn, the purity of the melt depends on the quality of the melt stock and on impurities coming from a partial dissolution of the crucible walls and the impurity concentration in the

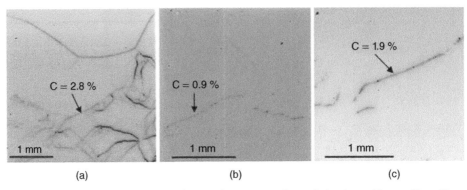

(a)   (b)   (c)

Figure 3.31   *EBIC contrast maps of recombining grain boundaries in multicrystalline silicon wafers. (a) Bottom of the ingot, (b) centre and (c) top of the ingot. Pizzini et al., 2005, [193]. Reproduced with permission from Trans Tech Publications*

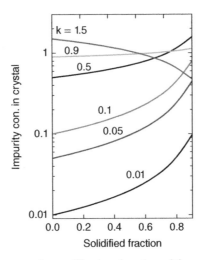

Figure 3.32   *Impurity concentration profiles as a function of the solidified fractions, for different values of the segregation coefficients of contaminating impurities*

grown crystal along the ingot height depends on their segregation coefficient k^o, following the Scheil equation (see Chapter 4, Section 4.2.6)

$$x_i^s = k^o x_i^o (1 - f)^{k^o - 1} \tag{3.36}$$

when the growth process is carried out close to thermodynamic equilibrium conditions, that is at very similar growth rates.

It comes out, see Figure 3.32, that for segregation coefficients lower than 1, the concentration of impurities increases with the increase in the solidified fraction f, that is in the distance from the bottom of the crucible, which corresponds to the fraction of the melt solidified first.

The electronic structure of GBs and the thermodynamics of segregation should therefore play the major role in the electrical properties of GB.

3.6.3 Structure of Grain Boundaries and Their Physical Properties

A grain boundary is a 2D defect which separates two grains of the same or different materials. It originates, formally, by a symmetry operation consisting in a rigid body translation *t*-around a rotation axis. The structure of a GB is entirely defined if the orientation θ, the translation *t* and the normal *n* to the boundary are known. When *n* is perpendicular to the rotation axis, the GB is conventionally called a tilt GB, when *n* is parallel it is called a twist GB, but mixed GBs are very common. When the median plane of a tilt GB is a symmetry plane of the crystal, the grain boundary is called symmetrical and the boundary plane has similar indices with respect to both grains. Thus the mutual orientation of the two grains can be described as a reflection with respect to the boundary plane or as a rotation of 180° about the normal of the boundary plane. These symmetrical GBs are called twins [194].

To give a phenomenological description of the structure of GBs and for their further classification the coincidence sites lattice (CSL) model could be first used [3, 5, 194–196]. This model assumes that the structure of a GB may be described as a short period superlattice rather than a simple contact between the surfaces of two neighbour grains and permits one to describe a GB as a function of the lattice sites that two neighbour grains might have in common, with the formation of a common sublattice when two neighbour crystals having the same structure are rotated around a common rotation axis.

With the CSL model a GB is a section, planar in particular cases, of a 3D superlattice constructed on common lattice sites of two neighbour crystals in physical contact.

If a coincidence lattice exists, which is a discontinuous function of θ, it can be defined by an index called the coincidence index Σ, which is the reciprocal density of coincidence sites and is always given by an odd integer.

Twins observed in natural and synthetic crystals always have a low coincidence index: for the <111> twin the coincidence index is 3, the lowest possible.

Details of the algebra involved in the determination of the atomic structure of GBs with the CLS model are outside the interest of this book, and can be find in the Honstra papers and in the excellent monograph of Fionova and Artemyev [5].

Tables of CLS for cubic, hexagonal and tetragonal materials are available in the literature, but this geometrical description is unable to give information on the energy of the GBs or to predict atomic relaxation at the interfaces leading to an equilibrium structure or the appearance of GB defects.

However, if the assumption is made that non-defective GBs arrange themselves uniquely this way and that the number of equilibrium structures depends only on these geometrical parameters, computer simulations using empirical potentials provide information on the energy of equilibrium GBs.

Energy calculation requires attention to atomic relaxation occurring in the GB superlattice in order to get minimum energy configurations. Atomic relaxation implies bond reconstruction with minimization of the number of dangling bonds in the structure with the set-up of arrays of five and seven member rings.

Geometrical structures derived by reconstruction were originally studied with the use of a simple stick and ball model [194, 195], which is actually able to predict the structure of a number of GB imaged by HRTEM studies.

Results of these computer modelling studies relative to GBs in the diamond structure are reported in Table 3.6 which shows, that the GB energy increases in the presence of

Table 3.6   *Calculated values of excess energy of close packed GB in silicon*

Coincidence index Σ	3	5	5	5ᵃ	7	7ᵃ	9
Plane orientation	{111}	{210}	{310}	{310}	{321}	{321}	{221}
Rotation axis	<111>	<100>	<100>	<100>	<111>	<111>	<110>
σ	—	1.118	0.7905	0.7905	—	—	1.5
Rotation angle (Θ)	60	36.9	36.9	36.9	38.2	38.2	39.9
Excess GB energy (mJ m^{-2})	30	770	1770	1980	2100	2600	640
Coincidence index Σ	9	9ᵃ	11	11ᵃ	13	13	15
Orientation	{411}	{411}	{331}	{331}	{320}	{510}	{521}
Rotation axis	<110>	<110>	<110>	<110>	<100>	—	<311>
σ	1.06	1.06	—	—	1.803	1.2745	—
Rotation angle (Θ)	38.2	38.2	50.5	50.5	2.6	2.6	48.2
Excess GB energy (mJ m^{-2})	2230	2580	1980	2580	2060	2180	2380

σ gives the relative area per one coincident site.
ᵃStructure with broken bonds.
Adapted from Fionova and Artemyev, 1993, [5] and from Artemyev, et al., 1990, [197].

dangling bonds and that GBs with the same value of Σ but with different orientation display different values of σ, and therefore a different density of coincident sites.

This leads to different values of GB excess energy, which is a marker of their chemical reactivity. Moreover, in structures like the Σ 7 {321}, Σ 9 {411} and Σ 11{311}, for which theory predicts the presence of broken bonds, higher values of excess energy are calculated.

Albeit GBs are not equilibrium defects in a strict thermodynamic sense, like other EDs, the phenomenology concerning their relative stabilities arising from the CSL model fulfils criteria of thermodynamic equilibrium, their excess energy increasing with the increase of every case of mismatch.

It is apparent, however, that the CLS model is unable to give any direct information on the true structure of GBs, which was revealed first by chemical etching procedures and then by HRTEM investigations. These last show [198] that GBs are essentially composed of regular line defects, simple dislocations, partial dislocations and SFs, depending on the mismatch. Frequently, several types of dislocations (edge and screw) are present in the same boundary and some may be dissociated into two partial dislocations that bound a SF. In many of these low-angle boundaries it is likely that reconstruction occurs with pairing of neighbouring unsatisfied bonds such that no dangling bonds remain. This involves formation of regular arrays of five- and seven-member rings on the boundary [194, 195].

Higher angle GBs are usually composed of distinct facets that are one or more nanometres long. These facets are subsections of the boundaries where bonding rearrangements have occurred. These configurations can usually, but not always, be predicted by the CSL theory. Not enough is known to predict whether or not all bonds are satisfied within the substructure of these facets, albeit the CLS model predicts that some of these structures present dangling bonds, as we have seen before.

Not enough is known about the electronic structure of GBs, although there is experimental evidence that the recombination activity of their minority carriers mostly depends on segregated impurities [199].

Non-equilibrium GBs are generated by interaction of point defects and impurities with GB, and non-equilibrium states may explain the variability of contrast along GBs observed in Figure 3.31.

3.6.4 Segregation of Impurities at Grain Boundaries and Their Influence on Physical Properties

Widespread and valuable information is available on this subject, which made significant advances in the last decades with the advent of spectrometric techniques of micrometric or submicrometric resolution, such as SIMS, μ-IR, X-ray fluorescence microscopy (μ-XRF), X-ray absorption spectroscopy μ(XAS) and μ-Raman spectroscopy, which allow one to map the impurity distribution in the proximity of or along a grain boundary. μ(XRF) and μ(XAS) experiments are performed with a synchrotron source beamline which delivers the necessary X-ray intensity.

As each technique has a specific spatial resolution range and individual sensitivity to a specific impurity family, the use of different techniques is necessary to determine the chemistry and the physics of the interaction, not a simple phenomenology.

An example is given in Figure 3.33 which displays the segregation patterns of metallic impurities in multicrystalline Si, as obtained with X-ray microfluorescence analysis by T.Buonassisi *et al.* [201–203]. It shows that, contrary to previous hypotheses, metallic impurities segregate as microprecipitates and not as homogeneously segregating unreacted impurities. It is, furthermore, easy to see that the localization of the impurities depends on the nature of the different metals, with Cu and Ni segregating at the GB, Fe at the GB and in its proximity, Ti always out of the GB.

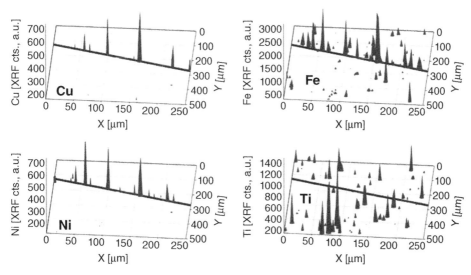

Figure 3.33 *Distribution of Cu, Ni, Fe and Ti impurities in the proximity of a GB of multicrystalline silicon using X-ray microfluorescence. Pizzini, (2009), [200]. Reproduced with permission from Springer Science and Business Media*

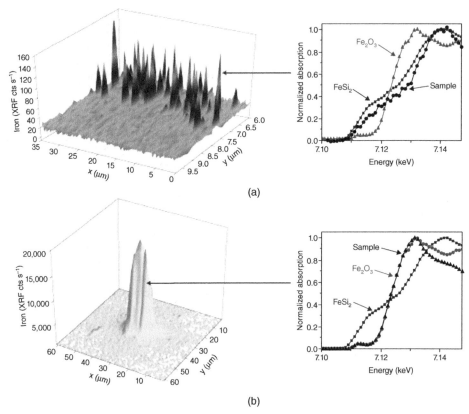

Figure 3.34 *Determination of the nature of the precipitates by combined X-ray microfluorescence (a) and X-ray absorption spectra (b) of the reference samples and the precipitates Buonassisi et al., 2005, [203]. Reproduced with permission from Macmillan Publishers*

The chemical nature of the precipitates can be and has been investigated using $\mu(XAS)$, taking advantage of the fact that every chemical compound has a unique $\mu(XAS)$ spectrum. A correspondence might be, therefore, found between the $\mu(XAS)$ spectra of standard samples and those of the microprecipitates [201–203].

An example is displayed in Figure 3.34 for the case of iron segregation in with exceptionally strong recombining GBs, which shows that both iron oxide and iron silicide precipitates may be detected, although only the $FeSi_2$ precipitates are homogeneously distributed along the GB (see Figure 3.34a) [201].

The indirect conclusion is that the chemistry of iron segregation at GBs is a complex process involving the partial oxidation of iron diffusing across an oxygen-rich interfacial layer, followed by the reaction of residual iron at active spots of the GB with the local formation of iron silicide. This conclusion fits with the results concerning the formation of iron silicide phases [204] grown as thin surface films by MBE on Si(111) substrates. Here, the heat treatment at 500 °C of the stoichiometric iron silicide film induces the formation of islands of the $FeSi_2$ γ-phase, with a $CaCl_2$ structure, for films thinner than 5 nm or of islands of the orthorhombic bulk β-phase for film thicknesses between 5 and 15 nm.

Another interesting example is that of copper, where the μ(XAS) spectrum of the Cu clusters, which segregate uniquely along the GB, matches well the spectrum of a standard Cu_3Si sample, although the reference μ(XAS) spectrum of CuO is also very close to that of the silicide.

Given the large difference in Gibbs energy of formation of Cu_3Si and CuO $(-194.25$ kJ mol^{-1} at 1000 K), the formation of the silicide is, however, thermodynamically favoured [202].

As a final example, Figure 3.35 displays the SIMS concentration profiles of B dopant and of C and O native impurities across a GB in a multicrystalline silicon wafer after annealing at 1150 °C [200]. It is possible to see that B and O segregate at the GB, whereas C segregates a few micrometres from it. The significant decrease in the Si signal of the GB indicates that a process involving the formation of B-O species and Si oxidation is probably occurring, the last with the ejection of silicon self-interstitials which allow the C precipitation. The distance between the GB and the C concentration profiles (about 25 ± 10 μm) could be taken as a measure of the diffusion length $L_D = \sqrt{Dt}$ of the self-interstitials at the annealing temperature. The resulting value of $6.9 \cdot 10^{-9}$ cm^2 s^{-1} is 6 orders of magnitude higher than that given in Table 2.7 [205] for the self-diffusion of silicon, but in this case the diffusion of self-interstitials occurs in the presence of a concentration gradient and a stress field which enhance the diffusion process.

3.7 3D Defects: Precipitates, Bubbles and Voids

3.7.1 Thermodynamic and Structural Considerations

Among EDs, precipitates, bubbles and voids play a significant role as potential scattering, recombination and light emitting centres in semiconductors, as well as sinks for atomic defects and impurities.

While bubbles (see Figure 3.36) [206] and voids are the result of a process of condensation and agglomeration of a supersaturated solution of gases or of supersaturated vacancies, the formation of precipitates, consisting of elemental substances or of binary or multinary compounds, occurs with the interaction/association of dissolved impurities and formation of an alloy or chemical compound.

Thermodynamic conditions for the segregation of a precipitate consisting of an elemental substance X occur when the supersaturation ratio $\frac{n_X}{n_X^0}$ of the chemical species X is larger than 1, and thus

$$RT \ln \frac{n_X}{n_X^0} > 0 \qquad (3.37)$$

where n_X^0 is the equilibrium solubility of the species X in solution and

$$\Delta G_{segr} = -RT \ln \frac{n_X}{n_X^0} \qquad (3.38)$$

is the excess Gibbs energy available for the precipitation process.

When the impurity X in the host crystal phase α reacts with the solvent or with another impurity and forms an alloy $X_m Y_n$ or a stoichiometric compound which segregates as a

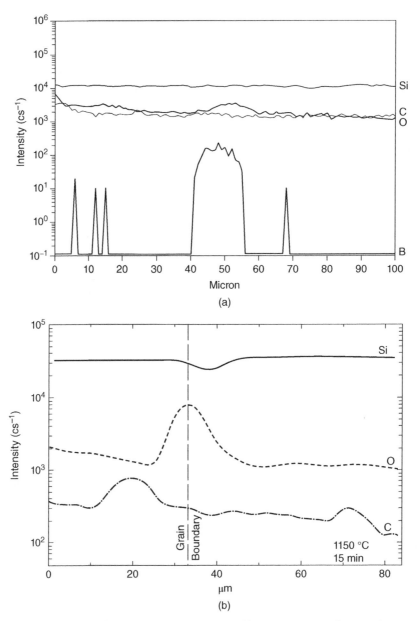

Figure 3.35   *(a) B, O and C SIMS concentration profiles across a GB. (b) C and O concentration profiles on an enlarged concentration scale*

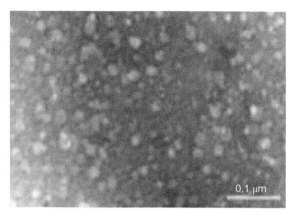

Figure 3.36   *Cross-section HRTEM image of a silicon sample implanted with He, after anneal-ing at 700°C to nucleate He bubbles. Leoni* et al., *2004, [206]. Reproduced with permission from Cambridge University Press*

second phase β, as would be the case for the formation of SiO_2, SiC or Si_3N_4 in supersat-urated solutions of O, C or N in Si, the driving force for the process is again given by the supersaturation ratio, where the n_X^o value is given by the equilibrium concentration of the species X in the host crystal phase α in the presence of the precipitate phase β.

Thermodynamic equilibrium conditions between the saturated solution and the second phase are satisfied by the sum

$$\sum_i (\mu_i^\alpha = \mu_i^\beta) \qquad (3.39)$$

for each of the system *i* components, but it should be mentioned that thermodynamic equi-librium is reached at the end of a process where the second phase formation occurs in the bulk of a solid matrix under lattice and volume misfits. In these conditions the driving force for the nucleation of the β phase is the supersaturation ratio and the barriers to nucleation are given by the strain energy ΔG_{str} arising from volume misfit between the nucleus and the host lattice and by the interfacial Gibbs energy of the nucleus/matrix interface ΔG_{int}.

The corresponding change in Gibbs free energy is given by

$$\Delta G_{nucl} = \Delta G_{segr} - \Delta G_{str} - \Delta G_{int} \qquad (3.40)$$

In the earliest precipitation stages nucleation occurs with lattice and volume misfit accommodation at the expense of elastic strain and interfacial energy. Further precipitate growth, in the case of large volume misfits, is only possible if plastic deformation occurs with dislocation emission or the excess or defect volume is accommodated by point defect ejection or absorption, as will be seen in Section 3.7.2.

Lattice misfit accommodation, in the case of low volume misfits, may occur by differ-ent routes. An example is given by the segregation of a compound presenting different stable polytypes, with different Gibbs energies of formation. In this case the thermody-namic equilibrium phase is not necessarily the most stable polytype, but that presenting the minimum volume- and lattice-misfit with the matrix, as the excess of thermodynamic energy is compensated by a lower elastic strain and interface energy.

Another condition to minimize the lattice misfit and the corresponding excess of interface energy is obtained when the precipitate grows inside the matrix crystal with a shape which favours the formation of the maximum number of coherent interfaces with the matrix, (which would imply that heteroepitaxial conditions are satisfied for these interfaces). The formation of octahedral {111} faceted voids in vacancy supersaturated CZ silicon [137] of amorphous silicon oxide octahedra in silicon [207, 208] and of silicide ($NiSi_2$) platelets in Ni saturated silicon, see Figure 3.37 [209] are typical examples of this.

In all these circumstances, under the assumption that the elastic strain is minimized thanks to small or negligible volume misfit and good structural compatibility, the driving force for the segregation remains the defects/impurities supersaturation ratio

$$f = RT \ \ln \ \frac{x_D}{x_D^o} \tag{3.41}$$

and the associated change in Gibbs energy permole of defects is

$$\Delta G = -f + \Delta G_{surf} = -RT \ \ln \frac{x_D}{x_D^o} + \lambda^{2/3} \tag{3.42}$$

where λ is a surface energy coefficient [137] which accounts for the excess of surface energy.

3.7.2 Oxygen and Carbon Segregation in Silicon

More complex situations are expected when the precipitate phase presents a large volume misfit with the host matrix and point defects injection or absorption at the precipitate interface occurs to compensate the volume misfit and the associated elastic strain.

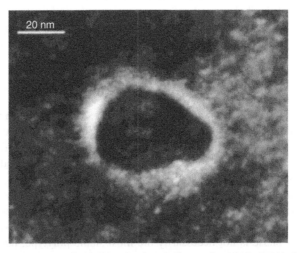

Figure 3.37 *HRTEM image of a NiSi₂ platelet. Seibt et al., 1999, [209]. Reproduced with permission from John Wiley & Sons*

Table 3.7 *Density, molar volumes and pressure stability limits of silicon and of different SiO$_2$ polymorphs*

Polymorph	Density (g cm^{-3})	Molar volume (cm^3)	Pressure (Pa)
α-quartz	2.65	22.67	Atmospheric
β-Tridymite	2.25	26.70	Up to 9.0×10^7
Cristobalite	2.33	25.78	Up to 1.5×10^8
Cohesite	2.90	20.07	2–4 GPa
Stishovite	4.28	14.03	9–12 GPa
Silicon	2.33	12.05	—

This is the case for SiO$_2$ precipitation in silicon [210]

$$Si_{Si} + 2O_i \rightarrow SiO_2(Si) \tag{3.43}$$

which occurs in supersaturated solutions of oxygen in silicon at temperatures above the brittle–ductile transition, which ranges between 900 and 1075 K in undoped silicon [48].

It can be seen in Table 3.7 that only one single SiO$_2$ polytype, the high pressure phase stishovite, has a molar volume comparable with that of silicon, while the low pressure phases have molar volumes more than double than that of silicon, leading to an average volume increase close to twice the volume of Si consumed to allow precipitation

$$\Delta V_{SiO_2} \approx 2\, \Delta V_{Si} \tag{3.44}$$

which must be compensated either by crystal deformation or by empty space formation.

The practicality of the first solution, at temperatures below the brittle–ductile transition, could be evaluated by computing the mechanical work to be spent in order to accommodate the oxide phase in the crystalline silicon matrix. From the equilibrium pressures reported in Table 3.7, the order of magnitude of the volume strain energy $\Delta G_{str} = P\Delta V$ under which the nucleation of spherical precipitates of amorphous cohesite could occur, neglecting the interfacial Gibbs energy contributions (see Eq. (3.40)), is around -400 kJ mol^{-1}, in rather good agreement with the result reported by Hu [148] (around 300 kJ mol^{-1}) for amorphous SiO$_2$.

Considering that the supersaturation ratio (about 10^4, see Section 2.3.4.4) results in a driving force amounting to 57.7 kJ mol^{-1} at 750 °K, that is below the brittle–ductile transition, the precipitation of SiO$_2$ by crystal deformation should be inhibited.

It is, therefore, accepted that the volume misfit is compensated by self-interstitials ejection and/or vacancy trapping, according to the exigent volume theory [147, 148] that postulates that the mass conservation law, displayed in Eq. (3.43) should be associated with a volume conservation law

$$V_{Si} + 2V_{O_i} + \delta V = V_{SiO_2} \tag{3.45}$$

and that the excess volume δV could be delivered by a process of self-interstitials ejection (and/or vacancies absorption) at the precipitate/matrix interface

$$(1 + y)\, Si_{si} + 2O_i \rightleftharpoons SiO_2(Si) + ySi_i \tag{3.46}$$

to allow the progress of the precipitate growth process.

In the case of SiO_2 precipitation in the form of one of the various low pressure silica polymorphs (see Table 3.7) the excess volume δV corresponds closely to the molar volume of silicon and, consequently, y in Eq. (3.46) is close to 1.

Taking for the Gibbs energy of formation of silicon self-interstitials a value of 3.8 eV, the calculated free energy of reaction (3.46) at 1000 K is largely less than zero

$$\Delta G_{3.46} = \Delta G_{SiO2} - \Delta G_I = -730.25 + 366.64 = -363.61 \text{ kJ mol}^{-1} \tag{3.47}$$

thus allowing the occurrence of the segregation process.

A local self-interstitials supersaturation is the consequence of this process, which leads to the nucleation of extrinsic self-interstitial types of dislocations, as already shown in Figure 3.20, which illustrates the formation of an expanding dislocation loop associated with the segregation of a SiO_2 microprecipitate.

Opposite conditions hold for the precipitation of SiC from a supersaturated solution of C in silicon, as in this case a volume decrease occurs, of about one atomic volume for each C atom incorporated in SiC.

The volume misfit may be accommodated by self-interstitial absorption

$$Si_{Si} + 2C_{Si} + \delta Si_i \rightarrow (1 + \delta) \text{ SiC} \tag{3.48}$$

if a local self-interstitial supersaturation is present or is mediated by the simultaneous precipitation of oxygen. Nevertheless, the segregation of carbon even from largely super-saturated solutions is a rather difficult process and it is also known that in oxygen-poor silicon the carbon segregation does not occur at all [211].

It is interesting to note that the exigent volume theory cannot explain why oxygen seg-regates much more easily than carbon, even when the thermodynamic conditions of seg-regation are satisfied. The metastability of supersaturated solid solutions of carbon can be, instead, understood by considering the influence of the interface energy σ_C of a SiC nucleus embedded in a Si host on the nucleation process. A quantitative estimate of the interface energies ruling the process of SiC nucleation may be obtained by using the fol-lowing equation

$$r_{crit} = \frac{2\sigma_C \Omega_{SiC}}{kT \ln\left[\left(\dfrac{x_C}{x_C^0}\right)\left(\dfrac{x_I}{x_I^0}\right)\right]} \tag{3.49}$$

which gives the critical precipitate radius r_{crit} as a function of the interfacial energy σ_C, of the atomic volume Ω_{SiC} of SiC and of the supersaturation ratios of carbon $\left(\dfrac{x_C}{x_C^0}\right)$ and of self-interstitials $\left(\dfrac{x_I}{x_I^0}\right)$. It can be seen that the nucleation at 900 °C of SiC with a critical radius of 50 nm occurs for values of σ_C around 2000 erg cm^{-2}, provided the supersatu-ration ratio of carbon is larger than 50, in the absence of self-interstitial supersaturation. Instead, for values of σ_C around 8000 erg cm^{-2}, nucleation occurs only if large carbon and self-interstitial supersaturations are present.

The conclusion is that the interfacial energy σ_C at a SiC/Si interface ranges between 2000 and 8000 erg cm^{-2}, much higher than that at a SiO_2 / Si interface (90 to 10 erg cm^{-2}) [211]. It has also been demonstrated that such large interface energies could be overcome, by co-precipitation of C and O and C–O complex formation [212].

3.7.3 Silicides Precipitation

Among metal silicides, major concern is devoted to copper silicides, because the introduction of copper interconnects in ULSI technology increases the danger of unintentional Cu contamination of silicon [213].

Three different copper silicide phases are stable, see Figure 3.38 and the $Cu_{19}Si_6$ phase, conventionally labelled Cu_3Si, has three polymorphs η, η', η'', differing slightly in composition, of which the η' results from a faulted stacking of the η one. The η'' phase is stable at room temperature and is the one most commonly found in Cu–Si thin film reactions [214] and is also the phase which fills the silicon voids, see Section 3.7.4.

Due to the large molar volume misfit between Cu_3Si ($\Omega = 29.71$ cm^3) and Si ($\Omega = 12.054$ cm^3 the Cu precipitation in Si as Cu_3Si is unfavourable and, when it occurs, the large volume difference should be accommodated by the ejection of self-interstitials [202]. Heterogeneous nucleation sites (SFs, dislocations, GBs) and high supersaturation levels are the necessary conditions for its precipitation.

The result of Cu precipitation is the formation of a planar array of (spherical) copper silicide colonies around a central precipitate bounded by an extrinsic dislocation loop [209] (see Figure 3.39.)

3.7.4 Bubbles and Voids

Bubbles of gases can be created by ion implantation of hydrogen or helium. Bubbles transform in faceted voids after a suitable thermal annealing, which causes the out-diffusion of

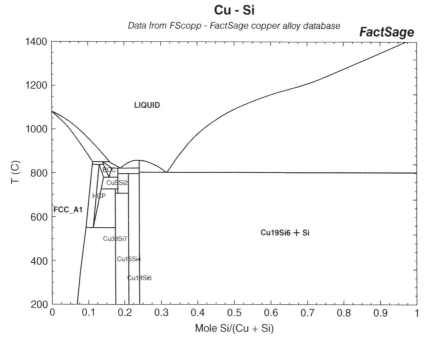

Figure 3.38 *Phase diagram of the Cu–Si system. Scientific Group Thermodata Europe (SGTE). Reproduced with permission from SGTE*

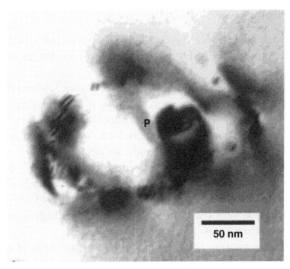

Figure 3.39   *TEM image of a copper silicide precipitate (P) in silicon, surrounded by spherical particles attached at a dislocation loop. Seibt, et al., 1999, [209]. Reproduced with permission from John Wiley & Sons*

the gases and the coalescence of irradiation-created vacancies. Void formation also occurs once vacancy supersaturation conditions are set up in a generic semiconductor during the growth cycle and, as such, may have a substantial influence on the physical and electronic properties of the material.

Voids in silicon formed by He implantation behave as deep sinks for metallic impurities, including Cu, Ni, Fe and Pt. As an example, copper chemisorbs at void walls with a binding energy of 2.2 eV [215] while interstitially dissolved copper interacts only slightly with silicon, leading to almost ideal Cu–Si solutions [213].

Copper interaction with voids has been used [216, 217] for the accurate determination of the total volume of native voids (cavities) in single crystal Si ingots used for the determination of Avogadro's number with sufficient accuracy (1 part in 10^8) and thereafter to allow the establishment of a physical standard for the kilogram-mass. It can be demonstrated that artificially created, less than 100 nm in diameter, voids in FZ silicon could be fully filled by Cu_3Si. Voids were created with a 30 keV helium implantation, followed by He out-diffusion and Cu diffusion from a surface source at 850 °C followed by a slow cooling (1 K min^{-1}). The Cu surface source was a 100 nm thick Cu layer, which transforms into a Cu_3Si layer during the annealing. As the void walls are stronger traps for Cu than the Cu silicide phase [215], redistribution of Cu within the silicide layer and the voids occurs during the annealing, until the voids are entirely filled.

The voids issue became, however, of crucial technological importance when it was discovered by low angle laser inspection procedures that large diameter (>200 mm) CZ grown silicon wafers presented a number of surface defects, called D-defects [111, 136, 137].

D-defects are different from other type of defects, the A-and B-swirls, which consist of self-interstitial aggregates and are also systematically found in silicon wafers. D-defects actually are voids and are generated by aggregation of supersaturated vacancies surviving during the growth cycle. The direct utilization of these defected wafers was discouraged by device manufacturers as these defects work as scattering and recombination centres, as heterogeneous nucleation sites for metallic impurity segregation, as shown before. Voids are also detrimental for the gate oxide integrity in MOS devices. The solution of the problem came from the discovery that the concentration of self-interstitials and vacancies in a silicon ingot during the growth process depends on the ratio of only two growth parameters, the growth rate V and the near axial temperature gradient G, whose suitable management (see Section 4.2.2) allows the control of vacancy or self-interstitial supersaturation conditions in a silicon ingot and also the growth of the so-called 'perfect' silicon.

It should, however, be noted that while in an interstitial type crystal the precipitation of oxygen is hindered and the segregation of carbon is enhanced, as shown in Section 2.6.2, in vacancy type crystals there is a competition between the process of voids formation and that of oxide precipitation.

The driving force for void nucleation is the vacancy supersaturation, while the driving force for oxide precipitation is both the vacancy and oxygen supersaturation, it being known that volumetric restrictions make the oxide precipitation possible only in the case of complementary vacancies supply or self-interstitials ejection.

This competition makes the control of voids and oxide particle formation quite complicated, because voids nucleate during the cooling of the crystals, and any long residence of the ingot in the range of temperatures typical of void nucleation must be prevented by suitable furnace design.

On the other hand, it is also known that in vacancy-type CZ crystals the oxide precipitation is enhanced if the residence time at $650 - 750\,°C$ is increased. This may be explained by considering that vacancies surviving to void formation facilitate the nucleation of the oxygen clusters in this temperature regime.

Concerning the hypothesis raised in the literature that germanium doping suppresses or limits void defect formation, it has been demonstrated [218] that actually the impact of Ge is only marginal, in good agreement with the reported weak influence of Ge-doping on the thermal equilibrium concentration of vacancies in silicon [219], given the small difference between the calculated binding energy of vacancy clusters V_n and the corresponding GeV_n clusters, as can be seen from the difference between the binding energy of di-vacancies $(1.58\,eV)$ and that of the GeV_2 cluster $(1.93\,eV)$.

References

1. Hirth, J.P. and Lothe, J. (1968) *Theory of Dislocations*, McGraw-Hill Book Company, New York.
2. J.P. Hirth, Dislocations, in *Physical Metallurgy*, Vol. **II** R.W. Cahn, P. Haasen Ed. pp 1223–1258, North Holland (1983).
3. J. Thiboult, J.L. Rouviere, A. Bourret, Grain boundaries in semiconductors, in *Materials Science and Technology, Electronic Structure and Properties of Semiconductors*, W. Schröter Ed., Vol. **4**, pp 321–378, Wiley-VCH Verlag GmbH, Weinheim (1991)
4. H. Alexander and H. Teichler, Dislocations, in *Materials Science and Technology, Electronic Structure and Properties of Semiconductors*, W. Schröter Ed, Vol. **4**, pp 249–322, Wiley-VCH Verlag GmbH, Weinheim (1991)

5. Fionova, L.K. and Artemyev, A.V. (1993) *Grain Boundaries in Metals and Semiconductors*, Les Editions de Physique, Les Ulis.

6. Hull, R. (1999) *Properties of Crystalline Silicon*, EMIS Data Review, vol. **20**, Inspec Publlications.

7. Nabarro, F.R.N., Hirth, J.P. and Kubin, L. (eds) *Dislocations in Solids*, vols. **10** (2001), **11** (2003), **12** (2005), **13** (2007), **14** (2008), **15** (2009), **16**, Elsevier, Amsterdam.

8. Kittler, M. and Reiche, M. (2011) Structure and properties of dislocations in silicon, in *Crystalline Silicon – Properties and Uses* (ed S. Basu), InTech, ISBN: 978-953-307-587-7, http://www.intechopen.com/books/crystalline-silicon%20-properties-and-uses/structure-and-properties-of-dislocations-in-silicon (accessed 22 December 2014).

9. Bullough, R. and Newman, R.C. (1963) The interaction of impurities with dislocations in silicon and germanium, in *Progress in Semiconductors*, vol. **7** (eds A.F. Gibson and R.E. Burgess), Heywood, London, pp. 100–134.

10. Bullough, R. and Newman, R.C. (1970) The kinetics of migration of point defects to dislocations. *Rep. Prog. Phys.*, **33**, 101–148.

11. Schröter, W., Scheibe, E. and Schoen, H. (1980) Energy spectra of dislocations in silicon and germanium. *J. Microsc.*, **118**, 23–34.

12. Sumino, K. (1989) Interaction of impurities with dislocations in semiconductors, in *Point and Extended Defects in Semiconductors*, NATO ASI Series, Series B: Physics, vol. **202**, Plenum Press, pp. 77–94.

13. Pizzini, S. (2002) *Defect Interaction and Clustering in Semiconductors*, Scitec Publications Ltd, Switzerland, ISBN: 3-908450-65-9.

14. Sumino, K. (1999) Impurity reaction with dislocations in semiconductors. *Phys. Status Solidi A*, **171**, 111–122.

15. Seibt, M., Abdelbarey, D., Kveder, V. *et al.* (2009) Interaction of metal impurities with extended defects in crystalline silicon and its implications for gettering techniques used in photovoltaics. *Mater. Sci. Eng., B*, **159-160**, 264–268.

16. Akatsuka, M., Sueoka, K., Katahama, H. and Adachii, N. (1998) Calculation of the slip length in 300 mm silicon wafers during thermal processes. Semiconductor Silicon 1998, Electrochemical Society Proceeding, vol. 98-1, p. 671.

17. Vanhellemont, J. (1998) Process induced mechanical stresses in silicon. Semiconductor Silicon 1998, Electrochemical Society Proceeding, vol. 98-1, pp. 997–1011.

18. Fischer, A., Richter, H., Kürner, W. and Kücher, P. (2000) Slip-free processing of 300 mm silicon batch wafers. *J. Appl. Phys.*, **87**, 1543–1549.

19. Cho, K.-C., Jeon, H.-T. and Park, J.-G. (2005) Slip formation in 300-mm polished and epitaxial silicon wafers annealed by rapid thermal annealing. *J. Korean Phys. Soc.*, **46**, 1001–1006.

20. Hirsch, P.B. (1980) The structure and electrical properties of dislocations in semiconductors. *J. Microsc.*, **118**, 3–12.

21. Blumenau, A.T. (2002) The modeling of dislocations in semiconductor crystals. Doctor degree thesis. University of Padeborn.

22. Hornstra, J. (1958) Dislocations in the diamond lattice. *J. Phys. Chem. Solids*, **5**, 129–141.

23. Alexander, H. (1986) in *Dislocations in Solids*, vol. **7** (ed F.R.N. Nabarro), North-Holland, Amsterdam.
24. Marklund, S. (1984) Structure and energy levels of dislocations in silicon. *J. Phys.*, **44**, C4–C25.
25. Ray, L.F. and Cockayne, D.J.H. (1971) The dissociation of dislocations in silicon. *Proc. R. Soc. London, Ser. A*, **A325**, 543–554.
26. Suzuki, K., Ichihara, M., Takeuchi, S. *et al.* (1992) Electron-microscopy studies of dislocations in diamond synthesized by a CVD method. *Philos. Mag. A*, **65**, 657–664.
27. Li, Z. and Picu, R.C. (2013) Shuffle-glide dislocation transformation in Si. *J. Appl. Phys.*, **113**, 0835191–0835197.
28. Frank, F.C. (1951) Crystal dislocations-elementary concepts and definitions. *Philos. Mag.*, **42**, 809–819.
29. Schröter, W. and Labusch, R. (1969) Electrical properties of dislocations in Ge and Si. *Phys. Status Solidi A*, **36**, 539–550.
30. Peierls, R. (1940) The size of a dislocation. *Proc. R. Soc. London, Ser. A*, **52**, 34–37.
31. Nabarro, F.R.N. (1947) Dislocations in a simple cubic lattice. *Proc. R. Soc. London, Ser. B*, **59**, 256–272.
32. Nabarro, F.R.N. (1997) Fifthy years study of the Peierls-Nabarro stress. *Mater. Sci. Eng., A*, **234**, 67–76.
33. Schoeck, G. (1999) Peierls energy of dislocations: a critical assessment. *Phys. Rev. Lett.*, **82**, 2310–2313.
34. Schoeck, G. (2010) Atomic dislocation core parameters. *Phys. Status Solidi B*, **247**, 265–268.
35. Schoeck, G. (2011) The Peierls stress in simple cubic lattice. *Phys. Status Solidi B*, **248**, 2284–2289.
36. Joos, B., Ren, Q. and Duesbery, M.S. (1994) Peierls-Nabarro model of dislocations in silicon with generalized stacking-fault restoring forces. *Phys. Rev. B*, **50**, 5890–5898.
37. Pizzagalli, L., Beauchamp, P. and Rabier, J. (2003) Undissociated screw dislocations in silicon: calculations of core structure and energy. *Philos. Mag.*, **83**, 1191–1204.
38. Bulatov, V.V., Yip, S. and Argon, A.S. (1995) Atomic modes of dislocation mobility in silicon. *Philos. Mag. A*, **72**, 453–496.
39. Bulatov, V. and Kaxiras, E. (1997) Semidiscrete variational peierls framework for dislocation core properties. *Phys. Rev. Lett.*, **78**, 4221–4224.
40. Joss, B. and Duesbery, M.S. (1997) The Peierls stress of dislocations: an analytic formula. *Phys. Rev. Lett.*, **78**, 266–269.
41. Nabarro, F.R.N. (1997) Theoretical and experimental estimates of the Peierls stress. *Philos. Mag.*, **A75**, 703–711.
42. Marchesoni, F. and Cattuto, C. (1998) On two conflicting estimates of the Peierls stress. *Philos. Mag. A*, **77**, 1223–1225.
43. Cai, W., Bulatov, V.V., Justo, J.F. *et al.* (2000) Intrinsic mobility of a dissociated dislocation in silicon. *Phys. Rev. Lett.*, **84**, 3346.
44. Cai, W. and Bulatov, V.V. (2004) Mobility laws in dislocation dynamics simulations. *Mater. Sci. Eng., A*, **387–389**, 277–281.
45. Pizzagalli, L., Beauchamp, P. and Jonsson, H. (2008) Calculations of dislocation mobility using nudged elastic band method and first principles DFT calculations. *Philos. Mag.*, **88**, 1–8.

46. Rabier, J. (2013) On the core structure of dislocations and the mechanical properties of silicon. *Philos. Mag.*, **93**, 162–173.

47. Duesberry, M.B. and Joss, B. (1996) Dislocation motion in silicon: the shuffle-glide controversy. *Philos. Mag. Lett.*, **74**, 253–258.

48. Tanaka, M., Maeno, K. and Higashida, K. (2010) The effect of boron/antimony on the brittle to ductile transition in silicon single crystals. *Mater. Trans.*, **51**, 1206–1209.

49. Rabier, J., Cordier, P., Demenet, J.L. and Garem, H. (2001) Plastic deformation of Si at low temperature under high confining pressure. *Mater. Sci. Eng., A*, **309–310**, 74–77.

50. Weber, E. and Alexander, H. (1979) EPR of dislocations in silicon. *J. Phys.*, **40**, C6-101–C6-106.

51. Schröter, W., Scheibe, E. and Schoen, H. (1980) Energy spectra of dislocations in silicon and germanium. *J. Microsc.*, **11**, 23–34.

52. Alexander, H. and Teichler, H. (1991) Dislocations, in *Electronic Structure and Properties of Semiconductors* (ed W. Schröter), Wiley-VCH Verlag GmbH, Weinheim, pp. 261–267.

53. Marklund, S. (1978) Electron states associated with the core region of the 60° dislocation in silicon and germanium. *Phys. Status Solidi B*, **85**, 673–681.

54. Marklund, S. (1979) Electron states associated with partial dislocations in silicon. *Phys. Status Solidi B*, **92**, 83–89.

55. Marklund, S. (1980) On the core structure of the glide-Set 90° and 30° partial dislocations in silicon. *Phys. Status Solidi B*, **100**, 77–85.

56. Marklund, S. (1983) Structure and energy levels of dislocations in silicon. *J. Physique, Coll. C4* **44**, 25–35.

57. Jones, R. (1977) Electronic states associated with the 60 degree edge dislocation in silicon. *Philos. Mag.*, **35**, 57–64.

58. Jones, R. (1977) Electronic states associated with the 60 degree edge dislocation in germanium. *Philos. Mag.*, **36**, 677–683.

59. Usmerski, A. and Jones, R. (1993) The interaction of oxygen with dislocation cores in silicon. *Philos. Mag.*, **A67**, 905–915.

60. Justo, J., Bazant, M.Z., Kaxiras, E. *et al.* (1998) Interatomic potentials for silicon defects and disordered phases. *Phys. Rev. B*, **58**, 2539–2550.

61. Engerness, T.D. and Arias, T.A. (1997) Multiscale approach to determination of thermal properties and changes in free energy: application to reconstruction of dislocations in silicon. *Phys. Rev. Lett.*, **79**, 3006–3009.

62. Schröter, W. and Labusch, R. (1969) Electrical properties of dislocations in Ge and Si. *Phys. Status Solidi B*, **36**, 539–550.

63. Labusch, R. and Hess, J. (1989) Conductivity of grain boundaries and dislocations in semiconductors, in *Point and Extended Defects in Semiconductors* (eds G. Benedek, A. Cavallini and W. Schröter), Plenum Press, New York, pp. 15–38.

64. Bigger, J.R.K., McInnes, D.A., Sutton, A.P. *et al.* (1992) Atomic and electronic structures of the 90° partial dislocation in silicon. *Phys. Rev. Lett.*, **69**, 2224.

65. Nunes, R.W., Bennetto, J. and Vanderbilt, D. (1998) Atomic structure of dislocation kinks in silicon. *Phys. Rev. B*, **57**, 10388.

66. de Araújo, M.M., Justo, J.F. and Nunes, R.W. (2004) Electronic charge effects on dislocation cores in silicon. *Appl. Phys. Lett.*, **85**, 5610.

67. Pizzagalli, L., Godet, J., Guénolé, J. and Brochard, S. (2011) Dislocation cores in silicon: new aspects from numerical simulations. *J. Phys. Conf. Ser.*, **281**, 012002.
68. Lehto, N. (1997) Shallow electron states of bounded intrinsic stacking faults in silicon. *Phys. Rev. B*, **55**, 15601–15607.
69. Kveder, V., Kittler, M. and Schröter, W. (2001) Recombination activity of contaminated dislocations in silicon: a model describing electron-beam induced current contrast behaviour. *Phys. Rev. B*, **63**, 115208-1–115208-11.
70. Castaldini, A., Cavalcoli, D., Cavallini, A. and Pizzini, S. (2005) Experimental evidence of dislocation related shallow states in p-type silicon. *Phys. Rev. Lett.*, **95**, 076401.
71. Sekiguchi, T. and Sumino, K. (1996) Cathodoluminescence study on dislocations in silicon. *J. Appl. Phys.*, **79**, 3253–3260.
72. Kveder, V., Badylevich, M., Steinman, E. *et al.* (2004) Room-temperature silicon light-emitting diodes based on dislocation luminescence. *Appl. Phys. Lett.*, **84**, 2106–2108.
73. Reiche, M., Kittler, M., Krause, M. and Ubensee, H. (2013) Electrons on dislocations. *Phys. Status Solidi C*, **10**, 40–43.
74. Radzimski, Z.J., Zhou, T.Q., Rozgonyi, G.A. *et al.* (1992) Recombination at clean and decorated misfit dislocations. *Appl. Phys. Lett.*, **60**, 1096.
75. Mankovics, D., Schmid, R.P., Arguirov, T. and Kittler, M. (2012) Dislocation-related photoluminescence imaging of mc-Si wafers at room temperature. *Cryst. Res. Technol.*, **47**, 1148–1152.
76. Pasemann, L. (1981) A contribution to the theory of the EBIC contrast of lattice defects in semiconductors. *Ultramicroscopy*, **6**, 237–250.
77. Donolato, G. (1983) Quantitative evaluation of the EBIC contrast of dislocations. *J. Phys.*, **44**, C-4-269–C-4-275.
78. Singh, J. (2001) *Semiconductor Devices, Basic Principles*, John Wiley & Sons, Inc., New York.
79. Seibt, M., Khalil, R., Kveder, V. and Schröter, W. (2009) Electronic states at dislocations and metal silicide precipitates in crystalline silicon and their role in solar cell materials. *Appl. Phys. A*, **96**, 235–253.
80. Kittler, M., Ulhaq-Bouilleta, C. and Higgs, V. (1995) Influence of copper contamination on recombination activity of misfit dislocations SiGe/Si epilayers: temperature dependence of activity as a marker characterizing the contamination level. *J. Appl. Phys.*, **78**, 4573–4583.
81. Castaldini, A., Cavalcoli, D., Cavallini, A. and Pizzini, S. (2005) Defect states in Czochralski p-type silicon: the role of oxygen and dislocations. *Phys. Status Solidi A*, **202**, 889–895.
82. Cavalcoli, D. and Cavallini, A. (2007) Electronic states related to dislocations in silicon. *Phys. Status Solidi C*, **4**, 2871–2877.
83. Schröter, W. and Cerva, H. (2001) Interaction of point defects with dislocations in silicon and germanium: electrical and optical effects, in *Defect Interaction and Clustering* (ed S. Pizzini), Trans Tech Publications Inc., Zürich, pp. 67–144.
84. Pizzini, S., Leoni, E., Binetti, S. *et al.* (2004) Luminescence of dislocations and oxide precipitates in Si. *Solid State Phenom.*, **95–96**, 273–282.

85. Pizzini, S., Acciarri, M., Leoni, E. and LeDonne, A. (2000) About the D1 and D2 dislocation luminescence and its correlation with oxygen segregation. *Phys. Status Solidi B*, **222**, 141–150.

86. Binetti, S., Pizzini, S., Leoni, E. *et al.* (2002) Optical properties of oxygen precipitates and dislocations in silicon. *J. Appl. Phys.*, **92**, 2437.

87. Bodnarenko, A., Vyvenko, O. and Isakov, I. (2011) Identification of dislocation luminescence partecipating levels in silicon by DLTS and Pulsed –CL profiling. *J. Phys. Conf. Ser.*, **281**, 012008-1–012008-7.

88. Simoen, E., Claws, F. and Vennik, J. (1985) DLTS of grown-in dislocations in p-and n-type high purity germanium. *Solid State Commun.*, **54**, 1025–1029.

89. Shevchenko, S. and Tereshchenko, A. (2010) Dislocation states and deformation induced point defects in plastically deformed germanium. *Solid State Phenom.*, **156–158**, 289–294.

90. Shevchenko, S.A. and Kolyubakin, A.I. (2013) DLTS study of plastically deformed copper doped *n* type germanium. *Semiconductors*, **47**, 849–855.

91. Arguirov, T., Kittler, M. and Abrosimov, N.V. (2011) Room temperature luminescence from Germanium. *J. Phys. Conf. Ser.*, **281**, 012021–012029.

92. Shevchenko, S. and Tereshchenko, A. (2007) Peculiarities of dislocation photoluminescence in germanium with quasi-equilibrium dislocation structure. *Phys. Status Solidi C*, **4**, 2898–2902.

93. Shevchenko, S. and Tereshchenko, A. (2008) Dislocations photoluminescence in silicon and germanium. *Solid State Phenom.*, **131–133**, 583–588.

94. Schöne, J., Spiecker, E., Dimroth, F. *et al.* (2008) Misfit dislocation blocking by dilute nitride intermediate layers. *Appl. Phys. Lett.*, **92**, 081905.

95. Amano, H.H., Sawaki, N., Akasaki, H. and Toyoda, Y. (1986) Metalorganic vapor phase epitaxial growth of a high quality GaN film using an AlN buffer layer. *Appl. Phys. Lett.*, **48**, 353–355.

96. Speck, S. and Rosner, S. (1999) The role of threading dislocations in the physical properties of GaN and its alloys. *Physica B*, **273-274**, 24–32.

97. Ruterana, P., Albrecht, M. and Neugebauer, J. (2003) *Nitride Semiconductors Handbook on Materials and Devices*, Wiley-VCH Verlag GmbH, Weinheim, pp. 337–345.

98. Ha, S., Skowronski, M. and Lendenmann, H. (2004) Nucleation sites of recombination-enhanced stacking fault formation in silicon carbide *p-i-n* diodes. *J. Appl. Phys.*, **96**, 393–398.

99. Zhang, Z. and Sudarshan, T.S. (2005) Basal plane dislocation-free epitaxy of silicon carbide. *Appl. Phys. Lett.*, **87**, 151913.

100. Chen, W. and Capano, M.A. (2005) Growth and characterization of 4H-SiC epilayers on substrates with different off-cut angles. *J. Appl. Phys.*, **98**, 114907.

101. Mion, C., Muth, F., Preble, E.A. and Hanser, D. (2006) Thermal conductivity, dislocation density and GaN device design. *Superlattices Microstruct.*, **40**, 338–342.

102. Friedrich, J., Kallinger, B., Knoke, I. *et al.* (2008) Crystal growth of compound semiconductors with low dislocation densities. Proceeding of the 20th International Conference on Indium Phosphide and Related Materials, IPRM 2008.

103. Müller, G., Schleswig, P., Birkmann, B. *et al.* (2005) Types and origin of dislocations in large GaAs and InP bulk crystals with very low dislocation densities. *Phys. Status Solidi A*, **202**, 2870–2879.

104. Kioseoglu, J., Kalesaki, E., Lymperakis, L. *et al.* (2011) Electronic structure of 1/6 (2023) partial dislocations in wurtzite GaN. *J. Appl. Phys.*, **109**, 083511-1–083511-6.

105. Wright, F. and Grossner, U. (1998) The effect of doping and growth stoichiometry on the core structure of a threading edge dislocation in GaN. *Appl. Phys. Lett.*, **73**, 2751.

106. Grzegory, I., Kurowski, S., Leszczynsi, M. *et al.* (2003) High pressure crystallization of GaN, in *Nitride Semiconductors, Handbook on Materials and Devices* (eds P. Ruterana, M. Albrecht and J. Neugabauer), Wiley-VCH Verlag GmbH, Weinheim, pp. 3–43.

107. Georgakilas, A., Ng, H.-M. and Komninou, P. (2003) Plasma assisted molecular beam epitaxy of III-C nitrides, in *Nitride Semiconductors, Handbook on Materials and Devices* (eds P. Ruterana, M. Albrecht and J. Neugabauer), Wiley-VCH Verlag GmbH, Weinheim, pp. 107–222.

108. Vetter, W.M. and Dudley, M. (2006) The character of micropipes in silicon carbide crystals. *Philos. Mag.*, **86**, 1209–1225.

109. Gudkin, M.Y., Sheinerman, A.G. and Argunova, T.S. (2009) Micropipes in silicon carbide crystals. *Phys. Status Solidi C*, **8**, 1942–1947.

110. Pirouz, P. (1998) On micropipes and nanopipes in SiC and GaN. *Philos. Mag. A*, **78**, 727–736.

111. Plekhanov, P.S., Gösele, U.M. and Tan, T.Y. (1998) Modeling of nucleation and growth of voids in silicon. *J. Appl. Phys.*, **84**, 718–727.

112. Cahn, J.W. (2013) Thermodynamic aspects of Cottrell atmosphere. *Philos. Mag.*, **93**, 3741–3746.

113. Higgs, V., Lightowlers, E.C., Norman, C.E. and Kightley, P.C. (1992) Characterisation of dislocations in the presence of transition metal contamination. *Mater. Sci. Forum*, **83–87**, 1309–1314.

114. Lelikov, Y., Rebane, Y., Ruvimov, S. *et al.* (1992) Photoluminescence and electronic structure of dislocations in Si crystals. *Mater. Sci. Forum*, **83–87**, 1321–1327.

115. Shevchenko, S.A., Ossipyan, Y.A., Mchedlidze, T.R. *et al.* (1994) Defect states in Si containing dislocation nets. *Phys. Status Solidi A*, **146**, 745–755.

116. Tajima, M., Tokita, M. and Warashina, M. (1995) Photoluminescence due to oxygen precipitates distinguished from the D lines in annealed Si. *Mater. Sci. Forum*, **196-201**, 1749–1754.

117. Shevchenko, S.A. and Izotov, A.N. (1995) Structure of the photoluminescence spectra in the vicinity of the lines D1 and D2 in plastically deformed Si. *Phys. Status Solidi A*, **148**, K1–K5.

118. Kveder, V.V., Steinman, E.A., Shevchenko, S.A. and Grimmeiss, H.G. (1995) Dislocation-related electroluminescence at room temperature in plastically deformed silicon. *Phys. Rev. B*, **51**, 10520–10526.

119. Sekiguchi, T. and Sumino, K. (1995) Cathodoluminescence study on dislocation-related luminescence in silicon. *Mater. Sci. Forum*, **196-201**, 1201–1206.

120. Sweinbjornsson, E.O. and Weber, J. (1996) Room temperature electroluminescence from dislocation-rich silicon. *Appl. Phys. Lett.*, **69**, 2686–2688.

121. Steinman, E.A. and Grimmeiss, H.G. (1998) Dislocation-related luminescence properties of silicon. *Semicond. Sci. Technol.*, **13**, 124–129.

122. Pizzini, S., Donghi, M., Binetti, S. *et al.* (1998) Luminescence from erbium-doped silicon epilayers grown by LPE. *J. Electrochem. Soc.*, **145**, L8–L11.

123. Cavallini, A., Fraboni, B., Binetti, S. *et al.* (1999) On the influence of dislocations on the luminescence of Si:Er. *Phys. Status Solidi A*, **171**, 347–351.

124. Pizzini, S., Binetti, S., Calcina, D. *et al.* (2000) Local structure of erbium-oxygen complexes in erbium-doped silicon and its correlations with the optical activity of erbium. *Mater. Sci. Eng., B*, **72**, 173–176.

125. Pizzini, S., Guzzi, M., Grilli, E. and Borionetti, G. (2000) The photoluminescence emission in the 0.7-0.9 range from oxygen precipitates, thermal donors and dislocations in silicon. *J. Phys. Condens. Matter.*, **12**, 10131–10143.

126. Pizzini, S., Binetti, S., Acciarri, M. and Casati, M. (2000) Study of radiative and non radiative recombination process at dislocations in silicon by photoluminescence and LBIC measurements. *MRS Symp. Proc.*, **588**, 117–122.

127. Emel'yanov, A.M., Sobolev, N.A. and Pizzini, S. (2002) Effect of surface state density on room temperature photoluminescence from Si-SiO2 structures in the range of band-to-band recombination in silicon. *Semiconductors*, **36**, 1225–1226.

128. Leoni, E., Martinelli, L., Binetti, S. *et al.* (2004) The origin of photoluminescence from oxygen precipitates nucleated at low temperature in semiconductor silicon. *J. Electrochem. Soc.*, **11**, G866–G869.

129. Kveder, V., Badylevich, M., Steinman, E. *et al.* (2004) Room-temperature silicon light-emitting diodes based on dislocation luminescence. *Appl. Phys. Lett.*, **84**, 2106–2109.

130. Steinman, E.A. (2005) Influence of oxygen on the dislocations related luminescence centers in silicon. *Phys. Status Solidi C*, **6**, 1837–1841.

131. Kveder, V., Badylevich, M., Schroter, W. *et al.* (2005) Silicon light-emitting diodes based on dislocation-related luminescence. *Phys. Status Solidi A*, **202**, 901–910.

132. Mchedlidze, T., Wilhelm, T., Arguirov, T. *et al.* (2009) Correlation of electrical and luminescence properties of a dislocation network with its microscopic structure. *Phys. Status Solidi C*, **6**, 1817–1822.

133. Kittler, M. and Reiche, M. (2009) Dislocations as active components of novel silicon devices. *Adv. Eng. Mater.*, **11**, 249–258.

134. Svensson, B.G., Jagadish, C., Hallen, A. and Lalita, J. (1997) Generation of vacancy-type point defects in single collision cascades during swift-ion bombardment of silicon. *Phys. Rev. B*, **55**, 10498–10507.

135. Schmidt, D.C., Barbot, J.F., Blanchard, C. and Ntsoenzok, E. (1997) Defect levels of proton-irradiated silicon with a dose of $3.6 \times 10^{13} cm^{-2}$. *Nucl. Instrum. Methods Phys. Res., Sect. B*, **132**, 439–446.

136. Falster, R., Voronkov, V.V. and Quast, F. (2000) On the properties of the intrinsic point defects in silicon: a perspective from crystal growth and wafer processing. *Phys. Status Solidi B*, **222**, 219–244.

137. Voronkov, V.V. and Falster, R. (1998) Vacancy-type microdefect formation in Czochralski silicon. *J. Cryst. Growth*, **194**, 76–88.

138. Hallam, B.J., Hamer, P.G., Wenham, S.R. *et al.* (2013) Advanced bulk defect passivation for silicon solar cells. *IEEE J. Photovoltaics*, **1**, 2156–3381, © 2013 IEEE.

139. Pizzini, S. (1993) Surface and interface segregation phenomena in oxide ceramics, in *Defects in Electronic Ceramics* (ed S. Pizzini), Trans Tech Publishers Inc., pp. 81–120.

140. Hirth, J.P. and Lothe, J. (1968) Equililibrium defect concentration, in *Theory of Dislocations*, McGraw-Hill Book Company, New York, pp. 443–483.

141. Lehto, N. and Heggie, M.I. (1999) Modelling of dislocations in silicon, in *Properties of Crystalline Silicon*, EMIS Data Review Series, vol. **20** (ed R. Hull), INSPEC, p. 357.

142. Ashcroft, N.W. and Mermin, N.D. (1976) *Solid State Physics*, HRM International Editions, Philadelphia, PA, p. 593.

143. George, A. (1999) in *Properties of Crystalline Silicon*, EMIS Data Reviews Series, vol. 20 (ed R. Hull), INSPEC, p. 104.

144. Seibt, M., Kveder, V., Schröter, W. and Voss, O. (2005) Structural and electrical properties of metal impurities at dislocations in silicon. *Phys. Status Solidi A*, **202**, 911–920.

145. Shen, B., Zhang, X.Y., Yang, K. *et al.* (1997) Gettering of Fe impurities by bulk stacking faults in Czochralski-grown silicon. *Appl. Phys. Lett.*, **70**, 1876.

146. Boldyreva, E. and Boldyrev, V. (1999) *Reactivity of Molecular Solids*, John Wiley & Sons, Ltd, Chichester.

147. Tan, Y. (1986) Exigent accommodation volume of precipitation and formation of oxygen precipitates in silicon. *MRS Symp. Proc.*, **59**, 269–279.

148. Hu, S.M. (1986) Oxygen precipitation in silicon. *MRS Symp. Proc.*, **59**, 249–267.

149. Binetti, S., Basu, S., Savigni, C. *et al.* (1997) Passivation of extended defects in silicon by catalytically dissociated molecular hydrogen. *J. Phys. III*, **7**, 1487–1493.

150. Lannoo, M. and Bourgoin, J. (1981) *Point Defects in Semiconductors I*, Springer-Verlag, Berlin, pp. 204–208.

151. Waite, T.R. (1957) Theoretical treatment of the kinetics of diffusion-limited reactions. *Phys. Rev.*, **107**, 463–470.

152. Paul, R. (1968) Statistical-mechanical analysis of the theory of diffusion-controlled chemical reactions. III. *J. Chem. Phys.*, **49**, 2806–2816.

153. Lehto, N. and Öberg, S. (1997) Interaction of vacancies with partial dislocations in silicon. *Phys. Rev.*, **B56**, R12706–R12709.

154. Justo, J.F., de Koning, M., Cai, W. and Bulatov, V.V. (2000) Vacancy interaction with dislocations in silicon: the shuffle-glide competition. *Phys. Rev. Lett.*, **84**, 2172–2175.

155. Antonelli, A., Justo, J.F. and Fazzio, A. (1999) Point defect interactions with extended defects in semiconductors. *Phys. Rev. B*, **60**, 4711–4714.

156. Carvalho, A., Johnes, R. (2015) Self-Interstitials in Silicon and Germanium, in *Silicon, Germanium and their alloys*, (eds G. Kissinger and S. Pizzini) Taylor and Francis, Boca Raton Fl, pp. 87–118.

157. Estreicher, S., Stavola, M. and Weber, J. (2015) Hydrogen in Si and Ge, in *Silicon, Germanium and their alloys*, (eds G. Kissinger and S. Pizzini) Taylor and Francis, Boca Raton Fl, pp. 217–254.

158. Pizzini, S., Acciarri, M., Binetti, S. *et al.* (1997) Recent achievements in semiconductor defect passivation. *Mater. Sci. Eng., B*, **45**, 126–133.

159. Stavola, M. (2003) Research on the Hydrogen Passivation of Defects and Impurities in Si Relevant to Crystalline Si Solar Cell Materials. Final Report NREL/SR-520-34821.

160. Spadoni, S., Acciarri, M., Narducci, D. and Pizzini, S. (2000) Surface microcharacterization of silicon wafers by the light-beam-induced current technique in the planar configuration and by attenuated total reflection spectroscopy. *Philos. Mag. B*, **80**, 579–585.

161. Nauka, K. and Kamins, T.I. (1999) Surface photovoltage measurement of hydrogen-treated Si surfaces. *J. Electrochem. Soc.*, **146**, 292–295.

162. Hartung, J. and Weber, J. (1993) Shallow hydrogen-related donors in silicon. *Phys. Rev. B*, **48**, 14161–14166.

163. Seager, C.H. and Ginley, D.S. (1979) Passivation of grain boundaries in polycrystalline silicon. *Appl. Phys. Lett.*, **34**, 337–339.

164. Makino, T. and Nakamura, H. (1979) The influence of plasma annealing on electrical properties of polycrystalline Si. *Appl. Phys. Lett.*, **35**, 551–553.

165. Johnson, N.M., Biegelsen, D.K. and Moyer, M.D. (1982) Deuterium passivation of grain-boundary dangling bonds in silicon thin films. *Appl. Phys. Lett.*, **40**, 882–884.

166. Hanoka, J.I., Seager, C.H., Sharp, D.J. and Panitz, J.K.G. (1983) Hydrogen passivation of defects in silicon ribbon grown by the edge-defined film-fed growth process. *Appl. Phys. Lett.*, **42**, 618–620.

167. Corbett, J.W., Linstroem, J.L., Pearton, S.J. and Tavendale, A.J. (1988) Passivation in silicon. *Solar Cells*, **24**, 127–133.

168. Corbett, J.W., Linstroem, J.L. and Perton, S.J. (1988) Hydrogen in silicon. *MRS Symp. Proc.*, **104**, 229–240.

169. Aucouturier, M. (1991) Hydrogen at semiconductor grain boundaries and interfaces. *Phys. B*, **170**, 469–480.

170. Heggie, M.I., Jenkins, S., Ewels, P. *et al.* (2000) Theory of dislocations in diamond and silicon and their interaction with hydrogen. *J. Phys. Condens. Matter*, **12**, 10263–10270.

171. Aberle, A.G. (2001) Overview on SiN surface passivation of crystalline silicon solar cells. *Sol. Energy Mater. Sol. Cells*, **65**, 239–248.

172. Duerinckx, F. and Szlufcik, J. (2002) Defect passivation of industrial multicrystalline solar cells based on PECVD silicon nitride. *Sol. Energy Mater. Sol. Cells*, **72**, 231–246.

173. Hallam, B., Tjaijono, B. and Wenham, S. (2012) Effect of PECVD silicon oxynitride film composition on the surface passivation of silicon wafers. *Sol. Energy Mater. Sol. Cells*, **96**, 173–178.

174. Ibuka, S., Tajima, M., Takeno, H. *et al.* (1997) Deep-level luminescence in Czochralski-grown silicon crystals after long-term annealing at 450°C. *Jpn. J. Appl. Phys.*, **36**, L494–L497.

175. Sauer, R., Kisielowski-Kemmerich, C. and Alexander, H. (1986) Dissociation-width-dependent radiative recombination of electrons and holes at widely split dislocations in silicon. *Phys. Rev. Lett.*, **57**, 1472–1475.

176. Izotov, A.N., Kolyubakin, A.I., Shevchenko, S.A. and Steinman, E.A. (1992) Photoluminescence and splitting of dislocations in germanium. *Phys. Status Solidi A*, **130**, 193–198.

177. Weronek, K., Weber, J. and Queisser, H.J. (1993) Passivation of the dislocation-hydrogen related D-band luminescence in silicon. *Phys. Status Solidi A*, **137**, 543–548.
178. Higgs, V. and Kittler, M. (1994) Influence of hydrogen on the electrical and optical activity of misfit dislocations in Si/SiGe epilayers. *Appl. Phys. Lett.*, **65**, 2804–2806.
179. Sekiguchi, T., Kveder, V.V. and Sumino, K. (1994) Hydrogen effect on the optical activity of dislocations in silicon introduced at room temperature. *J. Appl. Phys.*, **76**, 7882–7888.
180. Tulchinsky, D.A., Corbett, J.W., Borenstein, J.T. and Pearton, S.J. (1990) Exponential diffusion profile for impurity trapping at unsaturable trap. *Phys. Rev. B*, **42**, 11881–11883.
181. Ewels, C.P., Leoni, S., Heggie, M.I. *et al.* (2000) Hydrogen interaction with dislocations in Si. *Phys. Rev. Lett.*, **84**, 690–693.
182. Cavallini, A., Vandini, M., Corticelli, F. *et al.* (1993) Recombination at dislocations associated with oxygen precipitation. *Inst. Phys. Conf. Ser.*, **134**, 115–121.
183. Lightowlers, E.C. and Davis, G. (1996) Oxygen related luminescence centres created in Czochralski silicon, in *Early Stages of Oxygen Precipitation in Silicon* (ed R. Jones), Kluwer Academic Press, Dordrecht, pp. 303–318.
184. Harbeke, G. (ed) (1985) *Polycrystalline Semiconductors, Physical Properties and Applications*, Springer-Verlag, Berlin.
185. Polycrystalline semiconductors : grain boundaries and interfaces (POLYSE'88) : Proceedings of the International Symposium, Malente, Fed. Rep. of Germany, August 29- September 2, 1988 H.J.Moller, H.P.Strunk. J.H.Werner,
186. Moeller, H.J., Strunk, H.P. and Werner, J.H. (eds) (1989) *Polycrystalline Semiconductors, Grain boundaries and Interfaces*, Springer-Verlag.
187. Werner, J.H. and Strunk, H.P. (eds) (1991) *Polycrystalline Semiconductors II*, Springer-Verlag.
188. Strunk, H.P., Werner, J.H., Fortin, B. and Bonnaud, O. (eds) (1994) *Polycrystalline Semiconductors III*, Solid State Phenomena, vol. **37-38**, Scitec Publications.
189. Pizzini, S., Strunk, H.P. and Werner, J.H. (eds) (1996) *Polycrystalline Semiconductors IV*, Solid State Phenomena, vol. **51–52**, Scitec Publications.
190. Werner, J.H., Strunk, H.P. and Schock, H.W. (eds) (1999) *Polycrystalline Semiconductors V*, Solid State Phenomena, vol. **67–68**, Scitec Publications.
191. Bonneaud, O., Brahim, T.M., Werner, J.H. and Strunk, H.P. (eds) (2001) *Polycrystalline Semiconductors VI*, Solid State Phenomena, vol. **80–81**, Scitec Publications.
192. Fuyuki, T., Sareshima, T., Strunk, H.P. and Werner, J.H. (eds) (2003) *Polycrystalline Semiconductors VII*, Solid State Phenomena, vol. **93**, Scitec Publications.
193. Pizzini, S., Acciarri, M. and Binetti, S. (2005) From electronic grade to solar grade silicon: chances and challenges in photovoltaics. *Phys. Status Solidi A*, **202**, 2928–2942. doi: 10.1002
194. Honstra, J. (1960) Models of grain boundaries in the diamond lattice. II Tilt about <001> and theory. *Physica*, **26**, 198–208.
195. Honstra, J. (1959) Models of grain boundaries in the diamond lattice. I. Tilt about <110>. *Physica*, **25**, 409–421.

196. Kohyama, M. and Yamamoto, R. (1994) Theoretical study of grain boundaries in Si: effects of structural disorder on the local electronic structure and the origin of band tails. *Phys. Rev. B*, **50**, 8502–8522.

197. Artemyev, A.V., Polyak, L.E. and Fionova, L.K. (1990) Atomic and electron structure of grain boundaries in silicon. *Coll. Phys.*, **51**, C71.

198. Seager, C.H. (1985) Grain Boundaries in polycrystalline silicon. *Annu. Rev. Mater. Sci.*, **15**, 271–302.

199. Acciarri, M., Savigni, C., Binetti, S. and Pizzini, S. (1994) Effect of local inhomogeneities on the electrical properties of polycrystalline silicon. *Solid State Phenom.*, **219**, 37–38.

200. Pizzini, S. (2009) Bulk solar grade silicon: how chemistry and physics play to get a benevolent microstructured material. *Appl. Phys. A*, **96** 171–188.

201. Buonassisi, T., Istratov, A.A., Heuer, M. *et al.* (2005) Syncrotron-based investigations on the nature and impact of iron contamination in multicrystalline silicon solar cells. *J. Appl. Phys.*, **97**, 074901.

202. Buonassisi, T., Marcus, M.A., Istratov, A.A. *et al.* (2005) Analysis of copper-rich precipitates in silicon: chemical state, gettering, and impact on multicrystalline silicon solar cell material. *J. Appl. Phys.*, **97**, 063503.

203. Buonassisi, T., Istratov, A.A., Marcus, M.A. *et al.* (2005) Engineering metal-impurity nanodefects for low-cost solar cells. *Nat. Mater.*, **4**, 676–679.

204. Miglio, L. and Malegori, G. (1993) Electronic structure and total energy of $FeSi_2$ pseudomorphic phases. MRS Proceedings, Vol.320/1994 p. 119 DOI: http://dx.doi.org/10.1557/PROC-320-109

205. Tang, M., Colombo, L., Zhu, J. and Diaz de la Rubia, T. (1997) Intrinsic point defects in crystalline silicon: tight-binding molecular dynamics studies of self-diffusion, interstitial-vacancy recombination, and formation volumes. *Phys. Rev. B*, **55**, 14279–14289.

206. Leoni, E., El Bouyadi, R., Martinelli, L. *et al.* (2004) Structural and optical characterization of a dispersion of nanocavities in a crystalline silicon matrix. *Eur. Phys. J. Appl. Phys.*, **27**, 89–93.

207. Borghesi, A., Pivac, B., Sassella, A. and Stella, A. (1995) Oxygen precipitation in silicon. *J. Appl. Phys.*, **77**, 4169–4244.

208. Sassella, A., Borghesi, A., Geranzani, P. and Borionetti, G. (1999) Infrared response of oxygen precipitates in silicon: experimental and simulated spectra. *Appl. Phys. Lett.*, **75**, 1131–1133.

209. Seibt, M., Hedemann, H., Istratov, A.A. *et al.* (1999) Structural and electrical properties of metal silicide precipitates in silicon. *Phys. Status Solidi A*, **171**, 301–310.

210. U. Goesele and T.Y. Tan, Equilibria, nonequilibria, diffusion and precipitation. *Electronic Structure and Properties of Semiconductors*, Materials Science and Technology, Vol. **4**, pp. 197–247, W. Schröter Ed Wiley-VCH Verlag GmbH, Weinheim (1991).

211. Taylor, W.J., Tan, T.Y. and Gösele, U. (1993) Carbon precipitation: why is it so difficult? *Appl. Phys. Lett.*, **62**, 3336.

212. Binetti, S., Ferrari, S., Acciarri, M. *et al.* (1993) New evidences about carbon and oxygen segregation processes in polycrystalline silicon. *Solid State Phenom.*, **32-33**, 181–190.

213. Istratov, A.A., Flink, C., Hiselmair, H. *et al.* (2000) Diffusion, solubility and gettering of copper in silicon. *Mater. Sci. Eng., B*, **72**, 99–104.
214. Wen, C.-Y. and Spaepen, F. (2007) *In-situ* electron microscopy of the phases of Cu_3Si. *Philos. Mag.*, **87**, 5581–5599.
215. Myers, S.M. and Follstaedt, D.M. (1996) Interaction of copper with cavities in silicon. *J. Appl. Phys.*, **79**, 1337–1350.
216. Wen, C.Y. (2005) Precipitation of copper silicide in voids in silicon single crystals. PhD thesis. Harvard University.
217. Wen, C.-Y. and Spaepen, F. (2007) Filling the voids in silicon single crystals by precipitation of Cu_3Si. *Philos. Mag.*, **87**, 5565–5579.
218. Vanhellemont, J., Zhang, X., Xu, W. *et al.* (2010) On the assumed impact of germanium doping on void formation in Czochralski grown silicon. *J. Appl. Phys.*, **108**, 123501–123504.
219. Chroneos, A., Grimes, R.W. and Bracht, H. (2009) Impact of germanium on vacancy clustering in germanium-doped silicon. *J. Appl. Phys.*, **105**, 016102-1–016102-3.

4

Growth of Semiconductor Materials

4.1   Introduction

The production of microelectronic and optoelectronic devices (transistors, solid state memories, light emission devices (LEDs), lasers, Micro Electro-Mechanical Systems (MEMS), chemical sensors, radiation detectors and photovoltaic cells) depends on the availability of large amounts of physically and chemically homogeneous, bulk or layered semiconductor single crystals. They should be almost free of structural defects (dislocations, grain boundaries, precipitates) and of unwanted impurities, but may have added dopant- and/or compensating-impurities.

In several cases, polycrystalline elemental or compound semiconductors are also used, typically but not exclusively, in thin film solar cells and sensors.

As is well known, silicon is still the material of choice for microelectronic, power electronics, MEMS and photovoltaic applications and is commercially available in several hundred thousand tons/year amounts, as polycrystalline feedstock, crystalline or multicrystalline[1] ingots and wafers. These last are obtained by wire saw cutting Czochralski (CZ), float zone (FZ) or multicrystalline grown ingots. In the case of CZ silicon, wafers could have diameters up to 450 mm in the near future.

Compound semiconductors and their alloys are strategic for optoelectronic applications (lasers and light emission diodes in the full visible spectrum, thin film solar cells) as well as for high frequency applications in mobile phones and radiation detectors. They represent today a growing fraction of the overall semiconductor value chain.

Group IV semiconductor devices, typically silicon-based, but with an increasing role for Si-Ge, use single crystal wafers as substrates. Silicon solar cells are manufactured starting from either single crystal or multicrystalline (mc) silicon wafers.

Compound semiconductor devices are typically manufactured starting from thin, single crystalline layers homo- or hetero-epitaxially grown on single crystalline wafers, which

---

[1] Multicrystallinity is the conventional attribute of the microstructure of a polycrystalline silicon consisting of large, oriented columnar crystals.

Physical Chemistry of Semiconductor Materials and Processes, First Edition. Sergio Pizzini.
© 2015 John Wiley & Sons, Ltd. Published 2015 by John Wiley & Sons, Ltd.

behave as mechanically active substrates. Compound semiconductor solar cells are, instead, based on the use of thin polycrystalline layers deposited on metallic or polymeric substrates.

The aim of this chapter is to consider the physico-chemical backgrounds of a number of these growth processes, including the semiconductors of Group IV (diamond, Si, Ge) as well as of Group III–V and II–VI compound semiconductors. Specific technical details concerning the growth steps directly involved in device manufacturing processes will be kept to a minimum or totally omitted because they are amply available in the literature [1–10].

The next four sections of this chapter will focus on the growth of bulk and layered single crystal semiconductors from the liquid and/or vapour phase and on the correlations between specific growth process issues and the final structural and physical properties of the material. This class of semiconductor materials finds its main application as active substrates in microelectronic and optoelectronic applications. Major emphasis will be devoted to silicon and germanium as well as to Group III nitrides, not only for their technological importance, but also for the vast amount of physical chemistry that has been involved in their development.

The physical and chemical backgrounds of different processes, including liquid phase (CZ, FZ, directional solidification) and vapour phase processes (molecular beam epitaxy (MBE) sublimation, plasma deposition, chemical vapour deposition (CVD) and atomic layer deposition) will be discussed, in order to understand the challenges and potential of the different techniques.

Thin film growth will be discussed in the last section of the chapter, from the viewpoint that thin film semiconductors represent today a vast class of materials of current use in many applications, as thin film transistors, chemical and physical sensors and thin film photovoltaic cells, based on silicon, germanium and compound semiconductors. In this frame, the growth of silicon nanowires (SiNWs) will also be considered and used as a typical example of the growth of semiconductor nanowires and nanorods.

4.2   Growth of Bulk Solids by Liquid Crystallization

Ingot growth of semiconductors by liquid crystallization processes is of relevant technological interest in a wide range of different industrial applications, such as the growth of single crystals of silicon, germanium, gallium arsenide, indium phosphide and silicon carbide for microelectronic or optoelectronic applications and that of multicrystalline silicon ingots for photovoltaic applications.

Bulk growth may be carried out by liquid crystallization processes provided the precursor melt is thermodynamically stable at the melting temperature, the vapour pressures of the melt components are sufficiently low and the melting temperature is compatible with the mechanical, thermal and corrosion properties of materials of possible use as crucible material. The molten charge, in turn, can consist of an elemental or a compound semiconductor.

As an alternative to liquid crystallization, bulk crystal growth from the vapour phase should be selected when the material of potential interest is thermodynamically unstable at the melting temperature or when the melting temperature is so high, that crucible

Table 4.1 *Properties of few elemental and compound semiconductors*

Material	Melting point (K)	Heat of fusion (kJ mol^{-1})	Vapour pressure at MP (atm)	Density at 298 K (g cm^{-3})	Energy gap (eV)
Ge	1210	34.7	10^{-6} at 760 °C, 1 at 1330 °C	5.3267	0.66
Si	1690	50.21	10^{-6} at 900 °C, 1 at 1650 °C	2.328	1.12
3C-SiC	3100	n.a.	$\cong 10^{-5}$	3.166	2.3
InSb	800	25.5	4×10^{-8}	5.775	0.16
GaSb	985	25.1	10^{-6}	5.61	0.73
GaAs	1510	87	1	5.32	1.43
InP	1333	62.7	27,5	4.81	1.34
GaP	1730	117.6	32	4.14	2.26
GaN	2573	—	60 kbar	6.15	3.23 Cubic 3.45 Wurtzite
HgSe	1063	n.a		8.245	—
HgTe	943	n.a	12.5	8.63	—
CdSe	1623	44.8 (2)	0.3	5.81 (wurzite)	1.71
CdTe	1314	50.2	0.65	5.86 (zinkblende)	1.47
ZnSe	1373	56	0.5	5.26	2.82
ZnTe	1513	51.00	0.6	5.65	2.39

Data from [11–15].

materials of tolerable properties are not available or do not exist. As shown in Chapter 1, most compound semiconductors can be grown as single crystals only using vapour phase processes, which will be considered in Section 4.4.

It is, however, a fortunate circumstance that the vapour pressure of elemental and of several compound semiconductors of industrial interest is sufficiently low at their melting temperatures (see Table 4.1) [11–15] to allow their growth directly from a stable melt under normal pressure conditions or under technologically acceptable pressurization. In the case of melts spontaneously decomposing at or below the melting temperature, the applied hydrostatic pressure should be sufficiently high to stabilize the melt.

When bulk growth is carried out from a liquid charge in a crucible, the interaction of the melt with the crucible, if the melt wets the crucible walls, almost inevitably causes at least its partial dissolution, with unwanted contamination of the melt and, thus, with subsequent contamination of the ingot. To avoid contamination, crucible-less techniques may be adopted but these are not always economically or technically feasible.

Recrystallization from a liquid is also employed as a powerful purification technique in semiconductor technology. It is routinely used for compound semiconductors in the zone refining configuration [2], but also for the purification of impure, metallurgical grade feedstocks to be converted into solar grade silicon [16].

Additional constraints of an intrinsic nature influence the bulk growth of compound semiconductors of defined composition, such as the existence of a limited range of solubility of the elemental components, the possible non-ideality of both the liquid and solid solution and the absence of stable stoichiometric or quasi-stoichiometric phases.

For this reason, bulk growth and growth of most of the active sections of optoelectronic devices based on compound semiconductors is carried out using gas phase or vapour phase homo-or heteroepitaxial deposition processes on dedicated substrates where, however, lattice mismatch could play a crucial role.

The discussion of these topics, with relevant examples concerning the control of the defectivity of the grown solid and of lattice strain in the heteroepitaxial layers will be presented in Section 4.4.3.

4.2.1   Growth of Single Crystal and Multicrystalline Ingots by Liquid Phase Crystallization

The growth of single crystal ingots by liquid crystallization processes represents the best, when technically feasible, solution for the preparation of a material with a limited, or negligible, amount of structural defects (grain boundaries (GBs), dislocations, point defect clusters and precipitates).

The drawback, however, remains of a systematic impurity contamination due to the high process temperatures, which make chemical interaction processes of the liquid charge and its vapours with the crucible and furnace materials thermodynamically unavoidable and kinetically favoured.

Several elemental and compound semiconductors can be grown as single crystal ingots using seeded liquid crystallization techniques in a suitable furnace, which should be maintained under a protective atmosphere to avoid environmental degradation or contamination of the melt and of the ingot during the growth. The ingot has a cylindrical shape when grown with the CZ or FZ techniques [17, 18] (see Figure 4.1) [14] or a square section when grown with variants of the directional solidification (DS) technique [19–21] directly in the crucible (see Figure 4.2). The CZ process is today the main source of silicon single crystal ingots for the microelectronic industry [18]. Since 2010, with the successful achievement at lab scale of conversion efficiencies higher than 24% with n-type solar cells manufactured on CZ silicon substrates, large amounts of CZ silicon are also used in the photovoltaic market.

CZ and FZ processes are variants of the liquid phase epitaxy, where a suitable liquid, consisting of a molten elemental or compound semiconductor, is contacted with a single crystalline seed, on which the growth occurs under a mild thermodynamic driving force at a rate which allows the exact reproduction of the seed surface. CZ growth is carried out in the process configuration illustrated schematically in Figure 4.1a. The growth system consists of a rotating fused silica crucible, which serves as the container for the molten charge, a susceptor, generally manufactured with graphite and heaters, generally consisting of shaped graphite resistors. The silicon ingot, which rotates counterclockwise, is slowly extracted from the melt in the vertical direction, starting the growth with an oriented single crystal seed, which determines the orientation of the single crystal ingot. In turn, the seed orientation is selected with regard to the final application of the grown material. Necking [22, 23] (also known as Dash necking after the originator of the process) is a specific and necessary step of the CZ process in order to grow dislocation-free crystals. It consists of reducing the diameter of the originally dislocation-free seed crystal after its immersion in the melt to eliminate the dislocations generated in the seed by the thermal shock arising at the moment of contact. The growth axis is taken perpendicular or oblique with respect to the (111) slip plane, in the case of silicon and germanium, to allow dislocations to glide out of

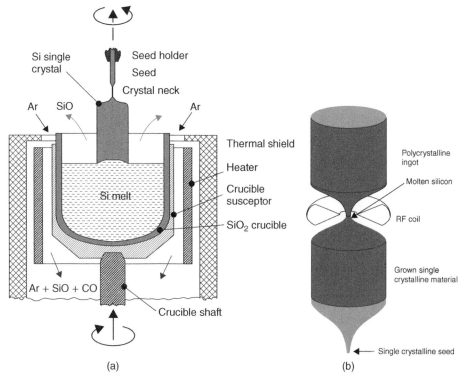

Figure 4.1 *(a) The hot chamber of a Czochralski furnace. Reproduced with permission from Professor Föll. Copyright (2014) H. Föll. (b) Growth configuration of a float zone furnace. Pfann, 1952. Reproduced with permission from OneMine, Inc*

the crystal as the ingot temperature is within the plastic regime of the crystal ($>0.5\ T_m$ (K) where T_m is the melting point). A dislocation-free, heavily B-doped Si crystal can, however, be grown without necking provided the B-concentration is of the order of 10^{18} at cm^{-3} [24]. Although no explanation is given concerning this apparently anomalous phenomenon, it could be supposed that the high concentration of B in this melt favours the formation of B–I and B–O complexes [25, 26] (see Section 2.3.4.3), thus inhibiting the homogeneous nucleation of dislocations from self-interstitials supersaturation arising from local thermal stress relief.

Counterclockwise rotation of the crucible and of the ingot ensures the stirring of the melt and allows control of the thermal and compositional profile of the liquid charge and the quality of the growth interface. Heat losses by irradiation from the free liquid surface are limited by the use of properly shaped metallic cover cones.

The growth process is carried out under a reduced pressure of argon to avoid oxidation of the liquid charge and the ingot. CZ growth is used also for germanium (under hydrogen as the cover gas), for silicon-germanium alloys and for the growth of GaAs and InP single crystals. In the case of III–V compound semiconductors, the liquid encapsulated configuration (LEC), see Figure 4.3, is frequently used [27], where a suitable oxide cap, molten boron oxide in the case of GaAs, prevents volatilization of high vapour pressure components.

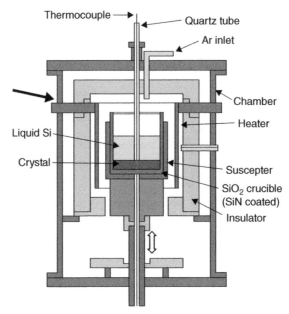

Figure 4.2 *Directional solidification (DS) furnace for the growth of multicrystalline silicon. Dhamrin et al., 2009, [21]. Reproduced with permission from Elsevier*

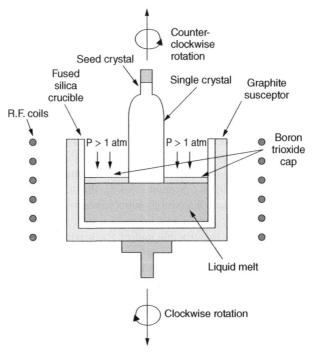

Figure 4.3 *Growth of GaAs crystals by liquid encapsulated CZ (LEC) growth [27]. Barron, [27]. Courtesy of A. Barron*

The fusion heat extraction (see Table 4.1 for the effective values) occurs spontaneously by heat conduction along the volume of the solid ingot and by surface radiation on the water-cooled stainless steel walls of the growth chamber.

The main problem with CZ growth processes has been, in the past, the achievement of a secure and accurate experimental knowledge of the heat flows within such a complex system in order to allow its complete modelling, necessary for the design of a software for a fully computer-controlled growth.

The growth rate, v, depends on the rate of extraction of the fusion enthalpy ΔH_f, that, in turn, depends on the temperature gradient in the solid G_s and in the liquid G_l phases and on the thermal conductivities of the solid k_s and of the liquid k_l

$$v = \frac{k_s G_s - k_l G_l}{\rho_s \Delta H_f} \tag{4.1}$$

where ρ_s is the density of the solid. It should be noted that the temperature gradients in the solid and in the liquid phases are complex functions of their surface emissivities, that of the solid ingot being rather inhomogeneous along its height, with important consequences for the final distribution of dopants, defects and microprecipitates, which will be discussed in Section 4.2.2.2.

Therefore, the irradiation losses from the liquid and the solid surfaces during the evolution of the growth process have been computer modelled with a global mathematical model to allow optimization of the thermal configuration of the furnace.

In modern CZ furnaces the entire growth process is computer controlled and is carried out at a pulling rate of several centimetres per hour, depending on the crystal diameter.

FZ is a crucible-less process carried out in the growth configuration illustrated in Figure 4.1b, using as the charge a polished cylindrical polycrystalline rod, which is zone melted starting from its bottom. Melting is done by induction heating of a limited zone of the poly-ingot and the growth of a single crystal ingot occurs by melt scanning the entire ingot.

As in the case of a CZ process, necking should be done to grow a dislocation-free crystal. It is carried out on a seed brought up from the bottom of the furnace to make a contact with the drop of melt formed at the tip of the poly rod. The neck is then allowed to increase in diameter up to the desired diameter of steady-state growth. The great advantage of the FZ process over CZ is that repetitive melting scans can improve drastically the quality of the crystal, due to impurity segregation processes, as will be seen later in this section. In addition, being a crucible-less process, impurity contamination is limited and the purity of any FZ material is far superior to a CZ one.

A different process, the Kyropoulus one (see Figure 4.4), is mainly used for the growth of alkali and alkaline earth halides and of large sapphire crystals, these last being used today, as an alternative to SiC and Si, as substrates for layered II–VI and III–V semiconductors growth from the vapour phase. The main difference from the CZ process is that the ingot, after the seed is dipped in the melt, grows directly in the melt, forming a hemispheric boule which grows in diameter until the crystal approaches within a few millimetres of the crucible walls. At this point, if the growth is computer controlled, the crystal is automatically extracted from the melt and the growth is further continued until the crucible walls are approached again, and so on.

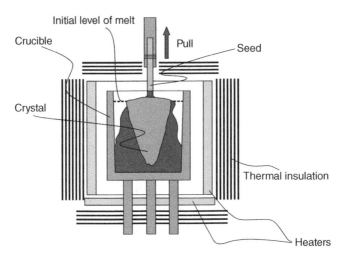

Figure 4.4 *Growth of sapphire single crystals by the Kyropoulos process. Akselrod & Bruni, 2012, [28] Reproduced with permission from Elsevier*

The advantage of the submerged growth is that it allows the crystal to grow in a shallower thermal gradient, an issue which is important for thermally sensitive ionic and semiconductor crystals that are prone to fracture and allows the growth of crystals of very high quality, with dislocation densities around 10^3 cm^{-2}. The other, and not least, advantage of the Kyropoulos method is that the size of the grown crystal is limited only by the size of the crucible, so that the boules may be as big as 300 mm in diameter and more, with a weight of 80 kg or more.

As with the CZ process, the selection of the crucible material is crucial as it will determine the impurity contamination of the ingot.

The use of mc semiconductors grown with variants of the DS technique could be an alternative to that of single crystal materials, when the carrier recombination at grain boundaries is minimized by the columnar growth of large crystals in a multicrystalline matrix [28]. The growth of large section, columnar crystals with low defect density occurs spontaneously, provided the process is carried out at moderate growth rates (≤ 2 cm h^{-1}) and the fusion heat is unidirectionally extracted from the bottom of the crucible, limiting the heat losses from the lateral walls to a minimum. This technique is in common industrial use for the growth of large (up to 800 kg) mc-Si ingots but may also be used for the growth of a number of other semiconductors where single crystallinity is not a requirement. Multicrystalline silicon for photovoltaic applications is, today, a unique example of a semiconductor material grown in hundreds of thousands of tons per year quantities with the DS technique, which has the advantage, with respect to CZ, of being cheaper with only a relatively limited decrease in the photovoltaic efficiency [29].

A directional solidification furnace is illustrated schematically in Figure 4.2, where one can see the ceramic crucible embedded in a graphite susceptor, the graphite heaters and the external insulation. Irradiation losses from the free liquid surface are limited by a graphitic cover on the top.

Heating may be carried out using lateral heaters, as in the figure, or by top and bottom heaters. Also in the case of DS, electromagnetic induction heating may be used.

Depending on the growth furnace configuration, solidification may be carried out by gradually removing the crucible from the hot zone or by modifying the temperature profile of the hot zone, while the fusion heat is extracted from the bottom of the crucible to ensure preferential growth of columnar crystals along the z-direction.

To this end, below the graphite pedestal on which the crucible sits, one can note the heat extraction system, consisting of a graphite block, whose cooling is carried out by irradiation on the water-cooled walls of the furnace or by the use of water or inert gases as coolants. As the growth rate is directly proportional to the rate of fusion heat removal, the cooling rate control tools are the key. Quasi-adiabatic lateral walls of the crucible would greatly improve the proper control of the lateral heat losses, which are detrimental for the columnar growth as they induce parasitic growth from the walls. Lateral induction heating could be an appropriate solution.

Recently, seeded growth has also been carried out with this process to get at least partially single crystal ingots, conventionally labelled mono-cast, for which commercial suppliers of mono-cast wafers claim an increase in cell efficiency of up to 1% absolute [30–32].

Avoiding or minimizing the melt contamination from the gaseous environment is mandatory for every process of crystal growth. Silicon is grown under a reduced pressure of Ar, germanium preferably under a hydrogen atmosphere or under vacuum and GaAs and InP under hydrogen as the covering gas.

The use of hydrogen when a semiconductor is kept molten in a silica crucible would lead, however, to a small, but not negligible silicon doping of the ingot [33], due to the reaction of silica with hydrogen

$$SiO_2 + 2H_2 \rightleftharpoons Si + 2H_2O \tag{4.2}$$

with, at the melting point of Ge, a Gibbs energy of reaction $\Delta G° = -26.125$ kJ mol^{-1} SiO$_2$.[2] This is not a real problem for Ge grown in a H$_2$ atmosphere as Si is for Ge a neutral substitutional impurity, but for GaAs and InP, silicon is a doping impurity.

Avoiding or minimizing the melt contamination arising from its interaction with the crucible walls is another necessary practice, which finds practical solutions with the use of suitable ceramic materials and with the control of the wettability of the inner crucible walls by suitable coatings, as will be deeply discussed in the following sections.

Elemental semiconductors, typically Si and Ge, do not present other critical thermodynamic constraints for their growth from a liquid charge, but this is not the case for binary and multinary alloys of elemental and compound semiconductors. For these materials, the growth under specific structural and chemical homogeneity conditions is dominated by the features of their specific phase diagrams, as will be shown for Si-Ge alloys and the tellurides.

Additional problems arise with compound semiconductors, where the liquid charge composition may change during the entire process (preliminary melting, charge homogenization, growth stages) due to the preferential evaporation of high vapour pressure components, as has already been shown in Chapter 1 for selenides, tellurides, arsenides and phosphides. Deviations from the stoichiometry may, therefore, be expected, with potential degradation of electronic properties. Charge losses from sublimation processes

---

[2] Most of the thermodynamic data reported in this chapter are obtained using the TLP Library of the University of Cambridge, which is acknowledged.

may, however, be at least partially suppressed by applying covering gas overpressures and/or convenient molten caps.

4.2.2   Growth of Single Crystals or Multicrystalline Materials by Liquid Crystallization Processes: Impact of Environmental Interactions on the Chemical Quality

The chemical quality of the as-grown materials depends on the impurity content of the melt from which the ingots have been grown, which, in turn, depends on

- the impurity content of the raw materials used to prepare the charge
- the partial dissolution of the crucible and of the crucible-impurities in the liquid charge
- the transfer of volatile impurities from the metallic or graphite heaters used to melt the charge and to keep it molten, from the thermal shields of the furnace chamber and from the current feedthroughs normally manufactured using copper
- the effective values of the segregation coefficients of the impurities in that particular melt, which will eventually determine the impurity content of the grown crystal.

Impurity contamination is a thermodynamically irreversible process as it causes a decrease in the free energy of the system by an increase in its configurational entropy

$$\Delta S = \sum_{1}^{n} x_i \ln x_i \tag{4.3}$$

where k is the Boltzmann constant and x_i is the atomic fraction of the ith impurity dissolved in the solid phase.

For crystals grown from a liquid phase in a ceramic crucible this problem may be particularly severe. Impurity contamination from the crucible walls and bottom implies that, as a consequence of a mechanical interaction with the liquid charge (convection fluxes are always active due to the presence of thermal gradients) or a chemical reaction, a matter flux from the crucible walls to the liquid charge will occur until thermodynamic equilibrium conditions are achieved.[3] Fortunately, this event, generally, is much slower than the growth process.

FZ-growth is a crucible-less technique and so is not subjected to impurity contamination arising from liquid charge–crucible interactions. Therefore, FZ processes will be ignored in the following section.

To prevent massive contamination, the crucible material should be suitably selected to ensure the best thermodynamic, chemical (and mechanical) compatibility with the liquid phase. This issue is particularly critical for the CZ-growth of semiconductor grade silicon from a charge of polycrystalline silicon produced with the Siemens process [16], which is today the purest synthetic material ever produced, with a typical metallic impurity content in the parts per trillion or sub-parts per billion range (As, Sb, Co, Ag, Au $\leq$ 1 ppt, B, Cu, Fe, Zn, P $\leq$ 0.1 ppb).

For this reason, a deep insight into crucible/semiconductor silicon interactions is particularly instructive and the conclusions thereafter obtained are straightforwardly applicable to other semiconductors, with limited differences.

---

[3] Consisting in the saturation of the melt with dissolved impurities.

4.2.2.1   Crucible Compatibility Problems

High purity fused silica crucibles are used as containers for molten silicon for both the CZ and DS growth processes and have the best compatibility with liquid silicon. Graphite, silicon carbide (SiC) and silicon nitride could be used, albeit graphite reacts severely and large size SiC or silicon nitride crucibles are not yet available at reasonable cost.

The compatibility of refractory materials with liquid silicon depends primarily on their wettability and chemical reactivity, as well as on the occurrence of phase transformations in the course of a duty cycle.

In the pure physical wetting case, the contact angle θ depends on the interfacial tension terms σ_{ij}

$$\cos\theta = \frac{(\sigma_{SV} - \sigma_{SL})}{\sigma_{LV}} \tag{4.4}$$

with subscripts S, L, V indicating solid, liquid and vapour respectively, and on the work of adhesion W

$$W = \sigma_{SV} + \sigma_{LV} - \sigma_{SL} = \sigma_{SV}(1 + \cos\theta) \tag{4.5}$$

When physical wetting is associated with a chemical reaction, as occurs at a Si/SiO_2 or Si/C interface, with the formation of SiO and SiC, respectively, the work of adhesion should be expressed by the sum

$$W = W_{equil} + W_{react} \tag{4.6}$$

where W_{react} represents the contribution of the chemical reaction to the work of adhesion. This last contribution essentially consists of the interfacial tension term σ_{SL}, because the chemical reaction is taking place at the solid/liquid interface. In the case of an interfacial reaction leading to the formation of a new solid phase, the decrease in the interfacial tension σ_{SL} is given by [34]

$$\sigma_{SL} = \sigma_{SL}^0 + \frac{d(\Delta G_R)}{d\Omega_{SL} \cdot dt} - \Delta\sigma_{SL} \tag{4.7}$$

(with $\Delta G_R < 0$), where σ_{SL} is the interfacial tension term at the time t and σ_{SL}^0 is the interfacial tension in the absence of reaction contributions. The second term of Eq. (4.7) is the contribution of the Gibbs energy of reaction ΔG_R per unit of interfacial surface Ω_{SL} and per unit of time. The third term is a measure of the difference in the interfacial tension at the liquid/solid interface when the pristine interface is substituted by the reacted interface. It is apparent that the actual values of ΔG_R and $\Delta\sigma_{SL}$ depend on the nature of the system considered, and should be experimentally or theoretically determined [35]. The wettability features of silica by liquid silicon have been recently critically reviewed [36] on the basis of literature results of contact angle θ measurements. These results show that the contact angle for silica ranges around 85–90° at 1723 K, decreases with time and is influenced by the reaction of silica with silicon, which leads to SiO formation and sublimation, as will be seen below.

The problem of silica wetting by liquid silicon is worsened by the α-quartz-crystobalite phase transformation, which occurs at 1145 °C, during the heating of the silicon charge and compromises the mechanical integrity of the crucible.

To prevent this last problem, fused silica crucibles are routinely used, which are manufactured with fused silica powder instead of α-quartz. As fused silica softens at temperatures above 1100 °C, the crucibles are physically embedded in a graphite susceptor to avoid their collapse during the growth process.

In spite of its superior performance with respect to other materials, fused silica reacts with liquid silicon, with the formation of SiO vapour

$$SiO_2^S + Si^L \rightleftharpoons 2SiO^V \tag{4.8}$$

$(\Delta G_{4.8} = -375.125 \text{ kJ mol}^{-1}$ (SiO$_2$) at the melting point of Si), which partly sublim out of the melt and collect as amorphous SiO on the cold walls of the furnace [37], while a fraction dissolves in the liquid silicon charge and dissociates to silicon and oxygen

$$SiO_{Si} \rightleftharpoons Si^L + O_{Si}^L \tag{4.9}$$

in the liquid charge [37]. As SiO$_2$ has a small, but not negligible solubility in liquid silicon at the melting temperature, see Figure 4.5 [38], the direct SiO$_2$ dissolution reaction might also occur

$$SiO_2 \rightleftharpoons Si + 2O_{Si}^L \tag{4.10}$$

[39], which couples with reaction (4.9).

A complete thermodynamic assessment of the Si–O system, with a detailed discussion about oxygen solubility in solid and liquid silicon and oxygen segregation features, may be found in the excellent paper of Schnurre [37], but here it is to be noted that the dissolution

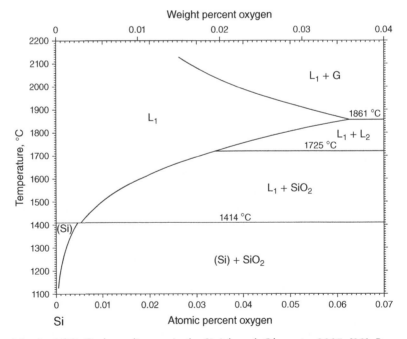

Figure 4.5   *Partial Si–O phase diagram in the Si-rich end. Okamoto, 2007, [38]. Reproduced with permission from Elsevier*

of oxygen may be associated also to a carbon contamination process. SiO vapours may react with the hot graphite parts of the furnace (susceptor and heaters) to give CO

$$SiO^V + C^S \rightleftharpoons CO^g + Si \qquad (4.11)$$

($\Delta G_{4.11} < 0$ only at $T > 2000$ K). Eventually, traces of oxygen in the argon covering atmosphere will contribute to the formation of CO while reacting with the high temperature graphitic parts of the furnace

$$O_2 + 2C \rightleftharpoons 2CO \qquad (4.12)$$

and CO will be brought towards the liquid silicon surface by the argon flow, which is injected into the furnace from a top inlet and discharged by a vacuum pump. There, it dissolves and dissociates

$$CO_g \rightleftharpoons C_{Si} + O_{Si} \qquad (4.13)$$

with consequent formation of oxygen and carbon defects, once incorporated in the ingot, as illustrated schematically in Figure 4.6a [40] for a CZ growth process, and in Figure 4.6b, for DS growth.

The basic chemistry of these processes is common to both CZ and DS growth, as well as the thermodynamics of the further distribution of oxygen, carbon (and impurities) between the liquid and solid phase during the solidification process, as will be discussed in the next sections.

Fused silica is, instead, well compatible with liquid germanium, thanks to the lower melting temperature of Ge and to the fact that the reaction

$$Ge + SiO_2 \rightleftharpoons GeO_2 + Si \qquad (4.14)$$

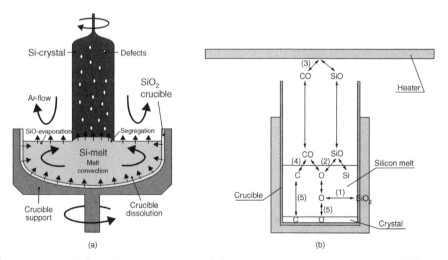

Figure 4.6 (a) Schematic representation of the oxygen transport processes in a CZ furnace. Müller et al., 1999, [40], Reproduced with permission from Elsevier. (b) Schematic view of the reactions and transport processes occurring in a directional solidification furnace for m.c. silicon growth. [41]. Reproduced with permission from Elsevier

is strongly shifted to the left, with a Gibbs free energy of reaction of $+345.87$ kJ mol^{-1} SiO_2 at the melting point of Ge (1210.55 K).

The equilibria involving germanium monoxide (GeO) are less understood than those for SiO. It is, however, known that the equilibrium constant for the reaction

$$2GeO + Si \rightleftharpoons SiO_2 + 2Ge \tag{4.15}$$

is about 5×10^{-9} at the melting point of Ge[4] [42], while the Helmholtz free energy of the GeO formation from a mixture of Ge and GeO_2

$$Ge + GeO_2 \rightleftharpoons 2GeO \tag{4.16}$$

has been calculated to be $+18.57$ kJ mol^{-1} (GeO) [42], making the role of this process negligible in practice [43].

Germanium reacts, instead, with graphite crucibles, leading to carbon contamination of the liquid and thus of the grown ingot [44]. The equilibrium carbon concentration ranges between $1 \cdot 10^{14}$ and $4.5 \cdot 10^{15}$ at cm^{-3}, depending on the position of the ingot, and is compatible with an effective segregation coefficient of carbon $k_{eff} = 1.85$.

Aside from silica, titania (TiO_2), alumina (Al_2O_3) and magnesia (MgO) (see Figure 4.7) are thermodynamically stable in the presence of silicon, but experiments show detectable ingot contamination, which is particularly dangerous in the case of Al and Ti which are deep recombination centres in silicon.

A number of metallic oxides are thermodynamically unstable in the presence of liquid silicon, as is the case for FeO

$$Si + 2FeO \rightleftharpoons 2Fe + SiO_2 \tag{4.17}$$

and most transition metal oxides (see Figure 4.7), for which the reduction would occur spontaneously at the melting temperature of silicon (1685 K), with consequent metallic impurity contamination of the melt, and then of the solid ingot.

Therefore, if the quartz sand from which the silica crucible is manufactured contains even minute amounts of metallic oxides (see Table 4.2), those thermodynamically less stable than SiO_2 (see again Figure 4.7) will slowly dissolve and contaminate the liquid silicon charge.

It is also easy to see from the diagram of Figure 4.7 that silicon would react with oxygen gas at any process temperature (i.e. during and after the growth), with the formation of SiO_2 at the surface of the liquid

$$Si + O_2 \rightleftharpoons SiO_2 \tag{4.18}$$

which then interacts with Si to give SiO according to reaction (4.8). The use of a deoxygenated covering gas is therefore mandatory in CZ and DS silicon growth, just to prevent, at least partially, its interaction. Because nitrogen also reacts with molten silicon, nitrogen gas cannot be used as a cover gas, and a reduced-pressure argon atmosphere is systematically used in every silicon growth process.

---

[4] $\Delta G_{4.15} = -37.36$ kJ mol^{-1} SiO_2.

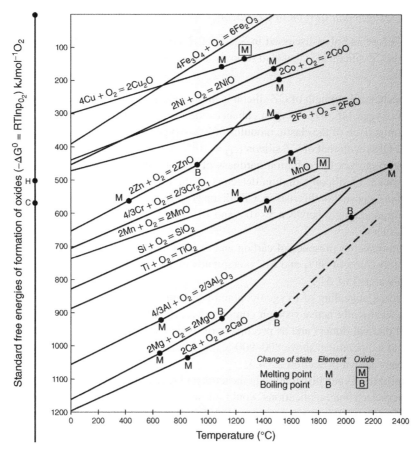

Figure 4.7 *Temperature dependence of the Gibbs free energy of formation of selected oxide systems. Ellingham Diagrams, Reproduced with permission from DoITPoMS, University of Cambridge*

Table 4.2 *Typical metallic impurities content in a powder – and rock – sample of a natural quartz (concentrations in ppmw)*

Impurity	Quartz powder	Quartz rock
Fe	0.08	4.80
Cr	<0.01	n.r.
Ti	0.70	0.50
Cu	0.01	n.r.
Al	11.0	20.50
Ca	0.09	0.40
Mg	0.02	n.r.
B	0.10	n.r.
K	0.41	1.60
Na	1.60	4.50
Li	0.540	2.00
Mn	<0.01	n.r.
Zr	<0.01	n.r.

4.2.2.2   Oxygen and Carbon in CZ Silicon

As seen before and also discussed in Section 2.3.4.3, oxygen is not only an unavoidable impurity in CZ-silicon [45], but also strongly affects its mechanical, chemical and physical properties.

- The mechanical strength of CZ silicon is improved by a dislocation locking mechanism due to oxygen [46, 47]. Oxygen enhances the elastic modulus E of silicon, as shown by a comparison of the elastic modulus of low-oxygen FZ silicon ($c_O = 10^{16} \text{cm}^{-3}, E = 103.78$ GPa) with that of CZ silicon ($c_O = 10^{18} \text{cm}^{-3}, E = 201$ GPa) [48].
- Oxygen improves the radiation hardness of silicon [49] and allows, for example, the employment of silicon radiation detectors in the high luminosity hadronic collider of CERN in Geneva. For this reason, CZ silicon is preferred to FZ silicon, where the oxygen concentration ranges between $5 \cdot 10^{15}$ and 10^{16} cm^{-3}, 2 orders of magnitude lower than in CZ silicon.
- The simultaneous presence of carbon and oxygen may be very detrimental consequent to the formation of SiO_2 and SiC precipitates in the bulk or at the extended defects of silicon wafers [50, 51].
- By thermal annealing of oxygen-contaminated silicon in the 450–650 °C temperature range, electrically active oxygen complexes are generated, which behave as (and are called) thermal donors and as precursors of oxygen precipitates which eventually segregate at temperatures above 800–900 °C [52], inducing the presence of gap states [53], as already shown in Section 2.3.4.3.
- The uniform oxygen distribution in the axial and radial directions of silicon ingots used for microelectronic applications would allow to perform optimized internal gettering processes at oxygen microprecipitates [54–58], as will be discussed in the next chapter, while the presence of oxygen defects severely affects the electronic properties of silicon substrates. The engineering of this process, as will be shown later in this section, is extremely difficulty.

There was, therefore, and still remains, the compelling need to address dedicated studies to the investigation and modelling of the oxygen transport and incorporation phenomena occurring during the silicon growth processes [37, 40, 59–63], which make oxygen the most studied impurity in silicon.

Several experimental methods are available [64] for the accurate determination of O and C in silicon, including IR spectroscopy for the determination of the concentration of interstitial oxygen and an electrochemical method for the direct measurement of the (total) oxygen distribution in the molten silicon [40]. These analytical techniques are complementary to the mathematical modelling of oxygen (and carbon) mass transport/incorporation processes, which is needed for a thorough optimization of the CZ process itself.

The oxygen mass distribution process between the liquid and solid phase relies on the oxygen concentration at the growth interface which, in turn, depends on the dynamic equilibrium between the oxygen fluxes into the melt, originated by the silica crucible dissolution and by the SiO sublimation fluxes, neglecting parasitic contributions from oxygen impurities in the cover gas. This process has been simulated [61] with the development of a mathematical model of the conjugated inward and outward oxygen transport.

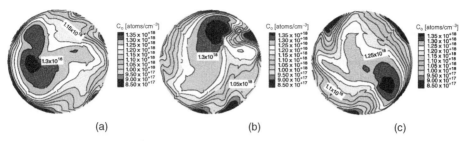

Figure 4.8 *Oxygen concentration distribution at the crystallization front (a) at t = 0, (b) at t = 900 s (c) at t = 1800 s. Smirnov and Kalaev, 2008, [61]. Reproduced with permission from Elsevier*

Boundary conditions for the oxygen incorporation in the melt are chosen in this way:

1. at the crystallization interface oxygen is incorporated into the solid with a segregation coefficient $k_O^{Si} = \frac{x_O^S}{x_O^L} \approx 1$.[5] As the amount of solid crystallized with respect to the mass of the liquid is relatively small at low pulling rates, the change in the oxygen concentration in the melt will be negligibly small

$$\frac{\delta C_o}{\delta n} = 0 \qquad (4.19)$$

   thus providing zero flux conditions as the boundary

2. in turn, the oxygen concentration C_o in the liquid depends on the crucible walls' dissolution reaction (4.10) and can be experimentally determined

3. at the melt/argon cover gas interface, the SiO sublimation reaction occurs

$$Si^L + O_{Si}^L \rightleftharpoons SiO_{Si}^V \qquad (4.9)$$

   and the boundary condition is the partial pressure of SiO at the equilibrium temperature, given by the equation

$$p_{SiO} = K_p^{SiO}(T)\, \gamma_o x_o \qquad (4.20)$$

   which depends on the equilibrium constant K_p^{SiO} of Eq. (4.9) and on the activity $a_o^L = \gamma_o x_o$ of the dissolved oxygen.

At all gas/solid boundaries the simplified condition of zero SiO mass fluxes is applied, and the oxygen transport in the melt was calculated using

$$\frac{\delta \rho C_O}{\delta t} + \nabla(\rho V C_O) = \nabla(D_{eff} \nabla C_O) \qquad (4.21)$$

where ρ is the density, V is the velocity vector and D_{eff} is the dynamic diffusivity, while the conservation of molar fluxes of oxygen in the melt and of SiO in argon is the additional condition.

The results of the simulation are displayed in Figure 4.8.

---

[5] Literature values of the segregation coefficient of oxygen in silicon range between 0.25 and 2.5, depending strongly on the experimental methods used for its determination. For the CZ process a value $k_O \approx 1$ is generally used [37].

Independently of the absolute numerical values which, however, are consistent with the experimental values of oxygen concentration obtained using Fourier transform infrared (FTIR) spectroscopy [61] and electromotive force (EMF) measurements [40], it is evident that the uneven distribution of oxygen in liquid silicon at the growth interface would not allow its further, uniform distribution in the crystal.

To obviate, at least partially, this problem, which depends substantially on the spontaneous melt convection fluxes in the crucible and on fluctuating growth rates [65, 66], the use of transverse magnetic fields has been successfully applied for the optimization of the oxygen distribution in the liquid [67], once it was demonstrated that magnetic fields stabilize the melt convection [68, 69].

Further problems concern the aggregation and segregation of point defects in a growing Si ingot, which accompany the oxygen segregation processes. These processes have as basic components the intrinsic point defects (vacancies V and self-interstitials I) [70, 71]

$$Si_{Si} \rightleftharpoons Si_{surf} + V_{Si} \tag{4.22}$$

$$Si_{Si} \rightleftharpoons I_{Si} + V_{Si} \tag{4.23}$$

which are generated in equilibrium conditions at the growth temperature.

Native defects aggregate in the form of microdefects (vacancy clusters (voids) and self-interstitial dislocation loops (swirls))

$$mD \rightleftharpoons P_{mD} \tag{4.24}$$

where D is an individual point defect and P_{mD} is a microdefect, as happens when a silicon ingot is cooled from the melting temperature of some hundred Kelvin [70]. Vacancies also react with oxygen, to form $V - O_n$ complexes

$$V + O \rightleftharpoons VO \tag{4.25}$$

$$VO + O \rightleftharpoons VO_2 \tag{4.26}$$

or cluster to polymeric oxygen-defect species [71–74].

$$P_{mO} + O \rightleftharpoons P_{(m+1)O} \tag{4.27}$$

The temperature dependence of the concentration of vacancies, self interstitials, oxygen and VO_2 complexes is reported in Table 4.3 [71, 75].

This makes the growth of a silicon ingot a unique example of a complex, non-isothermal thermodynamic process, where the interplay of vacancies and self-interstitials generation processes, of local vacancies and self-interstitials supersaturation and of oxygen and self-interstitials diffusion and clustering is at the origin of the uneven distribution of chemical and structural inhomogeneities. These are seriously prejudicial for the reliable and reproducible performance of technological processes to which silicon wafers are subjected during the device manufacturing and, therefore, the subject of dedicated studies aimed at the growth of a 'perfect silicon'.

A fundamental step towards the achievement of this objective was a study [72] that succeeded in demonstrating that the defect agglomeration phenomena [73, 74] depend on the V/G ratio, where V is the growth rate and G is the temperature gradient at the surface/melt interface.

Table 4.3 *Temperature dependence of the concentration of intrinsic defects, oxygen and oxygen–vacancies complexes in silicon*

$C_V = 7.52 \cdot 10^{26} \exp{-4.0}\,\text{eV})/kT$
$C_i = 6.1759 \cdot 10^{26} \exp{-4.0}(\text{eV})/kT$
$C_O = 9 \cdot 10^{22} \exp{-1.52}(\text{eV})/kT$
$C_{VO2} = 0.2 \cdot 10^{-22} \exp{-0.5}(\text{eV})/kT$

Data from [71, 75].

At large values of V/G the crystal is grown as a vacancy-type material while at low V/G values the crystal is grown as an interstitial-type of material, and a critical V/G ratio, the so-called Voronkov ratio ξ [72, 75]

$$\xi = \frac{E}{kT^2} \frac{(D_L C_L - D_V - C_V)}{C_V - C_L} \tag{4.28}$$

defines the condition at which one should observe the change from vacancy type to interstitial type of growth.

In Eq. (4.28) $E = \frac{(E_V + E_L)}{2}$ is the average free energy of formation of intrinsic defects, D is their diffusivity and C is their concentration at the growth temperature [70].

A value of $0.12\ \text{mm}^2\ \text{min}^{-1}\text{K}^{-1}$ is given as the critical value of ξ, using for the free energy of formation of vacancies and self-interstitials the values 4 and 4.6–4.8 eV, respectively [75].

Because the temperature at the growth interface is rather non-uniform, the axial temperature gradient G increases from the centre to the periphery, as does the critical growth rate $V_{cr} = \xi G$. It is, therefore, expected that the G/V values should vary radially across the surface.

The G/V contours may be indirectly observed on sections of the ingots by visualizing the contours of the domains of vacancy or self-interstitials excesses surviving after the ingot is cooled from the melting point of Si, using a copper decoration technique, which is very helpful for delineating defects in silicon.

Figure 4.9 [70] shows the section of an ingot whose growth rate was initially larger than V_{cr} and then slowed down. The copper decoration shows that the upper region, grown first, is a vacancy type whereas the lower section is an interstitial type. The effect of a non-isothermal growth surface is clearly identified by the shape of the vacancies and interstitials domain boundaries.

The growth of a virtually defect-free silicon ingot might be, therefore, conceived under the hypothesis of being able to improve the temperature uniformity at the growth interface and to grow the crystal at growth rates close to the Voronkov ratio ξ. In these circumstances the concentration of vacancies and of interstitials is very close and they will recombine before having the chance to interact with oxygen or to start a clustering process as the crystal cools.

Such an almost perfect material is today grown by different producers by proprietary methods and does not require post-growth engineering, such as annealing or epitaxy, before being used for IC processes. It is, however, now known [76] that this material still contains small defects consisting of oxygen precipitates, which can be evidenced by reactive ion etching (RIE) of the silicon wafer surface and which are, therefore, called reactive ion etching defects. As the presence of these defects in the near surface region is of prejudicial

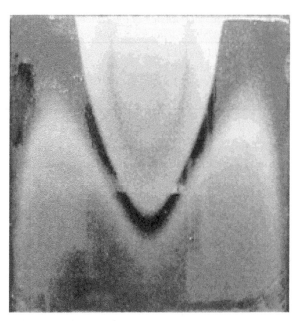

Figure 4.9   *Point defect contours in a silicon ingot showing the transition from a vacancy type (upper section) to an interstitial type of growth. Falster et al., 2000, [70]. Reproduced with permission from John Wiley & Sons*

to the proper performance of ULSI devices, R&D work is currently still being carried out to improve the growth process.

4.2.2.3   Oxygen in CZ Germanium

The behaviour of oxygen in germanium has attracted considerable attention since the historical utilization of Ge for semiconductor applications [77], in view of possible interferences with the transistor behaviour.

It has already been shown in the previous section that, differently from silicon, oxygen is not a dominant impurity in CZ Ge grown in fused silica crucibles. In fact:

- the chemical affinity of germanium with oxygen is lower than that of silicon at their respective melting points ($\Delta G°$ $SiO_2 = -610.875$ kJ mol^{-1}; $\Delta G°$ $GeO_2 = -347.625$ kJ mol^{-1})
- the chemical interactions of liquid germanium with the silica crucible are less than in the case of silicon
- when hydrogen is used as the cover gas for the growth of high purity (HP) Ge, the oxygen concentration of Ge grown in silica crucibles is very low ($\approx 5 \times 10^{13}$ at cm^{-3}) because of the occurrence of the reaction [78]

$$GeO + H_2 \rightleftharpoons Ge + H_2O \qquad (4.29)$$

high concentrations of oxygen ($10^{16} - 10^{17}$ at cm^{-3}) are only obtained by growing germanium in an oxygen-rich atmosphere [3]

- the lower segregation coefficient of oxygen in Ge ($k_O^{Ge} = 0.11; k_O^{Si} \approx 1.0$) favours the oxygen enrichment of the melt instead of the crystal.

As for silicon, however, when oxygen solubility conditions are overcome, GeO_2 precipitates segregate during thermal annealings.

4.2.2.4   *Oxygen, Carbon and Metallic Impurity Incorporation during the DS-Growth of mc-Si*

Different from the case of CZ growth, oxygen incorporation during the DS growth of mc-Si is associated with that of carbon [79], because the configuration of a DS furnace is different from that of a CZ furnace, as is shown in Figures 4.1 and 4.2. In the case of a CZ furnace, the cover limiting the irradiation losses from the melt, if present, is metallic, while in the case of a DS furnace the cover is graphitic and the crucible and its susceptor are inside graphite-based thermal shields, which necessarily generate a CO-rich environment and favour the occurrence of the processes described by Eqs. (4.11)–(4.13). Dissolved carbon in DS-grown mc-Si is an unwanted impurity, as it is the precursor of bulk SiC precipitates in silicon wafers used for the manufacturing of solar cells, which will work as p–n junction shunts and minority carriers recombination centres, thus seriously degrading the overall efficiency of the cell.

A schematic view of the reactions and transfer processes involving oxygen, carbon and silicon in a top-heated DS growth furnace is displayed in Figure 4.6b.

On that basis, and using comparable mathematics as used for the CZ growth [41], the oxygen and carbon incorporation during the growth of multicrystalline silicon could be modelled in the presence of an argon flow.

A result concerning the oxygen distribution in the liquid charge is reported in Figure 4.10, from which it is possible to observe the very inhomogeneous distribution of oxygen in the melt.

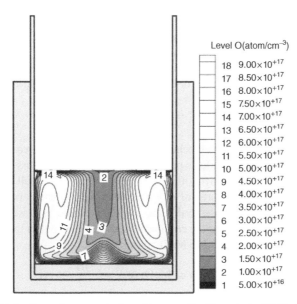

Figure 4.10   *Distribution of oxygen in the silicon melt. Gao et al., 2010, [41]. Reproduced with permission from Elsevier*

The whole system (liquid silicon as a carbon and oxygen sink, hot heaters as carbon sources and cold walls as SiO sinks) is typically not in thermal equilibrium conditions and the irreversible impurity transfer from the hotter impurity sources towards the melt should continue until a reaction interface exists or the liquid saturates. In the worse circumstances, supersaturation of an impurity in the melt may occur, with second phase segregation.

It is worth remembering here that in supersaturated C–Si solid solutions SiC segregates as a second phase only in the presence of a simultaneous segregation of oxygen as SiO_2, that behaves as the source of self-interstitials compensating the SiC volume deficit [80–82] (see Sections 2.3.4.3 and 3.6.2). This process is further evidence of the crucial role played by point defects (self-interstitials and vacancies) in oxygen and carbon segregation processes.

There is another important difference between the CZ and DS growth, which has not been discussed so far and which has relevant technological consequences for the growth of mc-Si.

In detail:

- with the DS-process the silicon ingot grows directly into the crucible and takes its shape, starting its solidification from the bottom. The volume expansion associated with solidification cannot be compensated without mechanical damage to both the ingot and the crucible unless both liquid and solid silicon do not wet the crucible walls.
- as liquid silicon actually wets fused silica, solid silicon will stick onto the crucible walls after solidification and the difference between the linear expansion coefficients of fused silica ($5.5 \cdot 10^{-7}$) and silicon ($2.6 \cdot 10^{-6}$) brings about the destructive breakage of both the crucible and the ingot.

The solution of this problem was originally proposed by the author of this book in the early 1980s. This involved coating the crucible walls with a thick layer of Si_3N_4, which was shown to work well as an anti-wetting agent, but only recently has the wettability of silicon nitride with silicon been critically discussed [36], allowing a quantitative conclusion to be reached.

It is demonstrated, in fact, that the wetting behaviour of silicon nitride is very complex and depends straightforwardly on the partial pressure of oxygen in equilibrium with the Si/Si_3N_4 system, as can be seen in Figure 4.11 [83], where biphasic Me/MeO_n systems are used to buffer the partial pressure of oxygen, according to the equilibrium reaction

$$MeO_n \rightleftharpoons \frac{1}{2}O_2(P_{eq}) + Me \qquad (4.30)$$

A transition from the non-wetting behaviour at high partial pressures of oxygen,[6] when the material is actually an oxynitride, to partial wetting conditions occurs at about 10^{-9} atm, consequent to the formation of a layer of SiO_2. A further transition to full wetting behaviour occurs at very low partial pressures of oxygen, when the system is brought to vacuum conditions. It is interesting to observe that the oxygen pressure in these experiments was controlled using different metals as the crucible materials, which buffer the oxygen pressure thanks to the formation of surface Me/MeO mixtures

$$Me + \frac{1}{2}O_2(p_{O_2}) \rightleftharpoons MeO; \Delta G^{\circ} = -\frac{RT}{2} \ln p_{O_2} \qquad (4.31)$$

---

[6] Contact angle 180°.

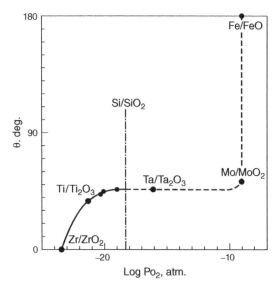

Figure 4.11   *Dependence of Si/Si$_3$N$_4$ contact angle on the partial pressure of oxygen at 1450°C. Li and Hausner, 1992, [83]. Reproduced with permission from Elsevier*

a well-known process, adopted industrially in getter pumps, which use metallic getters for the achievement of extremely high vacuum conditions coupled to very low oxygen pressures.

Last, but not least, impurity contamination of the solid ingot by solid state diffusion processes may also occur during and after the solidification process in the case of DS processes, where the hot crucible walls coated with often impure Si$_3$N$_4$ behave as sources of mobile impurities.

As an example, local iron and nitrogen supersaturation of a silicon ingot was observed during the cooling stage of a Si casting process, with second phase separation, due to iron and nitrogen release from the silicon nitride-coated crucible walls [84]. Here, the rate determining step of the contamination process is the solid state diffusion of the impurities, driven by a concentration gradient. Segregated impurities are normally confined in a restricted region close to the ingot walls and bottom.

4.2.3   Growth of Bulk Solids by Liquid Crystallization Processes: Solubility of Impurities in Semiconductors and Their Segregation

It has been shown in the last section that contamination of a semiconductor charge molten in a refractory crucible, and thus that of the further grown-in crystal, is unavoidable. As temperature–solubility relationships rule the behaviour of impurities during the growth and post-growth processes, their specific features, that depend on the nature of the system considered, should be considered in details to govern the technological processes.[7]

In the case of a system where the impurity has a negligible solid state solubility, at any temperature within the stability range of the host phase, the formation of an alloy may

---
[7] Here only physical processes are considered. Chemical processes involving reactive species would require a different approach.

be neglected and the concentration of the impurity dissolved in the solid phase takes its maximum value (the solubility) when the segregation of the corresponding second phase occurs. This is typically the case for carbon and oxygen in silicon displayed in Figures 2.17, 2.19 and 2.20, where the equilibrium second phases are SiC and SiO_2 and the equilibrium conditions are given by

$$\mu_C^{Si} = \mu_C^{SiC} : \mu_O^{Si} = \mu_O^{SiO_2} \tag{4.32}$$

However, the case of an impurity involved in a system characterized by extended solubility regions is different.

Looking, for example, at the Al–Ga phase diagram of Figure 4.12, the solidus and liquid lines give the solubility of Ga in the liquid Al-Ga alloy in equilibrium with the solid Al-Ga alloy, respectively, and the equilibrium conditions are expressed by the equation

$$\mu_{Ga}(Al_{1-x} - Ga_x)^L = \mu_{Ga}(Al_{1-y} - Ga_y)^S \tag{4.33}$$

The impurity solubility features obviously depend on the thermodynamic properties of the phases in which the impurities are dissolved, which may have ideal, regular or non-regular behaviour. It has been shown in Chapter 1 that the regular solution model is able to predict the behaviour of condensed phases in a vast number of cases, for which the excess Gibbs free energy of mixing is given by the product

$$\Delta G_{mix} = \Delta H_{mix} = x_A x_B \Omega \tag{4.34}$$

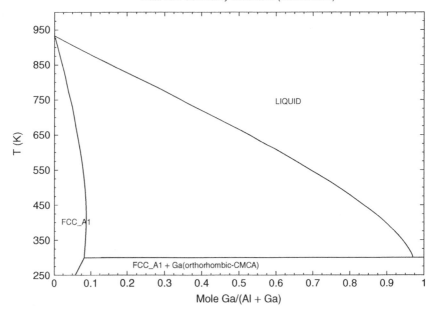

Figure 4.12 *Phase diagram of the Al-Ga system. Scientific Group Thermodata Europe (SGTE). Reproduced with permission from SGTE*

where Ω is an interaction coefficient which accounts for the interaction energies within the solution partners. Thereafter, the compositions of the phase in equilibrium, and thus the solubilities, may be calculated following the arguments discussed in Sections 1.5 and 1.6.

As also shown in Chapter 1, in eutectic type systems the solubility maximum is not necessarily found at the eutectic temperature: retrograde solubility should be observed, at least for regular systems exhibiting large positive values of the interaction coefficient Ω (and of the enthalpy of mixing) with the establishment of a solid/liquid equilibrium [85, 86]. When the solubility values of specific impurities for a multinary, liquid or solid, system are unknown and difficult to determine experimentally, or their behaviour cannot be extrapolated in a more extended temperature range to that experimentally available, or the simple regular solution model is not applicable, advanced modelling schemes should be applied.

To model a generic multicomponent solution and get its free energy functions, and hence the activity coefficients of the components by suitable differentiation procedures, it is convenient to use Eq. (1.107) which was given in Chapter 1 for the case of the $Ga_xIn_{1-x}P$ system, and used by Kattner [87] and Tang *et al.* [88] for the solution of impurities in silicon, but which can be generalized for any kind of multinary liquid and solid solutions.

Equation (1.107) can be rewritten for a liquid or solid solution of $(n - 1)$ impurities in a suitable solvent

$$G = \sum_{i=1}^{n} x_i G_i^{\circ} + RT \sum_{i=1}^{n} x_i \ln x_i + G^{exc} \qquad (4.35)$$

where the second term is the contribution of ideal mixing and the G^{exc} term is given by a sum of $G_{i,j}$ terms

$$G_{i,j} = x_i x_j \sum_{n=0}^{k} L_{i,j}^{k}(x_i - x_j)_{i,j}^{k} \qquad (4.36)$$

and is the measure of the excess free energy of mixing due to all the binary interactions, under the assumption that the solution is sufficiently diluted to ignore the ternary ones.

The sum $\sum_{n=0}^{k} L_{i,j}^{k}(x_i - x_j)_{i,j}^{k}$, the so-called Redlich and Kister polynomial [89], where the $L_{i,j}^{k}$ terms are interaction coefficients or convenient model parameters.

It can be noted that in the case of a binary solution the $\underline{G^{exc}}$ term of Eq. (4.36) corresponds to the excess free energy of mixing given by Eq. (4.34), taking only the first term of the polynomial.

By suitable differentiation of Eq. (4.36), for the case of a diluted solution of impurities in a solvent S, one obtains the activity coefficient of each solute

$$\gamma_i \approx \exp \frac{\sum_{n=0}^{i} L_{S-i}}{RT} \qquad (4.37)$$

while the activity coefficient of the solvent $\gamma^S \approx 1$ is considered unitary. Once the interaction coefficients are calculated or obtained from a database, the behaviour of any specific impurity may be modelled.

Using this approach [88], the temperature dependence of the solubility of several impurities in solid silicon can be calculated. The results reported in Figure 4.13 appear in good

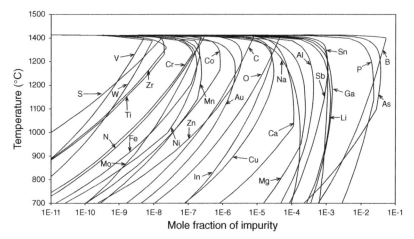

Figure 4.13   *Temperature dependence of the solubility of impurities in silicon. Tang, 2009, [88]. Reproduced with permission from Springer Science and Business Media*

agreement with the literature values and demonstrate the potential of the technique for different applications.

In an analogous way, the influence of a second impurity on the solubility of a main impurity could also be evaluated. This is the case for the results reported in Figure 4.14 [88] for carbon in silicon, which were obtained using the following equation

$$\ln x_C = -\ln K^{\circ}_{SiC} - \ln \gamma_C + \sum_i x_i (\ln \gamma_{Si\text{-}C} - \ln \gamma_{Si\text{-}i} - \ln \gamma_{C\text{-}i}) \qquad (4.38)$$

where K°_{SiC} is the equilibrium constant for the segregation of SiC from a C-saturated solution and the activity coefficients are evaluated using Eq. (4.37) with the appropriate interaction coefficients $L_{S\text{-}i}$.

It can be seen that the carbon solubility increases in the presence of Zr, P, B, Zn, As and Al, while the opposite is true for the other impurities, as a consequence of the different values of the interaction strength of the impurities with carbon and silicon. It should be noted that the maximum decrease in solubility occurs in the case of the interaction of carbon with oxygen, in good agreement with the experimental results and the arguments discussed in Section 2.3.4.2.

Better results could be obtained by an *ab initio* evaluation of the model parameters, but the evidence is already given that solubility relationships of impurities in semiconductors are intimately related to the chemistry of the impurity–solvent and impurity–impurity interactions, and to the stability of complex species in solid solution.

4.2.4   Growth of Bulk Solids by Liquid Crystallization Processes: Pick-Up of Impurities

Impurities in semiconductors play a crucial role, as they may work as dopants (donors and acceptors) or as minority carriers recombination centres [90]. In compound semiconductors they have to be used also to compensate the electrical activity of growth-related defects and

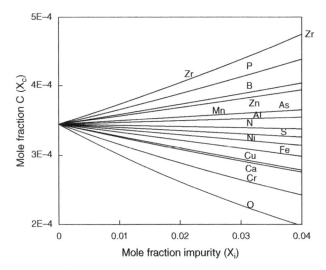

Figure 4.14 *Effect of a second impurity on the solubility of carbon in liquid silicon. Tang, 2009, [88]. Reproduced with permission from Springer Science and Business Media*

to reconduct the semiconductor to semi-insulating (SI) properties, as is the case for GaAs, where N, Al, In, Sb, B and P are the impurities of possible use. Other impurities, such as N in silicon, are used to improve the mechanical properties of the material

Knowledge of the behaviour of impurity segregation processes among a liquid and solid phase during a solidification process is therefore of immediate interest for the design of optimized growth of a doped crystal or a purification process for impure feedstocks, with a view to obtaining crystalline ingots with the desired properties.

A typical example is the purification of metallurgical silicon feedstocks for the production of solar grade silicon [16, 91], for which recrystallization processes are used as the final technological step.

Another critical example is the refining of the elemental components of compound semiconductors and of compound semiconductors themselves.

In the general case of a liquid charge contaminated by or with the deliberate addition of specific impurities, the impurity pick-up during the crystallization process is quantitatively determined by their segregation coefficients (see Section 1.6.2), which, in thermodynamic equilibrium conditions, reads

$$k_i^o(x, T) = \frac{x_i^S}{x_i^L} = \frac{\gamma_i^S}{\gamma_i^L} \exp{-\frac{\mu_i^{o,S} - \mu_i^{o,L}}{RT}} \tag{4.39}$$

where x_i^S is the atomic fraction of the impurity i in the solid phase, x_i^L is the atomic fraction of the impurity i in the liquid, $\gamma_i^S \gamma_i^s$ and γ_i^L are the activity coefficients of the impurity i in the solid and liquid phase and $\mu_i^{o,S} - \mu_i^L$ is the Gibbs free energy of fusion of the impurity i at the temperature of the growth process.

For a system consisting of a dilute solution of a *single* impurity in an elemental semiconductor, the process temperature closely corresponds to the melting temperature of the pure semiconductor $T_{growth} \sim T_m^\circ$. According to Eq. (4.39), the *equilibrium* segregation coefficient depends on the activity coefficients of the impurity i in the liquid and solid solutions having as the solvent the elemental semiconductor and as the solute the impurity i and on the Gibbs free energy of fusion of the impurity at the growth temperature.

For very dilute solutions of a single impurity, the activity coefficients may be taken equal to 1.

However, in the presence of several impurities and, again, of a dilute solution of them, the hypothesis that each of them ccan be treated independently and assigned with a unitary activity coefficient is not always tenable. We have, in fact, shown in Section 4.2.3, Figure 4.16 and in Chapter 2, that impurities interact chemically with other impurities, with the formation of complexes, even in the case of very dilute solutions. Therefore, their activity coefficients and their segregation behaviour will be influenced by specific interaction patterns.

There is also the experimental evidence that the segregation coefficients depend strongly on the growth rate (see Figure 4.15). Actually [92], the equilibrium segregation coefficient can be regarded as the ratio of the rate constants of two exchange processes occurring at the growth interface, one of which refers to the input flow of solute atoms into the crystal and the other to the corresponding output flow.

True equilibrium conditions may, therefore, be effectively achieved at growth rates approaching zero and, therefore, the equilibrium segregation coefficients k_{eq} are usually determined under very small growth rates. This is the case for CZ or FZ processes, where the condition of complete mixing in the liquid solution, which can be therefore considered homogeneous and isothermal, is close to being practically realized by thorough stirring.

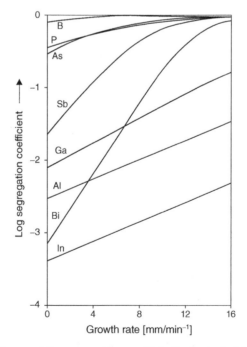

Figure 4.15 *Dependence of the segregation coefficient of selected impurities in silicon on the growth rate. Reproduced with permission from Professor H. Föll*

As industrial growth processes are generally carried out at a rate which would not allow the pick-up of impurities under true equilibrium conditions, the distribution of impurities among the liquid and the solid would depend on their *effective* segregation coefficients k_{eff}. These last depend on the growth rate and sensibly deviate from the equilibrium ones, as is shown in Figure 4.15, for the case of segregation of impurities in silicon.

Effective segregation coefficients (k_{eff}) need to be experimentally determined, because it is, usually, not very easy to predict their values.

Nevertheless, assuming that the incorporation rate depends on the diffusion of the impurities in a static (diffusion) layer of thickness δ at the liquid/solid interface, within which the fluid motion is laminar and non-mixing, and that diffusion in the solid is negligible [92, 93], the following equation describes the dependence of the effective segregation coefficients (k_{eff}) on the growth rate v

$$ k_{eff} = \frac{k_{eq}}{k_{eq} + (1 - k_{eq}) \exp\left(-\frac{v\delta}{D_i}\right)} \tag{4.40} $$

where D_i is the diffusion coefficient of the impurity in the melt and δ is the thickness of the diffusion layer, whose evaluation might be complicated because it depends on the convection fluxes in the liquid charge.

It is easy to see from Eq. (4.40) that the effective segregation coefficients should be virtually independent of the growth rate and equilibrium conditions will prevail at the growth interface for high diffusivity impurities. It is also apparent that the effective segregation coefficients should tend to a value corresponding to k_{eq} at very low growth rates and under good stirring conditions, for which $\delta \rightarrow 0$. Conversely, k_{eff} should tend to 1 for large values of the growth rate and poor stirring conditions.

The literature values of the segregation coefficients of impurities of impurities of technological importance in Si, Ge, GaAs and InP are reported in Table 4.4.

It is immediately obvious that the absolute values of the segregation coefficients of impurities depend on the nature of the host matrix (liquid and solids) in which they are distributed. Therefore, a direct comparison between the segregation coefficients of impurities in silicon or germanium and those in compound semiconductors is a physical nonsense.

Considering metallic impurities in silicon, it is immediately clear from Table 4.4 [33, 94–112] that the absolute values of the segregation coefficients for a FZ-growth are systematically lower than those pertaining to CZ- or DS- growth and even smaller than those quoted as the equilibrium values [94]. This result evidences that the FZ process has the best potential to grow extremely pure materials in comparison with CZ and DS, not only because it is a crucible-less technique but also because it can work close to true equilibrium conditions.

Concerning the basic physical and chemical factors that determine the absolute values of the equilibrium segregation coefficient, an experimental relationship has been found between the segregation coefficients and the solubility for a number of impurities in silicon and germanium [113] (see Figure 4.16). It gives a purely empirical, but useful correlation tool for laboratory practice. In fact, once the solubility or the segregation coefficient of a specific impurity is known, the corresponding value of the segregation coefficient or solubility could be estimated.

This correlation finds its qualitative rationale in the dependence of the segregation coefficient of impurities on their covalent radii in the solid at the melting point of silicon or germanium [104], as is shown in Figure 4.17.

Table 4.4 Segregation coefficients of donor, acceptor and deep-level impurities in Si, Ge, GaAs and InP

Impurity	Cz-Si [94]	CZ-Si	FZ-Si	DS-Si	Al-Si (1073 K)	Ge	GaAs	InP
C	$5.8 \cdot 10^{-2}$	$5 \cdot 10^{-2}$	—	—	—	≥ 1.85 [95]	1.44 ± 0.08 [96]	—
O	0.5–1.3	0.25	—	—	—	0.11 [97]	0.3 [98], 0.1 [99]	—
N	$8 \cdot 10^{-4}$	—	—	—	—	—	—	—
B	0.8	0.73 [100]	—	—	$7.6 \cdot 10^{-2}$ [101]	20 [102]	—	—
Al	$2 \cdot 10^{-3}$	$3 \cdot 10^{-2}$	—	—	—	$7.3 \cdot 10^{-2}$ [103]	—	4 [106]
Ga	$8 \cdot 10^{-3}$	0.008 [104]	—	—	—	$8.7 \cdot 10^{-2}$ [105]	$\approx 10^{-6}$ [105]	$\approx 10^{-6}$ [106]
In	$4 \cdot 10^{-4}$	—	—	—	—	10^{-3} [103]	0.1–0.07[98]	1 [105, 107]
As	0.30	—	—	—	—	$2 \cdot 10^{-2}$ [103]	$\approx 10^{-6}$ [104]	1 [106, 107]
Sb	0.002–0.023	—	—	—	—	$3 \cdot 10^{-3}$ [103]	—	—
Ge	—	0.6	—	—	—	—	—	—
N	—	$7 \cdot 10^{-4}$ [108]	—	—	—	—	—	—
P	0.35	0.35 [100]	—	—	$8.5 \cdot 10^{-2}$ [101]	0.08 [103]	2–3 [98]	$\approx 10^{-6}$ [105]
Cu	$4 \cdot 10^{-4}$	$8 \cdot 10^{-4}$	$3 \cdot 10^{-7}$	$2 \cdot 10^{-3}$ [109]	—	1.3 [110]	—	—
Ni	$1.5 \cdot 10^{-5}$	$3.2 \cdot 10^{-5}$	$1.5 \cdot 10^{-7}$	$8 \cdot 10^{-4}$ [109]	—	$3 \cdot 10^{-6}$ [110]	$6 \cdot 10^{-4}$ [98]	—
Au	$2.5 \cdot 10^{-5}$	—	$5 \cdot 10^{-5}$	—	—	$3 \cdot 10^{-5}$ [110]	—	—
Fe	$8 \cdot 10^{-6}$	$6.4 \cdot 10^{-6}$	$5 \cdot 10^{-6}$	$1 \cdot 10^{-5}$	$1.7 \cdot 10^{-11}$ [101]	$3 \cdot 10^{-5}$ [110]	$\geq 10^{-3}$ e) [98]	—
Co	$8 \cdot 10^{-6}$	10^{-5}	10^{-7}	10^{-7}	—	10^{-6} [110]	—	—
Ti	$2 \cdot 10^{-6}$ [18]	$2 \cdot 10^{-6}$	$1.8 \cdot 10^{-6}$	$3.5 \cdot 10^{-5}$	—	—	—	>4 [111]
Cr	$1.1 \cdot 10^{-5}$ [18]	$1.1 \cdot 10^{-5}$	$2.5 \cdot 10^{-6}$	$3.7 \cdot 10^{-3}$ [108]	—	—	10^{-3}–$5.7 \cdot 10^{-4}$ [97]	—
Zn	—	—	—	—	—	—	≈ 1 [112]	—
Si	—	—	—	—	—	—	—	30 [33]

Coletti et al., 2011, [90]. Reproduced with permission from John Wiley & Sons.

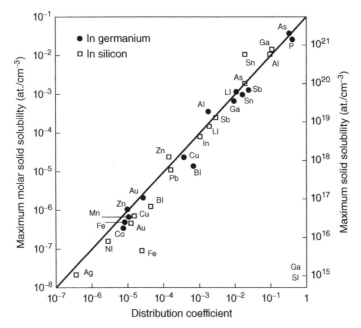

Figure 4.16   *Correlation between solid state solubility and segregation coefficients of impurities at the melting point of Ge and Si. Fischler, 1962, [113]. Reproduced with permission from AIP Publishing*

4.2.5   Constitutional Supercooling

Poor stirring conditions and high growth rates are favourable for the set-up of constitutional supercooling, a common phenomenon encountered in the directional solidification of metal alloys [114], but also known in the case of CZ- and DS- growth of elemental and compound semiconductors, which causes a conversion of the normal, planar growth to a cellular or dendritic growth.

To avoid the frequent confusion found in the literature concerning the set-up of constitutional supercooling, we shall note first that

1. the solidification of a binary alloy $A_{1-x}B_x$ (where the solute B is an impurity or the second component of an alloy) at high growth rates implies that the concentration of the rejected solute in the diffusion layer increases when $k_{eff,B} < 1$ or $\ll 1$ and, consequently decreases the freezing temperature of the solution
2. the same condition occurs when $k_{eff,A} > 1$ or $\gg 1$, causing an increase of the concentration of B in the diffusion layer
3. so far the freezing temperature of the solution in the diffusion layer is higher than the temperature of the melt outside, the situation is thermodynamically stable and normal growth continues. When the freezing temperature of the liquid inside the diffusion layer becomes lower than the temperature of the melt, the situation is unstable and conditions of supercooling occur, with local solution freezing, which will drive the beginning of a cellular or dendritic growth [115].

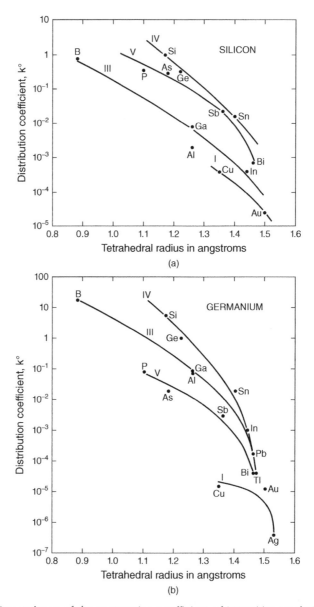

Figure 4.17 *Dependence of the segregation coefficient of impurities on their covalent radii in the solid at the melting point of (b) germanium and (a) silicon. Trumbore, 1960, [104]. Reproduced with permission from John Wiley & Sons*

4. supercooling would also cause second phase segregation when the temperature in the diffuse layer reaches the eutectic temperature. This could be the case of the binary B–Si system, on which we will return to in this section, which has a eutectic temperature (T_{eut} = 1394 °C) [116] very close to the melting point of silicon (see Figure 4.18).

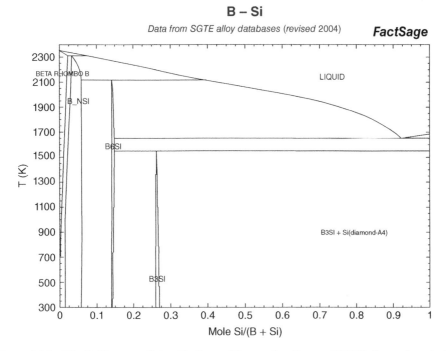

Figure 4.18 The B–Si system. Scientific Group Thermodata Europe (SGTE). Reproduced with permission from SGTE

The theoretical analysis of the constitutional supercooling was originally developed by Tiller *et al.* [118], with modest, subsequent modifications by other authors, and demonstrated to be applicable in a vast generality of cases.

According to Tiller, under the assumption of planar growth, negligible diffusion in the solid and negligible convection in the liquid, and a segregation coefficient independent of the growth rate,[8] the critical growth rate v_{crit} at which constitutional supercooling occurs, is given by the following equation, called also the Tiller stability criterion, and written adopting the nomenclature of this book

$$v_{crit} = DG_L \frac{k_{eq}}{mC_L(1 - k_{eq})} \qquad (4.41)$$

where D is the diffusion coefficient, G_L is the temperature gradient in the liquid normal to the solid/liquid interface, m is the slope of the liquidus curve and C_L is the impurity concentration in the bulk liquid.

It can easily be seen that by introducing into Eq. (4.41) the effective segregation coefficients of Eq. (4.40), the following equation could be obtained

$$v_{crit} = DG_L \frac{k_{eq}}{mC_L(1 - k_{eq})e^{-\Delta}} \qquad (4.42)$$

---

[8] These conditions are conventionally labelled 'normal freezing conditions' in physical metallurgy [114] and will be discussed in the next section.

where $\Delta = -\frac{v\delta}{D_i}$, which should adopted when the segregation coefficients depend on the growth rate.

Critical growth rates are, therefore, enhanced in the presence of fast diffusing species and of high temperature gradients at the growth interface, while constitutional supercooling is favoured by a high concentration of impurities.

The constitutional supercooling theory is not only one of the most important theories in physical metallurgy, but it has been systematically applied for the DS-growth of mc-Si, for the CZ- or DS-growth of heavily doped silicon, for Si-Ge alloys [119] and for compound semiconductor alloys, where often the Tiller's stability criterion is used to predict the critical value of the growth rate.

A recent numerical analysis [120] shows, however, that the stability criterion should be applied with caution in most practical circumstances.

Among the many reasons for the use of the Tiller's criterion giving erroneous results, the most important is that constant growth rate and a planar solidification front are not generally present in crystal growth processes, due to non-isothermal growth conditions, as already seen in Section 4.2.2.2.

As an example, some experiments on Bridgman growth of highly doped GaInSb semiconductor alloys, in agreement with previous numerical work, show significant variations of the growth rate and interface curvature during the solidification process. Interface curvature is systematically present in the directional solidification of mc-Si.

According to simulation results [120], the most important limitation of the stability theory is, however, the erroneous prediction of the liquidus profile, which shows a much lower slope than predicted.

In spite of these limitations, the theory of constitutional supercooling has been successfully applied to explain deviations from normal growth conditions in the directional growth of mc-Si from melts rich in metallic impurities [115, 121, 122] and also in the case of CZ-growth of heavily B-doped silicon, for B concentrations $> 3.10^{20}$ at cm^{-3} [123].

Constitutional supercooling phenomena are also relevant in the growth of III–V semiconductors (arsenides and phosphides) from non-stoichiometric melts, where the excess component has a negligible solubility in the solid ($k < 10^{-6}$). Precipitates formation due to supersaturation of impurities is also observed in single crystal InP, due to the segregation coefficient of the impurities being much less than 1 [124, 125].

4.2.6   Growth Dependence of the Impurity Pick-Up and Concentration Profiling

Once the values of the effective segregation coefficients for a specific solute (dopant or impurity) in a specific solvent are known, it is possible to foresee and optimize the solute profile along the ingot. This issue is of tremendous technological value because it makes it possible, for example, to grow a B-doped ingot and foresee the exact sections of the ingot which lie in a specific resistivity interval, considering that the B concentration p defines the concentration of electron holes in the material

$$p = \frac{1}{\rho q \mu_p} \tag{4.43}$$

where ρ is the resistivity, q is the electronic charge and μ_p is the holes mobility.

The same approach can be used in a purification process, to calculate the size of the ingot section suitably depleted of impurities.

To do this, we shall first consider that, during the crystallization process, the total amount of impurities in the crystal and in the melt remains constant and this condition can be written as

$$x_i^o = \int_0^f x_i^S dx + (1-f)x_i^L \tag{4.44}$$

where x_i^o is the initial concentration (in molar fraction) of the impurity i in the molten charge and f is the solidified fraction. After differentiation of Eq. (4.44) with respect to f one obtains

$$x_i^S + (1-f)\frac{dx_i^L}{df} - x_i^L = 0 \tag{4.45}$$

By definition $x_i^S = k_{eff}x_i^L$ and Eq. (4.45) becomes

$$\frac{dx_i^L}{x_i^L} = \frac{(1-k_{eff})}{(1-f)}df \tag{4.46}$$

By applying the boundary condition $x_i^L = x_i^o$ for $f = 0$, the solution of the Eq. (4.46) converts to the so-called *normal freezing equation* [14]

$$x_i^S = k_{eff}x_i^o(1-f)^{k_{eff}-1} \tag{4.47}$$

which gives the distribution of the impurity concentration in the ingot along the growth direction [101] and should be compared to the classical Scheil equation, valid for equilibrium segregation

$$x_i^S = k^o x_i^o(1-f)^{k^o-1} \tag{4.48}$$

where k^o is the equilibrium segregation coefficient.

Figure 4.19 [126] shows the distribution of impurities along the axial direction of a semiconductor crystal, calculated for different values of k_{eff}. For $k_{eff} < 1$, the solid phase contains less impurities than the liquid, so that the impurity concentration in the melt increases with the progress of crystal growth, as does the concentration of impurities in the solid. For $k_{eff} > 1$, the solid phase contains more impurities and the impurity concentration in the melt reduces with the progress of crystal growth and goes to zero when f approaches 1.

This behaviour is common to all crystals grown from the liquid phase, quantitative differences depending only on the values of the segregation coefficients and on the growth rate.

4.2.7 Purification of Silicon by Smelting with Al

Silicon refining through Al smelting is an interesting example of an impurity extraction process based on the segregation equilibria between a solid silicon phase and a Si-Al alloy [127].

The process is based on the circumstance that Al-Si alloys melt at considerably lower temperatures than silicon (see Al–Si diagram in Figure 1.35), that the solubility of Al in Si

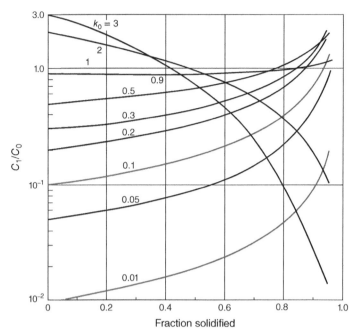

Figure 4.19 *Impurity concentration profiles as function of the segregation coefficients. Pfann, 1966*

is negligibly low and that the solubility of impurities in Al-Si alloys, irrespective of being liquid or solid [128], is higher than in silicon. As an example, at 1093 K the segregation coefficient of cobalt in Al is 10^4 times smaller than in silicon [129]. The segregation coefficients of Fe, Ti, B and P are also smaller,[9] making also the recovery of B and P possible [101]. The process is operated by first dissolving impure silicon in molten aluminium at temperatures ranging from 700 to 1100 °C to form a 20–55% Si–Al solution. Then, the solution is rapidly cooled to a temperature above the eutectic (577 °C), to segregate silicon as a solid phase and to establish an equilibrium between solid silicon and the molten alloy, with repartition of the impurities between the two phases, with a yield being defined by the respective segregation coefficients.

4.3   Growth of Ge-Si Alloys, SiC, GaN, GaAs, InP and CdZnTe from the Liquid Phase

Several elemental and compound semiconductors exhibit a continuous range of solid solutions. This is the case for GeSi alloys, nitride alloys and GaAs and InAs, as was discussed in Section 1.7.1.

The ability to grow a semiconductor alloy within its entire range of continuous, variable composition would be a strategic skill, because it would allow tailoring the width of the

---

[9] $K_{Fe}(1073\ K) = 1.7 \times 10^{-11}$, $K_{Ti}(1073\ K) = 3.8 \times 10^{-9}$, $K_B(1073\ K) = 7.6 \times 10^{-2}$; $K_B(1273K) = 2.2 \times 10^{-1}$; $K_P(1073\ K) = 4 \times 10^{-2}$; $K_P(1073\ K) = 8.5 \times 10^{-2}$.

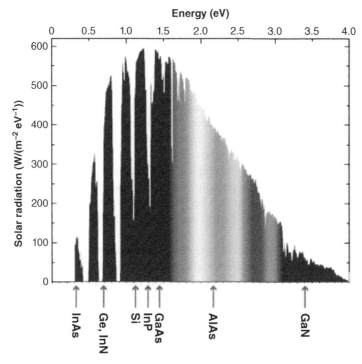

Figure 4.20 *Band gap of selected semiconductors of potential use in third generation solar cell. The solar radiation curve is the background. Brown & Wu, 2009, [130]. Reproduced with permission from John Wiley & Sons*

band gap as a function of the alloy composition. One of the important consequences of band gap tailoring capability would be, for example, the direct control of the wavelength of light emission of LEDs fabricated with direct gap semiconductors, like the arsenides and nitrides, or the optimization of the light absorption of solar cells, as shown in Figure 4.20 [130].

As most of the applications of this kind of alloys are based on thin film devices, the technological interest in bulk semiconductor alloys is addressed to the growth of single crystal boules, from which thick substrates could be obtained, to be used as a platforms for homoepitaxial or heteroepitaxial growth processes. The availability of bulk samples is, instead, mandatory for the experimental determination of their structural and physical properties.

4.3.1 Growth of Si-Ge Alloys

The Si-Ge alloys present a continuous variation of the fundamental, indirect band gap from that of pure Ge to that of pure Si [131], where the change of slope at $x = 0.15$ is due to a crossover of the minima of the conduction bands of Si and Ge [131, 132].

The defect physics of Si-Ge alloys is influenced by the alloy composition, making it possible, for example, to shift the energy levels of specific impurities within the band gap, as is the case of Tl, which is deep in Si and shallow in Ge [133].

Another intrinsic property of Ge-Si alloys is the partial suppression of the light induced degradation (LID effect) of the efficiency of silicon solar cells, already discussed in Section 2.3.4.3, due to the presence of B-O complexes. In the case of Si-Ge alloys the reduction of the LID effect increases with increase in the Ge content and is associated with the reduction in concentration of the O_{2i} species [134], a precursor involved in the formation of the B-O complexes.

Due to their indirect band gap, optoelectronic applications of these alloys are of modest interest, which is, instead, directed towards some different, but peculiar properties, such as their capability to extend towards the far-IR tail of the solar spectrum the sensitivity of Si-Ge solar cells and their high electronic mobilities.

Most of the more important technological applications of Si-Ge alloys are realized using thin layers grown with gas phase processes, whose features will be described in Section 4.5.2. Here, the discussion will be limited to the thermodynamics of the growth processes of bulk phases, in view of a full understanding of the limits and potential of this type of growth.

The thermodynamics of this system were discussed in Section 1.7.4, where it was shown that the solid solutions present small deviations from the ideal thermodynamic behaviour, with unmixing features at low temperatures. It was also shown that the liquid solutions are homogeneous mixtures of metallic- and covalently-bonded Si and Ge, still presenting a close to ideal behaviour.

It is, however, known already from the late 1950s, that only the growth of polycrystalline samples of $Si_{1-x}Ge_x$ alloys from the liquid phase can be carried out over the entire range of concentrations, using the Bridgman or the CZ techniques. Growth of single crystal ingots is possible only in the 0–20% and in the 90–100% (molar) range of Si in Ge, using pure germanium, silicon or Si-Ge seeds. In the intermediate compositions range only coarse polycrystalline alloys can be grown.

Constitutional supercooling is at the origin of degeneration of the planar growth and the onset of rough polycrystallinity in the $0.15 < x < 0.73$ composition range, due to the large segregation coefficient of Si in Ge ($k_{Si}^{Ge} \approx 5$) [135]. Consistent lattice mismatch of pure elements ($a = 0.5431$ nm for Si and 0.5658 nm for Ge at 300 K), a strong difference in the linear expansion coefficients α of the Si_xGe_{1-x} alloys for $x > 0.15$ ($\alpha = (2.6 + 2.55x)10^{-6}K^{-1}$ for $x < 0.15$ and $\alpha = (-0.89 + 7.73\ x)10^{-6}K^{-1}$ for $x > 0.15$) and the lack of $Si_{1-x}Ge_x$ single crystal seeds over the entire range of compositions are, instead, practical limitations to the seeded growth of single crystal $Si_{1-x}Ge_x$ alloys. As expected, the lattice mismatch is also at the origin of the presence of misfit dislocations in CZ-grown Si_xGe_{1-x} alloys, using silicon seeds.

Considering that, according to Eq. (4.41), the critical velocity above which constitutional supercooling occurs

$$v_{crit} = DG_L \frac{k_{eq}}{mC_L(1 - k_{eq})e^{-\Delta}} \tag{4.41}$$

increases with the increase of the temperature gradient G_L in the liquid phase, a criterion to reduce the impact of supercooling on the growth features of Si-Ge alloys would be that of inducing a large temperature gradient at the growth interface.

With temperature gradients up to 100 K cm^{-1} it was possible to grow [135] single crystalline Ge-Si alloys ingots up to 10% atomic Si, using a radiation heated, mirror-type vertical Bridgman furnace. Zone melting, with similarly large temperature gradients, of

Table 4.5 *Dependence of the segregation coefficients of dopants on the composition of Si_xGe_{1-x} alloys and silicon*

Dopant	Si_xGe_{1-x}			Si
	x = 0.8	x = 0.9	x = 0.95	
B	1.1–1.9	≈1.2	1.0–1.1	0.8
Ga	—	—	0.3	0.008
P	0.8	≈0.6	0.91	0.35

Yonenaga, 2005, [119]. Reproduced with permission from Elsevier.

an initially Ge-rich charge with a Ge seed resulted in a single-crystalline phase up to 50% atomic silicon.

A homogeneous dislocation density around 10^5 cm^{-2} is typical of these crystals, with the exception of the initial part of the ingot where the dislocation density is much higher.

Recent studies [136, 119] on the growth and fundamental properties of Si_xGe_{1-x} alloys confirmed, however, the earliest conclusions that single crystal boules of Si_xGe_{1-x} alloys could be successfully CZ-grown only in the $0 < x < 0.15$ and $0.73 < x > 1$ range, using necking procedures for the growth of Si-rich Si-Ge crystals with low dislocation densities. Constitutional supercooling was also definitely demonstrated to be responsible for the breakdown of single crystallinity in the intermediate composition range.

Due to the relatively large miscibility gap between the solid and the liquid (see Figure 1.15) and the composition-dependence of the segregation coefficient of Si in Si-Ge alloys, which ranges between 1 and 0.3 in Si-rich alloys and between 1 and 5.5 in Ge-rich alloys, the grown ingots also exhibit a composition profile along the solidification direction.

Several other problems arise with the growth of doped Si-Ge alloys. In fact, the composition dependence of the segregation coefficients of dopants (see Table 4.5) makes the doping procedure of Si-Ge alloys challenging.

The epitaxial growth of Si-Ge alloys of constant composition may be, instead, afforded without thermodynamic constraints, and over the entire range of compositions, using CVD, MBE or other gas phase techniques on suitable substrates [137]. In this case, however, lattice misfit problems with the substrate limit the useful range, as will be shown in Section 4.5.2.

4.3.2   Growth of SiC from the Liquid Phase

Direct seeded melt growth of SiC is not applicable, and even not possible, in view of its high melting temperature (3100 K) and of the difficulty in maintaining SiC stoichiometry at these elevated temperatures [138]. Its growth from saturated carbon solutions in silicon was, therefore, attempted, following a practice employed in the case of other compound semiconductors, as will be seen in the next section.

The drawbacks of this method applied to SiC are:

- the high temperatures needed to get acceptable carbon solubilities (the solubility of C ranges from 0.01% at the melting point of Si to 19% at 2830 °C),
- the high vapour pressures of silicon at these temperatures [139] (see Table 4.6) and
- the catastrophic degradation of the graphite crucibles used as containers, due to the permeation of silicon into the crucible pores with the formation of SiC.

Table 4.6 *Total vapour pressure of Si species ($\Sigma\ P_{Si}$) and of monomeric Si (P_{si}) over SiC*

T (K)	$\Sigma\ P_{Si}$ (MPa)	P_{Si} (MPa)
1700	$6.9 \cdot 10^{-8}$	$6.78 \cdot 10^{-8}$
1800	$3.2 \cdot 10^{-7}$	$3.17 \cdot 10^{-7}$
1900	$1.3 \cdot 10^{-6}$	$1.22 \cdot 10^{-6}$
2000	$4.6 \cdot 10^{-6}$	$4.53 \cdot 10^{-6}$
2100	$1.4 \cdot 10^{-5}$	$1.33 \cdot 10^{-5}$
2200	$3.9 \cdot 10^{-5}$	$3.83 \cdot 10^{-5}$
2300	$1.00 \cdot 10^{-4}$	$9.30 \cdot 10^{-5}$
2400	$2.36 \cdot 10^{-4}$	$2.17 \cdot 10^{-4}$
2500	$5.20 \cdot 10^{-4}$	$4.74 \cdot 10^{-4}$
2600	$1.08 \cdot 10^{-3}$	$9.73 \cdot 10^{-4}$
2700	$2.11 \cdot 10^{-3}$	$1.89 \cdot 10^{-3}$
2800	$3.95 \cdot 10^{-3}$	$3.35 \cdot 10^{-3}$
2900	$7.08 \cdot 10^{-3}$	$6.22 \cdot 10^{-3}$
3000	$1.22 \cdot 10^{-2}$	$1.06 \cdot 10^{-2}$
3100	$2.3 \cdot 10^{-2}$	$1.75 \cdot 10^{-2}$

The monomeric species is prevalent at every temperature140.Data from Sevast'yanov et al., 2010, [139].

In spite of these difficulties, it was possible [140] to grow SiC platelets from a Si melt at 1665 °C on a graphite tip [140], or SiC boules 20–25 mm in diameter and 20 mm long at a pulling rate of 5–15 mm h^{-1} in the 1900–2400 °C temperature range, under an Ar pressure of 100–120 bar [141].

This technology was, however, abandoned in view of the advantages presented by the PVT (physical vapour transport) method, which will be discussed in Section 4.4.1.

Liquid phase epitaxy of defective SiC wafers, a technique which works close to thermodynamic equilibrium conditions, has been, instead, proven to heal successfully micropipe defects, consisting of holes which penetrate deep in the crystal, and which are present in high density (1–100 cm^{-2}) in sublimation grown ingots [142], see Section 4.4.1.

4.3.3   Growth of GaN from the Liquid Phase

The availability of high quality, low cost GaN substrates would avoid the problems associated with lattice mismatch (see Table 4.7) and the consequent high dislocation densities found when using sapphire, SiC or silicon substrates for the heteroepitaxial CVD- or metal organic chemical vapour deposition (MOCVD) growth of blue and white Ga-In nitride LEDs, allowing homoepitaxial deposition conditions. This was the hint for the study of growth processes of single crystalline GaN, of which that from the liquid phase is discussed in the next sections.

4.3.3.1   Solution Growth of GaN

Among the possible, robust routes to grow large, high quality GaN boules, that based on the crystallization of GaN from a liquid phase was considered a very promising solution.

Table 4.7 *Lattice constants of GaN, sapphire, SiC and Si*

Material	Lattice constant (Å)
GaN	a 3.189
	c 5.185
Sapphire	a 4.785
	c 12.991
SiC	a 3.08
	c 15.12
Si	5.4307

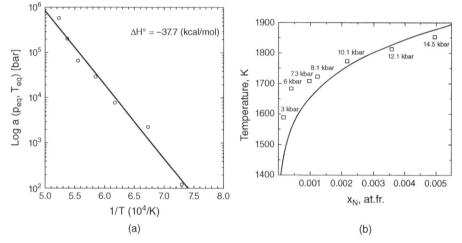

Figure 4.21 *(a) Equilibrium nitrogen pressures over GaN + Ga(l). Karpinski and Porowski, 1984, [146]. Reproduced with permission from Elsevier. (b) Solubility of nitrogen in liquid Ga (in molar fraction). Porowski and Grzegory, 1997, [144]. Reproduced with permission from Elsevier*

As in the case of SiC, the direct growth of GaN from a liquid GaN phase would be unfeasible due the critical temperature and pressure conditions involved [143–145] (see Table 4.1 and Figure 4.21).

A thermodynamically feasible and practically experimented solution is, instead, the growth of GaN from a saturated solution of GaN in liquid Ga [144, 145] or from a saturated solution of Ga and nitrogen in a liquid metal M, allowing operation at more convenient temperature/pressure conditions, according to the following equilibria

$$Ga^l + \frac{1}{2}N_2^g \rightleftharpoons GaN^S \tag{4.49}$$

or

$$Ga_{Ga_x - M_{1-x}} + \frac{1}{2}N_2^g \rightleftharpoons GaN^S_{Ga_x - M_{1-x}} \tag{4.50}$$

assuming that equilibrium occurs between the gaseous nitrogen and that dissolved in liquid Ga or in the Ga-M alloy.

The thermodynamics of this system have been extensively studied by Karpinski *et al.* [146, 147], who determined the experimental temperature dependence of the equilibrium pressure of nitrogen gas over GaN and Ga (see Figure 4.21)

$$\ln p_{N_2} = -\frac{\Delta H_{GaN}}{RT} \tag{4.51}$$

from which the standard free energy of formation of GaN was deduced.[10]

It should be noted that the solution kinetics of diatomic nitrogen in liquid Ga is determined by its dissociation energy, which amounts to 9.76 eV (-941.7 kJ mol^{-1}), making molecular nitrogen the strongest bound molecule in nature and its dissociation to atomic nitrogen difficult. The theoretical study of the dissociation kinetics at the surface of liquid Ga and of other metals has, however, shown that the metal surfaces have a strong catalytic effect, with energy barriers towards the formation of Me-N bonds (3.32 eV for Al, 3.4 eV for Ga and 5.6 eV for In) much lower than the dissociation energy [145].

The process involving only the reaction between nitrogen and liquid Ga, called high pressure solution growth (HPSG) was investigated at temperatures around 1800 K and 1200 MPa [144], without and with intentionally seeding, using multizone vertical furnaces. In this range of temperatures and pressures the solubility of nitrogen in liquid gallium is below 1 at% [144] (see Figure 4.21b), and supersaturation conditions needed to nucleate the crystals were obtained by imposing a vertical temperature gradient. In the absence of intentional seeding, in the upper and cooler part of the furnace a polycrystalline GaN layer segregates at the surface of the liquid. It acts as a nucleation site for larger, platelet-like crystals, crystallizing with the wurtzite structure. Small GaN single crystals were, instead, grown, seeding the growth with a needle-shaped seed.

To mitigate the extreme conditions of HPSG and to carry out the growth under more moderate process conditions, Ga-metal alloys have been considered as alternative reaction media.

Among a number of metals of potential use, Na has the advantage of not being dangerous impurity when dissolved in solid GaN. In addition, it offers the possibility to use sodium azide (Na_3N) as the GaN precursor

$$NaN_3 + Ga \rightarrow GaN + Na + N_2 \tag{4.52}$$

Actually GaN could be synthesized [148], using NaN_3 and Ga as the precursors, in stoichiometric amounts and reacted in a closed stainless steel tube at 750 °C for 100 h in a N_2 atmosphere. The practical advantage of this method is that sodium azide is stable at room temperature and decomposes only above 300 °C, providing the required amount of (active) nitrogen to react with Ga. With this method it is possible to grow prismatic crystals (a few millimetres in size), platelets or granules, depending on the Na/(Na-Ga) molar ratio.

It was also demonstrated, as a result of an investigation concerning the role of sodium in the nitrogen solubility in the Ga-Na system [149], that Na promotes the dissolution of nitrogen in Ga-Na solutions at tempperatures definitely lower than those used in the HPSG

---

[10] $\Delta G^0_{GaN} = -37.710^3 + 32.43\ T$ cal mol^{-1}, with $\Delta H_{GaN} = -37.7$ kcal mol^{-1} and $\Delta S = -32.43$ cal mol^{-1} K^{-1} as the standard enthalpy and entropy of GaN formation, respectively.

process. This favours the growth of GaN at temperatures and pressures definitely lower than those employed with pure Ga melts.

The physical chemistry of Na-enhanced nitrogen dissolution might be questioned, as it is known that the solubility of nitrogen in liquid sodium at atmospheric pressure is negligible [150] and that also the reactivity of nitrogen with sodium is negligible. The practical conclusions are, however, important, as a process based on the use of the Na-Ga alloy was effectively developed, as will be seen at the end of this section.

More insight on the Na–Ga–N system comes from the results of its thermodynamic modelling [151] and from the comparison of the calculated and experimental phase diagrams of the binary Na–Ga and ternary Na–Ga–N systems (see Figure 4.22) between 600 and 750 °C.

It can be seen, as an example, that the Ga–Na system exhibits important deviations from ideality, as could be deduced from the presence of a wide liquid immiscibility region in the Na-rich corner of the phase diagram of the binary Ga–Na system.

Considering, instead, the calculated phase diagram of the ternary Ga–Na–N at 750 °C and 50 MPa, one observes the presence of a wide region where the GaN phase is stable with a single Ga-rich liquid, favourable for the growth of GaN from a liquid Ga–Na phase.

Working around this temperature, pressure and composition range, it was possible to grow 3 in diameter GaN boules from a solution of Ga in sodium, using GaN seeds, at a temperature of 870 °C and nitrogen pressure of 3.2 MPa [152]. Better quality crystals are expected to be grown at higher temperatures and pressures under solution stirring, but it is already apparent that the process may be brought to a production stage.

4.3.3.2 Growth of Bulk Gallium Nitride by the Ammonothermic Process

Different from processes described in the former sections, where the growth is carried out from a molten elemental or compound semiconductor or from a molten elemental component of a semiconductor compound, ammonothermal growth is carried out in liquid supercritical ammonia solutions.

The starting point of the ammonothermic growth of bulk GaN is a process originally developed for the growth of AlN [153]. This process is based on the reaction of aluminium with supercritical ammonia in the presence of potassium amide (KNH_2) which leads to the precursor molecule $KAl(NH_2)_4$

$$Al + KNH_2 + 3NH_3 \rightarrow KAl(NH_2)_4 + 1.5H_2 \tag{4.53}$$

which then decomposes to AlN

$$KAl(NH_2)_4 \rightleftharpoons KNH_2 + AlN + 2NH_3 \tag{4.54}$$

The reaction (4.54) is reversible and pressure/temperature controlled and it was demonstrated that at 600 °C and a pressure of 2 kbar the reaction is shifted to the right.

It is interesting to note that it was demonstrated that with ammonobasic processes using alkali amides as mineralizers, the solubility of nitride phases in supercritical ammonia decreases with increase in temperature [154]. Therefore, if the process is carried out in a two zone reactor, reaction (4.53) is left to occur in the low temperature zone and reaction (4.54) in the high temperature zone, where the crystallization of the nitride phase may spontaneously occur.

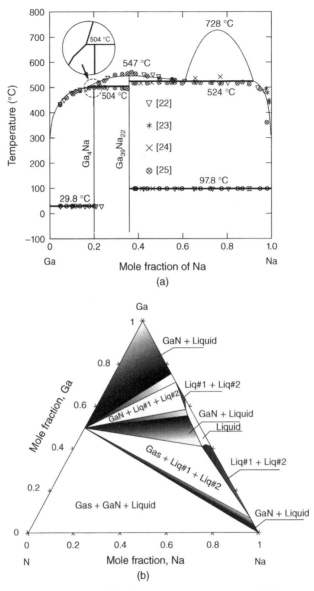

Figure 4.22 (a) Phase diagram of the binary Na-Ga system: solid lines represent the calculated values. (b) Calculated phase diagram of the ternary Ga-Na-N system at 750°C and 50 bar. Wang & Yuan, 2007, [151]. Reproduced with permission from Elsevier

In these conditions KAl $(NH_2)_4$ is transferred by convection from the low temperature region, where it is generated, to the high temperature region where it decomposes. As the driving force of the transport process is primarily given by the gradient of ammonia density, the hotter part of the autoclave where the AlN segregation occurs should be on the bottom side.

A similar process has been demonstrated to work for the production of GaN [155–157], using $NaGa(NH_2)_4$ as the precursor and sodium amide $NaNH_2$ as the mineralizer agent. The process was successfully tested [155, 156] in a Ni-Cr superalloy autoclave, at a pressure of 190–210 MPa, with the bottom region held at 700 °C, where seed crystals are also placed, and a top region held at 500 °C, where the precursors are housed. A metal baffle separates the two regions. After 82 days, a polyhedral crystal 4.5 mm in height and 5.7 mm in lateral dimensions was obtained, leading to an average growth rate of about 50 μm d^{-1}. The crystal presented an excellent microstructure and a reasonably low dislocation density ($N_D = 10^6 cm^{-2}$ on the Ga face and 10^7 on the N face).

The potential of the ammonothermic route to grow GaN crystals of excellent quality was confirmed by further experiments [157], that succeeded also in reducing the dislocations density down to $5 \cdot 10^3 cm^{-2}$.

Thanks to the demonstrated excellent microstructure of the crystals grown with this process, the (basic) ammonothermic route is of interest for commercial industrialization, considering its advantages with respect the HPVE process, which will be discussed in the next section, and the possibility of further improvements.

A variant of the basic ammonothermic process is carried out in the presence of acid mineralizers, such as NH_4Cl [158], with the advantage that the GaN solubility decreases with increase in temperature and that the process could be carried out with a more convenient configuration, having the low temperature section on the top of the autoclave [154]. The additional advantage is that the process can proceed at a reasonable rate at 500 °C and 120 kbar.

In this case, GaN is produced directly

$$Ga + NH_4Cl \rightleftharpoons GaN + HCl + \frac{3}{2}H_2 \tag{4.55}$$

and segregates from the solution when supersaturation conditions occur.

GaN recrystallization might then be easily carried out by a subsequent GaN transport process from the high temperature to the low temperature section of the autoclave [158].

Albeit the acid ammonothermal process has been successfully tested with the production of millimetre long GaN crystals, the severe corrosion of the autoclave construction materials in the presence of acidic mineralizers, with failure risks in the long term, discouraged industrialization of the process.

At the time of writing this chapter only small GaN bulk substrates grown by the ammonothermal process are commercially available, which are, furthermore, too expensive to be competitive with sapphire, SiC or silicon substrates.

4.3.4 Growth of GaAs, InP, ZnSe and CdZnTe

Before Nakamura's development of nitride-based blue, green and white LED in 1989 [159–164], GaAs and InP were the materials of choice for red and yellow LED and laser applications and ZnSe for blue-green LEDs. Today, GaAs and InP still represent an important fraction of the compound semiconductors value chain, and are also used as large diameter, dislocation-free substrates for homo- or heteroepitaxial applications. At the time of writing, there is an increasing interest in GaAs in view of its hybrid or monolithic integration on a silicon chip for 4G phones and for InP-based hybrid or monolithic photonic chips.

CdZnTe has, instead, a different and also strategic application in X-ray and γ-ray spectroscopy, being today the material of choice for the fabrication of radiation detectors used for security and medical applications. As the basic physical chemistry associated with their growth is common to a larger class of semiconductors, GaAs, InP and tellurides may be considered representative examples of compound semiconductors grown from the liquid phase.

One of the main problems encountered with the growth of compound semiconductors concerns the existence of an extended homogeneity region around the stoichiometric composition, as already seen in Chapter 1, whose width $\Delta\delta$ and shape not only influences the electronic properties of the material but also the practical growth procedures, as will be seen later in this section.

The width and shape of the homogeneity region are the quantitative signature of the deviations from stoichiometry of the compound semiconductor phase and depend on the formation energy of point defects (vacancies, interstitials, antisites), whose onset may be described with the following defect reactions, taking CdTe as an example

$$Cd_{Cd} \rightleftharpoons V_{Cd} + Cd^v \uparrow \tag{4.56}$$

$$Cd^L \rightleftharpoons Cd_{Cd} + V_{Te} \tag{4.57}$$

$$Te_{Te} \rightleftharpoons Te_i + V_{Te} \tag{4.58}$$

$$Te_{Te} + Cd_{Cd} \rightleftharpoons Te_{Cd} + Cd^v \uparrow \tag{4.59}$$

as already seen in Chapters 1 and 2, where the sign $\uparrow$ indicates a spontaneous sublimation process. One can see that the defect reaction (4.56) generates a Te excess, Eq.(4.57) a Cd excess, while the last two lead to the formation of Te interstitials and Te antisites, respectively.

The width of the homogeneity region of compound semiconductors is relatively narrow (see Figure 4.23 for the case of GaAs and CdTe) and consequently the concentration of point defects is small [165], due to the high values (>4 eV) of the formation energy of point defects [166, 167], reported in Table 4.7 for a number of selected semiconductors [165–167].

However, considering that the electronic structure of point defects is that of deep levels [168, 169], their influence on the electrical properties might be important, as already discussed in Chapter 2.

As the actual stoichiometry of a compound semiconductor, at the end of a growth process

- depends on the parameters of the growth and post-growth process steps (composition of the melt, growth temperature, growth rate, cooling rate after the end of the growth process) and
- determines the final amount of point defects and, thus, the electrical properties of the material

it is interesting to examine how far the details of a specific growth process could influence the final properties of the semiconductor.

Figure 4.24 shows four different routes that might be followed for the growth of an AB compound.

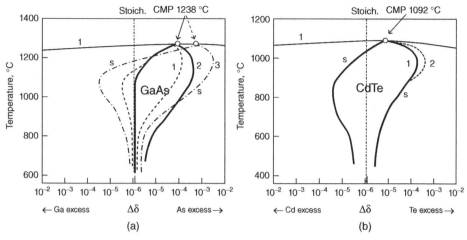

Figure 4.23 *Shape of the homogeneous region of (a) GaAs and (b) CdTe: (s) solidus lines (l) liquidus lines. Rudolph, 2003, [165]. Reproduced with permission from John Wiley & Sons*

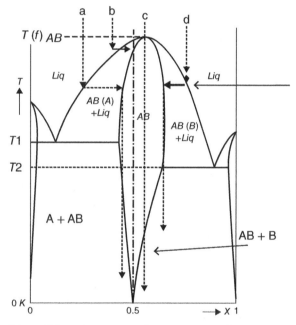

Figure 4.24 *Possible solidification routes for a non stoichiometric A_xB_{1-x} compound. Rudolph, 2003, [165]. Reproduced with permission from John Wiley & Sons*

Starting from an A-rich melt (route a), an A-rich AB phase will be grown, which, during a slow cooling towards room temperature decomposes in a two phase material, consisting of a (heterogeneous) mixture of the quasi-stoichiometric AB phase and precipitates of the α-phase.

Starting the growth with a melt at its equilibrium congruent melting temperature T_{cm} and composition (route c), a B-rich AB phase will be grown, which will decompose during its cooling towards room temperature in a two phase material consisting of an over-stoichiometric AB phase and of precipitates of the B-rich β-phase

Eventually, the growth from a B-rich melt (route d) will result in a B-rich AB phase, which on cooling will again decompose into a two phase material, with precipitates of the B-rich β-phase.

Only following route (b), starting with a melt in equilibrium with the stoichiometric solid, one should remain within the stoichiometric phase after cooling at room temperature. Phase decomposition may occur, in the long term, also in this case, due to melt enrichment by the rejected excess of A.

With the exception of case c, constitutional supercooling may also occur, due to a possible enrichment of the diffusion layer at the growth interface by the rejected excess of A or B.

It is therefore apparent that knowledge of the phase diagram topology, the segregation coefficients of the components and the homogeneity range of the non-stoichiometric semiconductor phase is mandatory for the appropriate design and conduction of a growth process.

The influence of point defects and extended (volume) defects on the properties of the grown crystals will not be considered in the following, unless mandatorily necessary, as the issue has already been considered in Chapters 2 and 3.

4.3.4.1   Growth of GaAs

GaAs crystals are grown from the melt in fused silica crucibles at 1240 °C, with the liquid encapsulated LEC-CZ technique (see Figure 4.3), using B_2O_3 as the encapsulating material, in order to prevent or minimize stoichiometry deviations due to the high vapour pressure of As over the stoichiometric melt.

The equilibrium vapour pressure of As (mostly consisting of As_4 and As_2 species) is 73 ± 4 kPa, [170], slightly below a value of 97.6 kPa quoted in the older literature [171].

As in the case of silicon, the compatibility of fused GaAs with ceramic crucibles plays an important role in the final properties of the grown crystal, because fused silica crucibles are slightly corroded by molten GaAs, with a consequent oxygen and silicon contamination of the melt and thus of the grown crystal.

Pyrolitic boron nitride (PBN) crucibles could be used instead of fused silica, with only modest boron and nitrogen contamination of the melt. B contamination of the melt and of the grown GaAs crystal is instead occurring as a consequence of the chemical interaction of the capping layer with molten GaAs and found to range between $5 \cdot 10^{17}$(1.7 ppmw) to $1.9 \cdot 10^{19}$(64 ppmw).

One of the main problems with the melt-grown GaAs, however, is that the composition of the melt at its maximum freezing point (see Figure 4.23a) is a little over-stoichiometric, leading to an As-excess in the crystal and, consequently, to a high density (about 10^{16} cm^{-3}) of EL2 deep donor defects. These defects have been supposed for a long time to consist of As_{Ga} antisites, albeit the most recent structure proposed consists of a three-center-complex made by the association of an arsenic vacancy (V_{As}), an arsenic antisite As_{Ga} and a gallium antisite (Ga_{As}) [172]. The donor activity of EL2 defects should be compensated by acceptor impurities (typically Cr) to obtain a semi-insulating (SI) material. This high concentration of EL2 defects limits also the charge collection efficiency of GaAs radiation detectors.

High purity, detector grade GaAs with a 10-fold lower concentration of the EL2 defects concentration has been, however, obtained from an ingot which was initially zone refined in a vertical zone melting furnace, using a PBN crucible [173]. This ingot was then zone levelled by passing a molten zone enriched with Ga through the entire ingot. To do this, the ingot was first machined to create an annular cavity in which ultrapure Ga was inserted.

Zone refining was shown to be very effective also in removing Zn impurities, which behave as acceptor impurities when sitting in Ga lattice sites (Zn_{Ga}).

As said above, the interaction of molten GaAs with the fused silica or the PBN crucible and with the B_2O_3 flux causes Si, O, N and B contamination of the liquid and then of the GaAs ingot during the growth process.

Among these impurities, Si is of most concern, as it is responsible for impurity-induced layer disordering of $Al_xGa_{1-x}As/GaAs$ quantum-well laser structures [174] and behaves as donor or acceptor depending on whether it sits on Ga or As sites. It could, therefore, at least partially, compensate the EL2 centres, when it presents acceptor properties [175], making more critical the management of compensating procedures with added acceptor impurities.

Severe oxygen contamination was demonstrated to be responsible for the presence [176] of a midgap level, labelled ELO, assigned to an oxygen sitting in an arsenic position (O_{As}) or to a ($O-V_{As}$) complex, falling virtually at the same level as the EL2 defect (825 ± 5 meV), but with a capture cross section ($\sigma = (4.8 \pm 0.6) \times 10^{-13} cm^2$) four times larger. This conclusion was recently confirmed by exo-electron emission spectroscopy measurements [177] on oxygen centres generated at the surface of GaAs samples. As the oxygen concentration in CZ-and Bridgman-grown GaAs is generally low, the ELO defect concentration is generally below that of the EL2 defects.

Most of the theoretical works dedicated to nitrogen doped-GaAs agree that nitrogen is expected to go to isoelectronic substitutional As sites, as a N_{As} species, although the stability of nitrogen interstitials is not excluded.

It should be noted that nitrogen and arsenic present relevant differences in size (N 75 pm, As 119 pm) and electronegativity (N 3.04, As 2.18). These differences lead to a significant local strain and also to a decrease in the band gap width of GaAs when added in amounts of a few percent [178], as in the case of the GaAsN dilute alloys, where N can occupy an interstitial position.

Concerning the effect of the interaction of a silicon-doped GaAs melt with the B_2O_3 flux, the results of a study on the equilibrium between Si-doped molten GaAs and B_2O_3 [179] may be interpreted by assuming that the following reaction occurs in the GaAs melt

$$3Si_{GaAs} + 2B_2O_3 \rightleftharpoons 3SiO_2 + 4B_{GaAs} \tag{4.60}$$

Thus, the B_2O_3 reaction with dissolved silicon reduces the concentration of Si in the melt with an increase in the B concentration and then in the grown GaAs ingot, as was indirectly demonstrated experimentally. A B content ranging between $2 \cdot 10^{17}$ and $2 \cdot 10^{18}$ at cm^{-3} (0.68–6.8 ppmw) was found for the LEC process with undoped GaAs, while the B content ranged between $5 \cdot 10^{18}$ and $3.8 \cdot 10^{19}$ cm^{-3} (17–128 ppmw) in the case of Si-doped GaAs.

B-doping of GaAs occurs also as the consequence of partial dissolution of the PBN crucible, when used. In this case, it was found that the B-doping depended on the partial pressure of nitrogen in the cover gas atmosphere and, therefore, also on the nitrogen content in the melt. The boron nitride dissolution in molten GaAs occurs, in fact, with the evolution of nitrogen

$$BN \underset{}{\overset{GaAs}{\rightleftharpoons}} B_{Ga} + \frac{1}{2}N_2(g) \tag{4.61}$$

The results of these measurements [179] are displayed in Figure 4.25. It should be noted that the formation of a substitutional B_{Ga} centre in solid GaAs will not influence the electronic properties as B_{Ga} is electroneutral. Due to the difference in the covalent radii of Ga (126 pm) and B (82 pm), the set-up of local strain may, instead, be expected.

4.3.4.2 Growth of InP

The growth of InP in the laboratory and in industry is carried out using either the LEC-CZ growth or the zone melting, Bridgman or vertical gradient freezing techniques [180, 181], that yield crystals with very low densities of dislocations (10^2–10^3 cm^{-2}), are also used although the seeded LEC-CZ-pulling gives the best results in terms of crystal quality.

Due to the high equilibrium pressure of P (27.3 atm) at the melting temperature (1059 °C), the growth of InP from stoichiometric melts is carried out under controlled pressure.

Crucible wetting from the melt, in the case of horizontal Bridgman growth, is known to be the origin of a high density of dislocations in the crystal [182], higher than 10^6 to 10^4 cm^{-2} when a silica boat is used and 10^3 cm^{-2} when a PBN crucible is used [183]. When the growth is carried out in fused silica boats in a vertical Bridgman furnace or in a

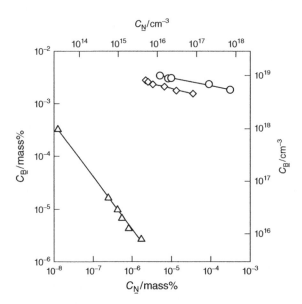

Figure 4.25 *Dependence of the B content in the GaAs melt equilibrated with BN on the nitrogen concentration Ref. [179] ○ and ◇ (this work) Δ calculated values from the literature. Yamada, et al., 2004, [179]. Reproduced with permission from The Japan Institute of Metals and Materials*

LEC-CZ furnace, wetting causes, similarly to the case of Si and GaAs, silica dissolution in the melt, which is associated with the loss of volatile indium monoxide

$$4In_{InP(L)} + SiO_2 \rightarrow Si_{InP(L)} + 2In_2O^V \uparrow \qquad (4.62)$$

to which corresponds a Gibbs free energy of reaction $\Delta G = 163\,920 + 1.460 \times 10^{-3}TInT + 0.19\,10^{-3}T^2 + 7.4510^4/T - 49.44T(\text{cal mol}^{-1})$. The reaction is sensibly shifted to the right, at the melting temperature of InP, if In_2O is allowed to sublime and condense on a cold section of the furnace [184]. As in the case of GaAs, Si-doping causes the formation of donor or acceptor centres, depending on its incorporation as a Si_{In} or a Si_P species.

It can be seen in the phase diagram of Figure 4.26a, that InP (together with InAs and InSb) presents almost negligible deviations from the stoichiometry, compared to GaAs, CdTe and CdSe (see Table 4.8) and that the almost stoichiometric phase can be grown (see Figure 4.26b) [185] over a wide range of In concentrations from low temperature, low pressure In-rich melts. However, In-inclusions [186] and small deviations from the stoichiometry are generally observed when the growth is carried out from In-rich melts, making the growth from stoichiometric melts the preferred solution.

LEC pulling from stoichiometric melts is, therefore, generally carried out in a pressurized furnace, operating in dry nitrogen gas at slightly higher pressures than the equilibrium pressure (about 3800 kPa) [187]. The loss of P from the hot ingot after its extraction from the melt is prevented by the presence of a thin layer of B_2O_3 which wets the ingot. Its freezing on the InP surface at lower temperature causes, however, the onset of local stress, with the generation of surface microcracks which could, however, be eliminated by suitable polishing.

The stoichiometry of InP crystals may be adjusted by a reconditioning process at low temperatures (800 °C), carried out under a P pressure of 810 kPa [188].

As an alternative to fused silica and PBN crucibles, glassy carbon and AlN crucibles could also be used [189, 190], with an almost negligible increase in the residual carrier concentration with respect to that occurring in InP ingots grown in silica crucibles. However, B and Al impurities were found in InP crystals LEC pulled in AlN crucibles, arising from a reaction between the crucible and the boron oxide encapsulant.

4.3.4.3 Growth of ZnSe, CdTe and Cadmium Zinc Telluride (CZT)

While the interest in bulk ZnSe, which is typically grown with a sublimation process [191, 192], has gradually decreased with the advent of nitride semiconductors, cadmium zinc telluride ($Cd_{1-x}Zn_xTe$, with $0.08 < x < 0.1$) is considered the material of choice for X-ray and γ-ray spectroscopy. It is unique in virtue of its large energy gap (1.6 eV) which allows its use at room temperature, different from the case of Ge detectors, which require cooling at liquid nitrogen temperature to limit the thermal noise. It is, therefore, strategic for security airport screening detectors, medical imaging and astronomical (space telescopes) applications. Depending on the energy of the incident radiations, the thickness of the detector should vary from 1 to 15 mm, for security and medical applications [193] and around 10 cm or more for astrophysical applications in the MeV range [194]. Bulk growth processes are therefore needed for its production, which are mostly carried out with modified variants of the vertical Bridgman [19] or with CZ furnaces using graphite or PBN crucibles. MBE and FZ growth has also been used [195].

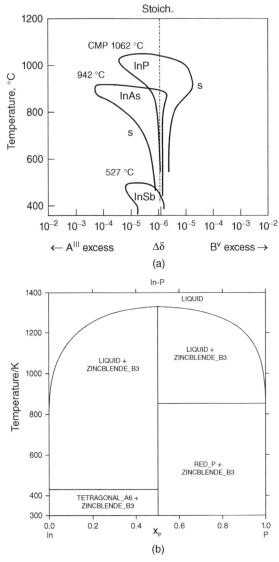

Figure 4.26 *(a) Existence domains of the non-stoichiometric phases of InP, InAs and InSb. Rudolph, 2003, [165]. Reproduced with permission from John Wiley & Sons. (b) Phase diagram of InP. MTDATA - Thermodynamics and Phase Equilibrium Software from the National Physical Laboratory; Davies et al., 2002, [185]. Reproduced with permission from National Physical Laboratory*

The thermodynamic properties of this system and of its precursor CdTe were discussed in Section 1.7.10, where it was shown that the solid solutions of ZnTe in CdTe behave very close to ideality. CZT being a diluted solution of Zn in CdTe, its growth may be treated as a slight variant of that of pure CdTe.

Table 4.8 *Maximum width $\Delta\delta_{max}$ of the homogeneity region for selected compound semiconductors [165], defect concentration and formation energies of Schottky pairs*

Material	GaAs	InP	CdTe	CdSe
$\Delta\delta_{max}$ (mole fraction)	$\sim 2 \cdot 10^{-4}$	$\sim 5 \cdot 10^{-5}$	$\sim 1 \cdot 10^{-4}$	$\sim 5 \cdot 10^{-4}$
Side of $\Delta\delta$ max	As rich	In rich	Te rich	Cd rich
Defect concentration (max)	$V_{Ga} \sim 4.4 \cdot 10^{18}$	$V_{P} \sim 2 \cdot 10^{17}$	$V_{Cd} \sim 1.4 \cdot 10^{18}$	$V_{Se} \sim 9.1 \cdot 10^{18}$
Formation energy of a Schottky pair (eV)	4.82	5.27	4.91	4

Data from [166, 167].

The growth of (semi-insulating) SI CdZnTe crystals, with spatially uniform charge transport properties required for radiation detector applications, poses, however, a number of severe challenges for crystal growers. In addition to the difficulties of growing large-volume single crystals with low dislocations-, precipitates- and inclusions-density, there is the challenge concerning the control of the stoichiometry of the material. The main problem of the CdTe growth from the liquid phase, and thus of $Cd_{1-x}Zn_xTe$ alloys is the high equilibrium vapour pressure of Cd, whose sublimation would induce a Te enrichment of the melt and of the crystal to a maximum value corresponding to its solid solubility. For undoped CdTe the maximum Te solubility is less than 10^{-3} in molar fraction (see Figure 4.23b) and around 10^{-3} in molar fraction for $Cd_{1-x}Zn_xTe$, with $0.01 < x_{Zn} < 0.1$ [196].

To partially suppress the Cd losses due to its sublimation, either B_2O_3 encapsulation may be used [197] with the consequence of a modest B contamination (800 ppma), or the growth is carried out in sealed quartz glass ampoules [85], and the Cd vapour pressure is maintained constant at its equilibrium value by a separate Cd vapours source.

The additional drawback is that the growth of a CdTe crystal from a stoichiometric melt or from a melt at its congruent melting composition yields a material with a slight ($\sim 10^{17}$ at cm^{-3}) Te excess [198] (see Figure 4.23b). To this excess corresponds an equivalent amount of point defects, consisting of tellurium antisites (Te_{Cd}) and secondary phases formation, consisting of tellurium inclusions or precipitates [165, 198]. The effect of point defects and extended defects on the electrical properties of II–VI compounds has already been discussed in Chapters 2 and 3, making reference to the excellent review of Neumark [169].

Here, it is important to note that there is a difference between inclusions and precipitates. Inclusions are droplets of liquid Cd or Te incorporated into the crystals grown from Cd-or Te-rich melts, as a consequence of the retrograde solubility of Cd and Te in the CdTe phase, for growth temperatures higher than the Cd-CdTe and CdTe-Te eutectics (321.06 and 449.57 °C), respectively [199]. Second phase precipitates are, instead, generated from solid state reactions occurring in CdTe phases containing an excess of Cd or Te above the solubility, at temperatures lower than the eutectic ones.

The hypothesis that a fast cooling of the ingot at the end of the growth process below the retrograde range could avoid the segregation of precipitates from a Te-rich melt was tested experimentally [85], with limited success, as only the edge of the ingot was preserved by Te precipitates.

To grow a stoichiometric crystal and to avoid the presence of inclusions and precipitates, a substantial excess of Cd in the melt is needed ($> 10^{-4}$ at%) (see Figure 4.23), to which corresponds a vapour pressure of Cd around 2×10^2 kPa.

This condition may be realized by using a separate Cd vapour source, consisting of a liquid Cd reservoir heated at $850\,^\circ$C, in order to maintain a constant Cd vapour pressure corresponding to the ternary liquid/solid/vapour equilibrium at the growth temperature [198, 200] and a thick silica glass ampoule capable of resisting the overpressure.

As the main contaminant of CZT grown using B_2O_3 as encapsulant is B, the influence of B-doping on the properties of CZT have been studied by Pavesi *et al.* [201], who demonstrated that B, unlike in the case of CdTe, where it behaves as an A-donor center ($B_{Cd}V_{Cd}$), is optically and electrically inactive in CZT.

4.4   Single Crystal Growth from the Vapour Phase

4.4.1   Generalities

Two main targets are at the forefront of these processes. The first is to grow large semiconductor boules to be used, after sectioning and polishing, as mechanical substrates on which the further growth of the active layers of microelectronic or optoelectronic devices could be carried out.

The second, is to deposit high electronic quality, dislocation-free, thin homo- or hetero-epitaxial layers on suitable single crystalline substrates which behave as templates and are selected with regard to their structural parameters, which should be compatible with those of the epilayer.

The rationale behind these last processes is that the thickness of the active layer of most microelectronic or optoelectronic devices, including radiation detectors and solar cells manufactured with direct gap semiconductors, is often in the micrometre or less range. Furthermore, the vapour phase operation conditions are often mild enough to grow layers of high chemical and structural quality.

This procedure allows, as an example, the use of single crystalline silicon wafers with oxygen precipitates and other extended defects as templates on which to grow high quality, defect-free homo-epilayers of silicon for IC applications.

These processes may be carried out with a variety of techniques, ranging from PVT to the CVD and MBE in almost all the configurations available today.

Only in a few cases have vapour phase processes and, very rarely, the CVD techniques, been used for the growth of single crystal ingots. This is mostly because of their very low growth rates, that are of the order of a few μm h^{-1}.

This is the case for SiC and GaN, for which the sublimation method and the HVPE [202], respectively, are used to grow SiC and GaN boules, and that of silicon, for which a few laboratory attempts to grow single crystal Si ingots have been successful in the past [203, 204].

While the homoepitaxial deposition processes, such as those used for silicon, SiC, GaAs and GaN, yield excellent results in terms of the structural, electrical and optical properties of the deposited layer, the heteroepitaxial deposition is inevitably faced with lattice mismatch and physical inhomogeneity problems (e.g. different thermal conductivity and thermal expansion coefficients). The quantitative impact of these issues depends on the nature of the substrate and leads, unless suitably managed, to poor epilayer quality. Only at the beginning of the deposition does a pseudomorphic-type of layer grow on the underlying substrate, characterized by the presence of a coherent interface.

A large elastic strain will accumulate in this layer, depending on the lattice misfit, as long the layer grows, until a critical thickness is reached, at which the elastic strain energy is totally or partially relaxed with the emission of misfit dislocations and plastic deformation. This issue will be discussed in Section 4.4.3 with reference to Si-Ge alloys, where it will be shown that strain in Ge and Si also induces remarkable, and beneficial, changes in their electronic structure. Strain relief with emission of misfit dislocations is favoured by high temperatures, but strain relieved layers are of limited or no interest as active sections of microelectronic or optoelectronic devices, as misfit and threading dislocations are powerful recombination centres for minority carriers, as shown in Chapter 3.

While the next section addresses issues dealing with the growth from the vapour phase of bulk Si, ZnSe, SiC and GaN, the following section will deal with the structural and physical chemistry aspects of homoepitaxial and heteroepitaxial processes.

4.4.2 Growth of Silicon, ZnSe and Silicon Carbide from the Vapour Phase

4.4.2.1 CVD Growth of Bulk Silicon

Single crystal silicon bars, 40 cm long and about 1 in in diameter were grown in a laboratory Siemens type of reactor using conventional $SiHCl_3$ (TCS) – H_2 mixtures and cylindrical single crystal seeds [203]. The TCS/H_2 molar ratio was kept in the range between 0.05 and 0.016 and the deposition temperature between 1200 and 1280 °C. With a temperature kept at 1250 °C and a TCS – H_2 flow rate of 600 l h^{-1}, the deposition rate reached its maximum value of 5.5 μm min^{-1}. Apparently, at $T > 1200$ °C the rate controlling step is the mass transport in the gas phase, while at lower temperatures the surface reactions are rate determining.

It was demonstrated that after an induction period whose duration depends on the temperature, the rod habit changes from cylindrical to hexagonal, as the growth rate along the <110> direction is higher than along the other directions, in good agreement with the relative order of rates given in the literature [205]. The attempt to grow silicon crystals using laser induced chemical vapour deposition and silane (133 mbar, without carrier gas) and single crystal wafers as substrates [204] succeeded only in growing, at temperatures in excess of 1500 K, single crystal rods with a hexagonal habit of a maximum diameter of only 200 μm.

Neither process found industrial applications.

4.4.2.2 Growth of Zinc Selenide (ZnSe)

Single crystal boules of ZnSe are typically grown using the PVT process without a seed crystal in a vertical furnace [191, 192]. The starting material, consisting of polycrystalline ZnSe lumps, is charged in a clean quartz ampoule with a top capillary (tip) which serves to initiate the crystallization. The quartz ampoule is then sealed under high vacuum, the charge is sintered and then the growth is carried out by bringing the system to the sublimation temperature. The sublimation reaction is dissociative

$$ZnSe_S \rightleftharpoons Zn_v + \frac{1}{2}Se_{2(v)} \qquad (4.63)$$

with a temperature-dependent equilibrium constant.

$$K_p = p_{Zn} p_{Se_2}^{1/2} \tag{4.64}$$

The difference between the values of the equilibrium constant with reference to the sublimation source and at the tip is the driving force of the process. In general, high quality crystals can be obtained at relatively high growth rates (0.41 mm h^{-1}) [192].

4.4.2.3   Growth of SiC Ingots

Vapour phase growth is the current process for the production of single crystal boules of SiC, which are commercially available since CREE, about 20 years ago, succeeded in bringing to industrial production a modified Lely process, which still takes the name of the inventor. Single crystal ingots of silicon carbide are not only important for the use of SiC itself in power electronic and high frequency applications, but also for providing the substrates for GaN heteroepitaxy, on which will be reported in a later section.

The Lely process and its subsequent modifications are based on the sublimation of SiC vapours in an Ar atmosphere, at a temperature ranging between 2200 and 2450 °C, from a source consisting of SiC powder or lumps and their deposition on a single crystal seed of SiC. The powder or the lumps charge is kept inside a graphite crucible. A vertical temperature gradient of about 20–40 °C cm^{-1} is applied over the length of the crucible [138, 206] to drive the mass flux.

The growth furnace consists of an outer double -walled or single-walled quartz tube [207] and a rigid graphite insulation frame which embeds the graphite susceptor and a graphite crucible which is the reaction chamber. On top of the reaction chamber is located the single crystal seed, consisting of a wafer of a Lely grown SiC polytype (3C, 4H or 6H). The furnace is induction heated.

The process starts with a number of degassing and backing stages of the growth zone. When the growth temperature is reached (2300–2450 °C in the case of 4H- and 6H-SiC), the Ar gas pressure is reduced to 5–30 mbar to favour the sublimation.

The first stage of the growth process is the dissociative sublimation of the SiC source, followed by gas phase transport of the vapour species, which eventually crystallize onto the seed. The vapour phase transport is driven by a temperature gradient, which also induces supersaturation and controlled growth at the seed surface.

On the basis of the experimental values of the partial pressures of the species present in the vapour phase equilibrium at the deposition temperature, see Table 4.9, [139], one can assume that the deposition reaction is ruled by the SiC$_2$ and SiC species and that the multiple equilibria involved can be described by the following reactions

$$SiC_2^g + Si_2C^g \rightarrow 3SiC^s \tag{4.65}$$

$$SiC_2^g + 3Si^g \rightleftharpoons 2\,Si_2C^g \tag{4.66}$$

$$Si_2C^g \rightleftharpoons C^s + 2Si^g \tag{4.67}$$

$$Si^g \rightarrow Si^L \tag{4.68}$$

noting, however, that reaction (4.67) is also a source of Si$_2$C molecules, as Si vapour can react with the graphite walls of the crucible, shifting the reaction to the left.

Table 4.9 *Partial pressure (in atm) of different equilibrium species over SiC*

T (K)	Si	SiC	SiC$_2$	Si$_2$	Si$_2$C	Si$_2$C$_2$	Si$_3$
2149	$2.1 \cdot 10^{-5}$	—	$1.9 \cdot 10^{-6}$	$3.8 \cdot 10^{-8}$	$1.4 \cdot 10^{-6}$	—	—
2168	$2.7 \cdot 10^{-5}$	—	$2.5 \cdot 10^{-6}$	$4.8 \cdot 10^{-8}$	$1.9 \cdot 10^{-6}$	—	—
2181	$3.3 \cdot 10^{-5}$	$2.2 \cdot 10^{-9}$	$4.2 \cdot 10^{-6}$	$6.7 \cdot 10^{-8}$	$2.6 \cdot 10^{-6}$	—	—
2196	$4.1 \cdot 10^{-5}$	—	$4.4 \cdot 10^{-6}$	$1.1 \cdot 10^{-7}$	$3.9 \cdot 10^{-6}$	$8.5 \cdot 10^{-9}$	—
2230	$6.5 \cdot 10^{-5}$	—	$6.5 \cdot 10^{-6}$	$1.6 \cdot 10^{-7}$	$5.1 \cdot 10^{-6}$	$1.6 \cdot 10^{-8}$	$3.2 \cdot 10^{-8}$
2247	$8.3 \cdot 10^{-5}$	$6.3 \cdot 10^{-9}$	$1.1 \cdot 10^{-5}$	$2.1 \cdot 10^{-7}$	$8.1 \cdot 10^{-6}$	—	—
2316	$2.0 \cdot 10^{-4}$	$9 \cdot 10^{-8}$	$3.1 \cdot 10^{-5}$	$7 \cdot 10^{-7}$	$2.2 \cdot 10^{-5}$	$7.5 \cdot 10^{-8}$	$1.6 \cdot 10^{-8}$

Data from Sevast'yanov et al., [139].

Figure 4.27 *A 6H-SiC boule grown with the seeded sublimation process. Yakimova et al., 1999, [207]. Reproduced with permission from Elsevier*

The major drawback of this process is the large number of structural defects (dislocations, stacking faults, inclusions, micropipes) present in the ingot and the polytype stability during the growth. Actually [142, 207], these problems are inter-related to a large extent. In fact, due to the close values of the heat of formation and the structural parameters of the 4H- and 6H-polytypes (see Table 1.9) defect formation occurs if local change of the stacking sequence occurs and consequently foreign polytype inclusions or stacking faults are formed.

Today, 3 and 4 in diameter 4H-SiC ingots and wafers are commercially available (see Figure 4.27), almost free of micropipe defects (see Section 3.3.1) and with dislocations densities around 10^4 cm^{-2}.

Doping of SiC can be done directly during the PVT growth or by ion implantation of the SiC wafers. n-type doped SiC, which is mainly used as the substrate for gallium nitride-based blue LED, is PVT grown with the addition of nitrogen, which is the doping impurity, to the inert argon atmosphere. N is assumed in the literature to occupy either a Si or C site, but it has been demonstrated recently [208] that the effect of substitution of C or Si with N is not comparable. In fact, in the case of formation of a substitutional nitrogen in a C site (N_C) no new states are introduced in the gap, while in the case of the formation of a N_{Si} species, new states associated with the N defect appear.

A significant side-effect of n-type doping is the reduction in density of basal plane dislocations, with a tremendous impact on the SiC-based devices [4]. The nitrogen doping itself is a relatively simple process, as it involves the sole control of the N_2/Ar ratio. In turn, the net donor concentration increases from $7 \cdot 10^{18}$ to $1.7 \cdot 10^{19}$ with an almost linear dependence on increasing nitrogen concentration in the gas in the 3–10% range [209]. Extremely high N-doping, however, induces polytype transformation. As an example [210], high resolution transmission electron microscopy (HRTEM) studies show that when extremely high nitrogen doping concentrations are applied on a 6H polytype, leading to a free carrier concentration around 10^{19} at cm^{-3}, evidence of 4H- and 15R-polytypes could be observed.

p-type, Al-doped SiC, used for the production of SiC-based high frequency and high power devices, is instead PVT grown, starting from a mixture of Al and SiC powder or by injecting vapours of an organometallic compound of Al (e.g. trimethyl-aluminium) into the growth chamber. The main problem associated with the use of Al/SiC mixtures as the source of both SiC and Al vapours is that the vapour pressure of Al is much larger than that of SiC (100 kPa at 2517 °C), with a consequent depletion of Al in the source and a doping gradient in the ingot. A solution which is adopted is to provide the furnace with an Al reservoir, separately heated, where a flux of argon works as a carrier gas for the Al vapours [211]. Using this variant it was possible to grow Al-doped SiC with an axial charge carrier profile of $2.5 \cdot 10^{15}$ mm^{-1}cm^{-1}, in comparison to a value of $7.5 \cdot 10^{15}$ mm^{-1}, in a crystal grown without the external Al supply. It should be noted that this solution is equivalent to that used in the case of CdTe and CZT to control the semiconductor stoichiometry, as shown in Section 4.3.4.3.

4.4.2.4   Growth of Bulk GaN by the Hydride Vapour Phase Epitaxy (HVPE) Process

The main advantage of this process [212] in comparison with the ammonothermic and the high pressure–high temperature ones, already discussed in the previous sections, is its high growth rate (> 200–300 μm h^{-1}) and the softer operational temperature/pressure conditions (see Table 4.10). Different from the other processes, it is basically a heteroepitaxial deposition process on sapphire substrates from which a freestanding GaN wafer can be separated by removing the sapphire substrate by mechanical means [213, 214].

Typically, the HVPE process is carried out at 1030 °C, using a 350 μm thick sapphire substrate in a horizontal quartz reactor operating at atmospheric pressure. The Ga precursor

Table 4.10 *Comparison between the process parameters of the GaN bulk crystals growth*

Process	Temperature range (°C)	Pressure range	Growth rate	N_D (cm^{-2})
HVPE	>1000	100 kPa	> 200–300 μm h^{-1}	n.d.
Ammonothermic	500–700	190–210 MPa	50 μm d^{-1}	10^6–10^7
High pressure	Max 1550	1000–2000 MPa	—	10^3–10^6
High pressure (Ga-Na)	700–870	3.2 MPa	38–46 μm h^{-1}	n.d.

The quality of the crystals, when available from literature data, is reported in terms of dislocation density N_D (cm^{-2}).

species is GaCl, which is synthesized in the cooler part of the reactor at 850 °C by reacting Ga with HCl or is prepared using metal organic precursors [214]. NH_3 is used as the source gas to allow the GaN formation reaction to occur

$$GaCl + NH_3 \rightarrow GaN + HCl + H_2 \qquad (4.69)$$

and nitrogen is used as the carrier gas [215]. In order to grow an almost dislocation-free deposit, the process temperature should be slightly higher than 1000 °C.

Once the epilayer reaches a thickness of 100 μm, the substrate is cooled to 200 °C and then brought again to the growth temperature and a 350 μm thick GaN layer is homo-epitaxially grown on the previously deposited layer, which behaves as a buffer layer. The sapphire substrate is then removed by diamond abrasive cloths and a crack-free, free-standing GaN substrate is obtained. It should be noted that during the cooling–heating cycle to which the sample is first subjected, the sapphire substrate cracks, not the epilayer. In the heteroepitaxial deposition of thick GaN layers on sapphire the presence of cracks is common, with severe consequences for the quality of the GaN layer.

More recently, Soitec used the smart cut technique, originally developed for the preparation of freestanding silicon epitaxial layers from a silicon on insulator (SOI) template, to remove the epitaxial layer from a sapphire reusable template, leading this technology towards industrial application. Nevertheless, the hybrid HVPE/ammonothermal process, Section 4.3.3.2, seems to be the best solution available today as it avoids the mechanical detachment step of the heterogeneous template and produces a freestanding substrate with a very low dislocation density.

4.4.3 Epitaxial Growth of Single Crystalline Layers of Elemental and Compound Semiconductors

4.4.3.1 Basic Issues in Epitaxial Growth

The spectacular development of Si-Ge, GaAs, InP, GaN-based micro- and opto-electronic devices, on which depend modern communication and computation systems, and the onset of solid state lighting for general illumination, was possible thanks to the growth of epitaxial thin semiconductor layers on suitable substrates.

The challenge was, and remains, compliance with the thermodynamics, lattice mismatch and strain with composition, thickness and defectivity of the active layers.

For the majority of the applications CVD, MOCVD and MBE were found to be the most appropriate solutions, although MBE is not yet suitable for mass production.

While homoepitaxial deposition processes run generally with limited problems, only a few single crystal substrates, with the appropriate mechanical, structural, thermal and chemical compatibility with the epitaxial layer and the growth atmosphere at the growth temperature, are available for heteroepitaxial depositions, even before considering cost issues.

As an example, silicon carbide, sapphire, GaN and silicon are the best choice, at least today, for Group III nitrides. Silicon is the more appealing candidate with respect to cost issues, and GaN on silicon products are already available on the market with a diameter of 6″ and more [216]. Although the efficiency of GaN-LEDs grown on sapphire has been, at least in the recent past, higher than those grown on silicon, the gap is closing very fast and

Table 4.11 *Structural and thermal properties of substrates used for the heteroepitaxial deposition of Si-Ge and nitride alloys*

Substrate	Epilayer	Melting temperature substrate (°C)	Lattice constant substrate (Å)	Lattice constant epilayer (Å)	Lattice mismatch (%)
Sapphire	InN	2040	a 4.785 c 12.991	a 3.545 c 5.7034	25.9
Sapphire	InGaN	2040	a 4.785 c 12.991	a 3.207 c 5.2516	7.6
Sapphire	GaAs	2040	a 4.785 c 12.991	5.6531	15.3
Sapphire	GaN	2040	a 4.785 c 12.991	a 3.189 c 5.185	24.4
α-SiC	GaN	>2650	a 3.08 c 15.12	a 3.189 c 5.185	3.4
GaN	InGaN	2500	a 3.1908 c 5.185	a 3.207 c 5.2516	1.2
AlN	InAlN	—	a 3.112 c 4.982	a 3.196 c 5.138 ($x = 0.205$)	3.1
Si	InN	1410	5.4307	a 3.536 c 5.709	4.7
Si	GaN	1410	5.4307	a 3.189 c 5.185	4.7
Si	$Si_{0.15} - Ge_{0.85}$	1410	5.4307	5.595	2.9

today their efficiency is only 10% less [217] and is expected to become negligible in the near future.

Table 4.11 reports the thermal and structural properties of the substrates used in the heteroepitaxial deposition of Si-Ge and Group III nitrides and nitride alloys [218].

The main problem with heteroepitaxial deposition is the lattice misfit between the substrate and the epilayer, defined by the ratio $\frac{a_{epi} - a_{sub}}{a_{sub}}$, where a_{epi} is the lattice constant of the epilayer and a_{sub} is the lattice constant of the substrate, calculated for the lattice planes allowing the best epitaxial conditions which, in the case of sapphire and GaN or AlN, are the (0001) planes.

Lattice misfit inevitably induces mechanical strain in the epilayer and has a crucial impact on its properties, independently of the kind of deposition process used.

In fact, a critical layer thickness h_c exists, above which the misfit strain relaxes with the generation of misfit dislocations, threading dislocations and plastic deformation. Lattice mismatch depends on the deposition temperature and varies during the after-deposition temperature ramps, inducing second-order effects on the resulting final strain.

It is a common experience [219] that for misfits larger than 4% the dislocation density in fully relaxed epilayers would be in the 10^9 cm^{-2} range. From inspection of Table 4.10 it can be seen that only a few substrate/epilayer combinations could stay below this figure.

On the other hand, strain could also induce important (and beneficial) changes in the electronic properties of the material. Taking, as a non-exclusive example, the case of Si and Ge, it is known that the electronic mobilities of strained Si and Ge are enhanced. It also known that the crossover from indirect to direct band gap may be obtained by the application of a 1.9% biaxial strain [220] in the case of Ge, due to the small difference in energy (0.18 eV) between the indirect and direct band gaps.

For this reason, strained layers of Ge are used as the active components of Si-Ge/Ge/multi-quantum well devices.

In the case of the heteroepitaxial deposition of an alloy, the critical layer thickness h_c is given by [221]

$$h_c = \frac{b}{8\pi f_m (1 + v)} \left(1 + \ln \left(\frac{2h_c}{q} \right) \right)$$
(4.70)

where b is the active component of Burger's vector, v is the Poisson coefficient of the strained deposited layer, q is the radius of the dislocation core and

$$f_m = \frac{a(x) - a_{sub}}{a_{sub}}$$
(4.71)

is a misfit parameter, where $a(x)$ is the lattice constant of the deposited alloy, which depends on its composition and a_{sub} is the lattice constant of the substrate.

It turns out from Eq. (4.70) that the critical thickness decreases with increase in the lattice misfit, favouring solutions which imply the use of the most compatible substrate.

Unless the properties of the strained layer before relaxation are interesting for device applications, as is the case of strained Ge and Ge-Si alloys, there is a general need to grow dislocation-free layers, thicker than the critical thickness. The solution of this problem, as we will see in what follows, is the use of intermediate layers, on which the misfit strain relaxes.

Lattice mismatch is important, but the surface compatibility is crucial. As an example, the lattice mismatch between silicon and InN is only 6% but the quality of InN deposited on Si is worse than that deposited on sapphire, because the InN film becomes polycrystalline when deposited on Si, as a result of the formation of an amorphous SiN_x layer on the Si substrate [222].

4.4.3.2   *Epitaxial Growth from the Vapour Phase of Si-Ge Alloys*

It has been seen already that the seeded growth of bulk $Si_{1-x}Ge_x$ alloys from the liquid phase suffers from thermodynamic (i.e. constitutional supercooling) and structural constraints, these last associated with lattice misfit problems with the silicon seed. The full relaxation of misfit strain is associated with the formation of misfit (threading) dislocations, whose density, in the case of the deposition of pure germanium on silicon, is of the order of 10^8–10^9 cm^{-2} [219]. Even if electronic devices can be engineered on substrates having comparable dislocation densities, these high dislocation densities limit their reliability, as dislocation interaction would occur under operational stresses with the nucleation of additional dislocations.

To deposit pure Ge or $Si_{1-x}Ge_x$ alloys on Si without excess dislocation problems, the first solution found was the creation of thin graded junctions, consisting of a sequence of epilayers with increasing concentration of Ge, grown by MBE or CVD at relatively high temperatures (900 °C), where the dislocations are not yet mobile [219]. Typically, for the deposition of a Si_xGe_{1-x} alloy with $0.1 < x < 0.53$ and a compositional gradient of 10% Ge/μm starting with a layer of $Ge_{0.1}Si_{0.9}$, one gets a series of fully relaxed intermediate layers and a dislocation free top layer which is fully strained if its thickness is lower than the critical thickness. A cap layer of the same composition deposited on top of the graded junction exhibits dislocation densities in the range 10^{-5} to 10^{-6} for x values ranging between 0.23 and 0.50, in comparison with a dislocation density of $\approx 2 \times 10^8$ for a 4 μm thick layer of $Ge_{0.32}Si_{0.64}$ abruptly grown on a silicon substrate [219, 223]. These graded buffer layers serve as virtual substrates for a number of applications. As an example, on these buffer layers multiwell structures can be deposited, consisting of a sequence of $Ge_{0.85}Si_{0.15}//Ge//Ge_{0.85}Si_{0.15}$ multiple layers. They suffer, however, from being grown several micrometres thick to optimize the surface layer quality and limit the threading dislocation density. Also, a considerable amount of precursor is consumed.

The availability of thinner virtual substrates would clearly be an advantage and among several attempts carried out with this aim, the process developed by von Känel *et al.* [224] seems to be particularly promising. It consists in depositing on a silicon substrate, after an HF dip to remove the native oxide layer, a 100–150 nm thick silicon buffer layer at 750 °C, using the low energy plasma enhanced chemical vapour deposition (LEPECVD), which will be dscussed in detail in Section 4.5.2 when discussing the issues concerning the growth of nanocrystalline silicon (nc-Si).

Different from the case of thermal CVD, where the deposition of $Si_{1-x}Ge_x$ graded buffers with Ge content $x > 0.5$ requires an intermediate chemical mechanical polishing (CMP) step to limit the surface roughening, such a step is not necessary when using the LEPECVD process [225], because the $Ge_{0.85}Si_{0.15}$ layers work as barriers and the strained Ge layer is the optically active layer.

More recently, an atypical seeded growth was developed [226, 227], which offers a reliable and effective control over defects, leading, for example, to a 2 orders of magnitude higher photoluminescence from germanium layers grown on silicon substrates.

In the example given in Figure 4.28, the seeds are silicon micropillars deeply micromachined into a (001) silicon substrate, 8 μm in height, 2 μm in base width, with inter-pillar spacing around 3 μm.

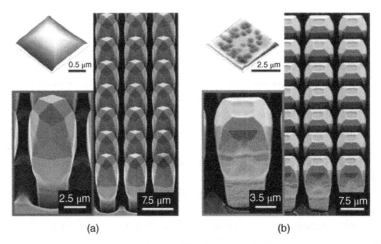

(a)	(b)

Figure 4.28   *Electron microscope micrographs of Ge-towers grown on (a) 2×2 μm^2 and (b) 5×5 μm^2 Si(001) pillars. The insets show atomic force images of the top surface of the Ge towers after selective etching. Pezzoli, et al., 2014, [227]. Reproduced with permission from the American Physical Society*

The germanium deposition was carried out using a low energy plasma CVD process from a silane source, at very high growth rate (4 nm s^{-1}) at 560 °C. The process is highly non-conformal and germanium grows in the form of towers, as the lateral expansion is self-confined and the growth occurs vertically. The horizontal square section of the towers is shown to depend only on the spacing between the silicon pillars. High resolution X-ray diffraction measurements showed that the crystal structure of the Ge towers is perfect, completely unstrained and dislocation- and crack-free. Crack formation is, apparently, prevented by the patterning features which inhibit the formation of a continuous layer and strain relief is completely elastic in view of the absence of dislocations. Misfit dislocations are confined to the bottom region of the towers and 60° dislocations originating in the strain relief process glide out of the crystal, favoured by the faceted structure of the towers.

Although having been successfully tested only for Ge and Ge-Si alloys, this process could provide the solution for the manufacturing of compound semiconductor devices monolytically integrated in silicon substrates, where thick active layers are required, as in the case of X-ray or radiation detectors.

4.4.3.3 Epitaxial Growth of Group III Nitride Semiconductors

Epitaxial growth of nitride semiconductors may be taken as a typical, and therefore, illustrative example of the problems encountered with the heteroepitaxial deposition of compound semiconductors.

As already shown in Chapter 1, nitride semiconductors are the materials of choice for the fabrication of LEDs and lasers emitting in a wide range of wavelengths (see Figure 4.29) high electron mobility transistors (HEMTs) and radiation sensors. Depending on the targeted wavelength, InGaN alloys are ideal for visible light emission, while InAlN alloys are ideal for HEMT devices [228], for UV emission and for white light emission using phosphors excitation.

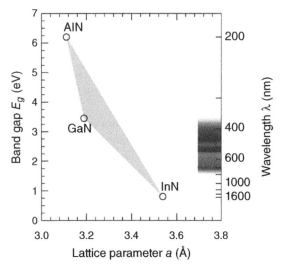

Figure 4.29 *Band gap energy versus lattice parameters for nitride alloys. Courtesy of D. Holec, Montanuniversität Leoben, Austria*

For years, the development of this sector was, however, inhibited by the intrinsic difficulties present in the fabrication of heterostructures based on GaN and InAlN, GaInN and GaAlN [229]. Difficulties which, in part, still remain [228, 230–232], in spite of the recent, spectacular development of the sector.

As GaN bulk or free-standing GaN substrates, which would permit the homoepitaxial deposition of nitrides, were not available until only recently, the problem was to find a solution to mismatch failures occurring (see Table 4.11) using the available substrates (sapphire, α-SiC and, more recently, silicon),[11] which would cause the presence in the active layer of both misfit and threading dislocations [233].

The mismatch is particularly important in the typical case of nitrides grown on the basal surface {0001} of hcp sapphire, with a growth direction normal to the {0001} basal plane [234–238]. Here, the epitaxial relationships satisfy the condition that the anion stacking (O^{2-} in the substrate) continues with the anions stacking (N^{3-}) in the epitaxial layer, while the cations switch from octahedral (Al^{3+} in Al_2O_3) to tetrahedral sites (Ga^{3+} in GaN).

Considering that in GaN the lattice constant a_{GaN} in the basal plane corresponds to the N–N distance, while in sapphire the O–O distance is $a_{Al2O3} \cos 30° = a' = 4.1432$ Å, the calculated lattice misfit is

$$f = \frac{a_{GaN} - a'}{a'} = 23.03\% \tag{4.72}$$

or

$$f = \frac{\sqrt{3}\, a_{GaN} - a_{sapph}}{a_{sapph}} = 16.09\% \tag{4.73}$$

if one accounts for a 30° rotation needed to accommodate the lattices superposition [230].

Silicon is considered now to be potentially the best substrate, due to its availability in large diameter size, with high quality, low cost and reasonably low lattice mismatch. The main problem with the growth of GaN on silicon is the large thermal expansion of Si (116% more than GaN), which inevitably causes cracks on small diameter wafers. When the deposition is carried out on large diameter silicon wafers (150 mm and more), the thickness of the wafer is sufficiently large to allow the nitride growth under mechanically induced compressive stress. Consequently, when brought to room temperature, the film is almost unstrained [239].

The main problem was to develop a process, suitable to be brought to the industrial scale, for the deposition of thin layers of n-and p-doped layers of GaN or of a nitride alloy, in order to realize np junctions or quantum well structures.

The first, viable solution was found by Nakamura *et al.* [159–164]. They used a GaN buffer layer deposited on a sapphire substrate with the MOCVD technique carried out at atmospheric pressure, on which InGaN/$Al_{0.15}Ga_{0.85}N$ heterojunctions and InGaN quantum wells were eventually grown. The role of buffer layers is to provide an intermediate substrate on which to grow layers of active material of the best crystallographic quality possible. For this reason, GaN is chosen for the deposition of InGaN alloys and InN or AlN for InAlN alloys (see Table 4.10) [218, 228, 234, 240].

MBE and ion beam assisted molecular beam epitaxy (IBA-MBE) [222] are also well suited for the deposition of GaN and nitride alloys, but despite past and ongoing research,

---

[11] These present the best structural relationships.

Table 4.12   *Thermodynamic properties of Group III nitrides*

	Density (g cm^{-3})	Melting temperature (K)	Decomposition temperature (°C) ($p < 10^{-6}$ mbar)	Activation energy for thermal decomposition (kJ mol^{-1})
AlN	3.29	3487	1040	414
GaN	6.07	2791	850	379
InN	6.81	2146	630	336

Data from Ambacher, 1998, [234].

a complete understanding of the physico-chemical problems, common to all processes involved in the Group III nitride vapour phase deposition, is, however, not yet available.

The main problem with Group III nitride deposition from the vapour phase, independently of the technique used (MOCVD, MBE or IBA-MBE), is the high N_2 decomposition equilibrium pressure of InN and GaN [241, 242]

$$2MN \rightleftharpoons 2M + N_2 \tag{4.74}$$

which limits the deposition temperature, for processes carried out at atmospheric pressure, to 550 °C for InN and 850 °C for GaN (see Table 4.12). Thanks to its high decomposition temperature, AlN can be directly prepared by reactive evaporation of Al in ammonia (NH_3) at 1400 °C [241].

Nitrogen losses from the nitride epilayer also affect the electronic properties of the semiconducting material, as nitrogen vacancies are donor defects and will influence the type and amount of conductivity, with serious impact on the device behaviour.

Concerning, specifically, MBE and IBA-MBE, which are processes carried out under vacuum at a pressure lower than 10^{-6} mbar, the main problem is the high thermodynamic stability of the nitrogen molecule ($\Delta G_{dec} = 942.661$ kJ mol^{-1}), which would result in low nitridation rates when used as the reactant. The nitrogen desorption during the growth process is a further problem, as it limits the growth temperature of GaN and InGaN alloys below the critical ones and, consequently, the structural quality of the material.

To reduce the impact of the limited nitrogen reactivity, nitridation in MBE or IBA-MBE processes is, therefore, carried out using radiofrequency excited atomic nitrogen and nitrogen radicals, or low energy ($E_{max} = 25$ eV) nitrogen ions [243], whose use allows to reach a growth rate of the order of 2 nm min^{-1}.

The indirect result of a study concerning the nitrogen effusion rates from nitride films prepared by MOCVD, measured under vacuum at $< 10^{-6}$ mbar, shows that stable nitride film growth occurs at deposition temperatures of about 700, 800 and 1400 °C for InN, GaN and AlN, respectively [242] in the presence of a source of nitrogen radical fluxes in the range $10^{15} - 10^{16}$ cm^{-2}s^{-1}, easily available by carrying out the deposition using IBA-MBE.

Concerning MOCVD processes, which are, currently, the principal methods used to grow thin films of Group III nitride semiconductors, the main problem is associated with a lack of definitive knowledge of the complex chemistry of the ammonia pyrolysis and the trimethyl-indium, -gallium and -aluminium used as In, Ga and Al precursors.

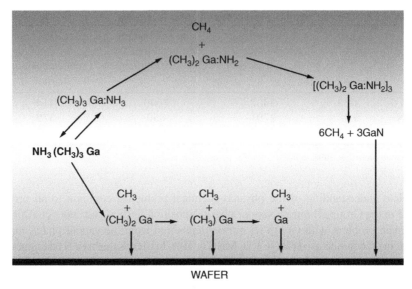

WAFER

Figure 4.30 *Trimethylgallium decomposition pathway (lower route) and adduct pathway (upper route). Parikh and Adomaitis, 2006, [244]. Reproduced with permission from Elsevier*

The overall process for the case of GaN production (see Figure 4.30), easily extended to InN and InGaN alloys, is, formally, the coupling of

1. the reaction of decomposition of trimethyl gallium to Ga on a surface site (bottom route in Figure 4.30) [244]

$$(CH_3)_3Ga \rightarrow Ga(s) + 3(CH_3^\bullet) \tag{4.75}$$

where $(CH_3^\bullet)$ is a radical species, followed by the formation of H_2, C_2H_6 and CH_4

2. the ammonolysis process, with the formation of an active N^* species

$$2NH_3 \rightleftharpoons 2N^* + 3H_2 \tag{4.76}$$

available for reacting with Ga released by reaction (4.75)

3. the direct, surface reaction of Ga(s) with the active nitrogen species

$$Ga(s) + N^* \rightarrow GaN \tag{4.77}$$

and

4. the adduct route (upper route in Figure 4.30), with the initial formation of a Lewis acid–Lewis base adduct

$$(CH_3)_3Ga + NH_3 \rightleftharpoons (CH_3)_3Ga : NH_3 \tag{4.78}$$

and the final formation of the $[(CH3)_2Ga : NH_2]$ adduct, which decomposes to methane and GaN

$$3[(CH3)_2Ga : NH_2] \rightarrow 6CH_4 + GaN(s) \tag{4.79}$$

The trimethyl gallium decomposition pathway (lower route in Figure 4.30) is well understood [244, 245], as well as that of the reactions involving trimethyl radicals and hydrogen [246].

The upper route has also been studied in detail [244] and the process parameters determined.

The studies of NH_3 pyrolysis show that the reaction pathway consists of 21 reactions with nine intermediate species, of which the most important are NH_2 (hydrazine), NH and N [247]. These studies show also that even at high temperatures and at pressures close to atmospheric, the ammonia conversion is very low (< 20% at 1400 K), and minimal at pressures employed with MBE. However, considering the large excess of ammonia used in MOCVD processes, the ammonia pyrolysis is not considered a critical pathway in the GaN process, as the required amount of active nitrogen, not necessarily consisting of atomic nitrogen, should always be available for the completion of the surface reaction (4.78).

Very little is known, however, about the actual pathways followed in the surface reaction, which involve chemisorption/desorption of active species from the gas phase, surface mobility of reaction partners and decomposition/recombination reactions. An example of a similar situation occurring in the course of the nc-Si plasma enhanced chemical vapour deposition (PECVD) is described in Section 4.5.2.

Almost independently of the growth process, the buffer layer (see Figure 4.31) and the active layer of GaN, InN or of nitride alloys contain large amounts of extended defects, among which the most common are threading (screw and edge) dislocations (up to $\approx 10^9 \ cm^{-2}$) and nanopipes, similar to those found in SiC, crossing the entire epitaxial layer [230]. Dislocations originate by a mosaic type of growth on top of a (0001) sapphire or SiC substrate, which leads to islands rotated mostly around the c-axis, bounded by edge dislocations [230]. This is not the general case, because there are literature reports [248] showing that a threading dislocations-free, 100 nm thick, $In_{0.1}Ga_{0.9}N$ epitaxial layer can be grown on a dislocated GaN buffer.

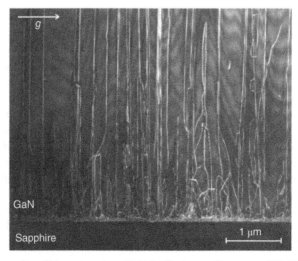

Figure 4.31 *Threading dislocations in a GaN buffer layer. Courtesy of D. Holec, Montanuniversität Leoben, Austria and J. S. Barnard, University of Cambridge, UK*

It could be expected, therefore, on the basis of the existing knowledge concerning other semiconductors (see Chapter 3) that extended defects would strongly influence the performance of nitride alloy-based devices.

Concerning nanopipes, their density is relatively low in the best epi-layers ($< 10^5$ cm^{-2}), such that their reciprocal distance is higher than the average minority carriers diffusion length (about 0.1 µm). Therefore, only threading dislocations should be considered as potential minority carrier killers in epi-layers [230], as their mean distance is comparable with the diffusion length of minority carriers [232]. Different from analogous situations in Si, Ge or Si-Ge alloys, where such a high threading dislocation density would lead to unusable devices, these high dislocation densities in nitride alloys are definitely tolerated. This issue would indirectly lead to the conclusion that most threading dislocations in GaN are electrically inactive.

Actually, the comparison of experimental and self-consistent simulations of the nitrogen K-edge spectra from the bulk, edge, screw and mixed intrinsic dislocations show that the cores of these dislocations do not give origin to band states [249]. A slightly different conclusion can be derived from the computational work carried out by Blumenau *et al.* [250], who show that full core (stoichiometric) screw dislocations induce deep band states, while open-core screw and edge threading dislocations are electrically inactive in their impurity-free intrinsic configurations (see Section 3.3.1 for the structure of full core and open core dislocations). Cores of screw and edge dislocations interact, however, with Ga vacancy–oxygen complexes [$V_{Ga} - O_N$], normally found in bulk GaN, but whose (calculated) formation energy is lower in the strain field of a dislocation [250, 251]. As these defects are electrically active, their trapping at dislocation cores may lead to the electrical activity of dislocations.

4.5   Growth of Poly/Micro/Nano-Crystalline Thin Film Materials

4.5.1   Introduction

Polycrystalline semiconducting films are, potentially, low cost alternatives to single crystal materials, especially if they could be grown as thin films by low temperature, vapour phase techniques on top of low cost, inorganic or polymeric substrates. Their microstructure would depend on the nature of the material and on the growth temperature, with variable grain sizes in the micrometres range.

The film thickness should be selected in relation to the specific application, as done for single crystal epi-layers used for the fabrication of microelectronic or optoelectronic devices. In the case of solar cells, which will be considered in major detail here below, the light absorption coefficient α ($h\nu$) and the diffusion length (L_n or L_h) of the minority carriers are the key functional parameters. Both depend strongly on the nature of the semiconducting film, as is shown in Table 4.13 and Figure 4.32 [252] for a number of semiconductors used today in thin film solar cells [252–255]. It is immediately apparent from Figure 4.32 that the absorption coefficient $\alpha(h\nu)$ of crystalline silicon (c-Si) is 1 order of magnitude lower than that of amorphous hydrogenated silicon (a-Si:H)[12] and copper indium gallium

---

[12] Amorphous silicon (a-Si) prepared with hydrogen-free techniques (e.g. sputtering) is a very defective material, with defect densities around 10^{20} cm^{-3}, unusable for electronic applications.

Table 4.13   *Electrical properties of Si, Ge, CdTe and CIGS (*) this book*

Material	ρ (Ω cm)	E_g (eV) (300 K)	α (cm^{-1}) ($hv > E_g$)	Diffusion length (μm)	Notes	References (*)
c-Si	360 to 1	1.12 1.124	$1 \cdot 10^4$	1150–280	Single crystals n-type	[252]
c-Si	400 to 15	—	—	977–120	Single crystals p-type	[252]
a-Si:H	10^{13} to 10^{-2}	1.7	$\approx 10^5$	0.1–2	Amorphous, hydro-genated	[253]
CdTe	$\approx 10^{-2}$	1.59	$> 5 \cdot 10^5$	0.6–2.8	Single crystals undoped	[254]
Copper indium gallium selenide (CIGS)	—	1.38	$\approx 10^5$	0.3–0.9	—	[255]

Data from [252–255].

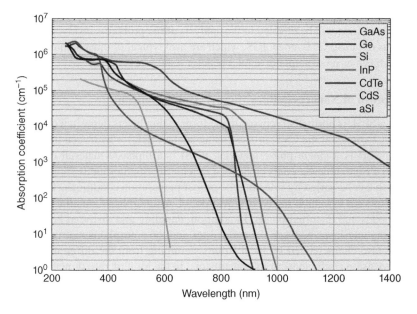

Figure 4.32   *Absorption coefficients of selected semiconductors. PVEducation.org*

selenide (CIGS) and almost 2 orders of magnitude lower than that of CdTe. For the last ones a 500 to 100 nm-thick active layer would absorb more than 90% of the incident photons having an energy $hv > E_G$, working as very efficient active substrates.

In addition, the energy gap of a-Si:H, CdTe and CIGS provide an optimum energy match with the solar spectrum [253, 256–258].

These features make a-Si:H and compound semiconductors ideal candidates for thin film solar cells, although the conversion efficiency of a-Si:H is limited by residual structural defects and by the presence of GBs as parasitic recombination centres of minority carriers in the other cases. GBs, as already seen in Chapter 3, are active recombination centres for minority carriers and their recombination activity is, possibly, enhanced by interaction with metallic impurities.

Hydrogenation of a-Si during a CVD growth process using silane as the precursor passivates the majority of defects and leads to a hydrogenated amorphous Si phase labelled a-Si:H.

Residual defects limit the conversion efficiency of a-Si:H single junction solar cells to values around 10%, too low to compete with c-Si and compound semiconductor solar cells.

Amorphous silicon is used, instead, with record efficiencies (23%) in Sanyo a-Si/c-Si heterojunction solar cells, consisting of a pin junction, where the thin, top p-type layer consists of a-Si:H, the intrinsic layer is undoped a-Si:H and the n-type section is crystalline silicon.

The GBs problem is particularly severe for polycrystalline silicon, which is, in fact, of no interest for solar cell applications. It has been overtaken, with a solution of still only partial success, by the use of microcrystalline silicon (μc-Si), which is a heterogeneous mixture of a-Si:H and nc-Si.

The GB problem is, instead, critical but soluble, for CdTe and CIGS, for which specific, and still empirical, GB passivation techniques are adopted [256, 258].

The next sections will be devoted to the analysis of the problems involved in the growth of thin films of μc-Si, CdTe and CIGS, which are also typical of other semiconductors and of SiNWs, taken as an example of semiconductor nanowires.

4.5.2   Growth of Nanocrystalline/Microcrystalline Silicon

nc-Si, like a-Si:H, is normally prepared in the form of thin layers deposited on a convenient substrate by variants of the CVD process using silane (SiH_4) − hydrogen mixtures.

While a-Si:H is a single phase material, nc-Si is a biphasic material consisting of a dispersion of silicon nanocrystals embedded in a matrix of amorphous silicon, whose volume fraction can be varied by selecting the proper preparation conditions. The amorphous matrix is expected to exhibit the typical disorder of an amorphous material that is residual dangling bonds and distorted bonds. In addition, dangling and distorted bonds should be present at the interfaces between the crystallites and the amorphous matrix. The biphasic material is also characterized by local strains stemming from the different densities and thermal expansion coefficients of its constituents. In the limit of zero or negligible amount of the amorphous phase, the material is called microcrystalline silicon (μc-Si).

Due to the presence of hydrogen in the vapour phase, nc-Si is also a hydrogenated material, conventionally labelled nc-Si:H.

Since the adverse effects of disorder can be, at least partially, suppressed by hydrogenation, nc-Si:H looks, potentially, very appealing for photovoltaic and optoelectronic applications. In fact, nc-Si:H mimics a dispersion of Si dots in a larger-energy-gap matrix of a-Si, which potentially induces quantum confinement effects, provided the crystallites are a few nm in diameter.

In most cases nc-Si:H is deposited by PECVD on glass substrates. Among several variants of this technique, capacitatively coupled radio-frequency (RF) plasmas are most commonly used, which allow the fragmentation of silane to SiH_3 which is the precursor species of Si, once absorbed on a silicon substrate, as will be discussed in detail later in this section. All these process suffer, however, from low growth rates and ion-induced damage of the deposited film.

For years, one of the main issues in this field has been how to achieve growth rates suitable for industrial production (several nm s^{-1}), while avoiding ion-induced damage on the surface of the growing films. An approach, pioneered by Shah and coworkers from the University of Neuchatel [259], was to increase the excitation frequency from the usual 13.5 MHz to typically 60 MHz or more [260].

The hot-wire (HW) CVD approach, developed at the University of Utrecht [261], is a truly ion-free deposition technique, based on the use of heated ($T > 1750\,°C$) W, Ta or TaC-coated graphite filaments, on which SiH_4 is catalytically decomposed to Si and atomic hydrogen.

$$SiH_4 \rightarrow Si + 4H \qquad (4.80)$$

The decomposition reaction is followed by a parasitic reaction leading to the formation of Si_2H_4

$$Si + SiH_4 \rightarrow Si_2H_4 \qquad (4.81)$$

which remains unreacted in the vapour phase and is lost for further processing, while atomic hydrogen goes through a secondary reaction with silane

$$H + SiH_4 \rightarrow SiH_3 + H_2 \qquad (4.82)$$

with the formation of SiH_3, which acts as a predominant nc-Si precursor, as will be seen below discussing the LEPECVD, which is also a technique capable of reducing the ion damaging effect of plasma CVD [262–264].

LEPECVD was originally developed for Si homoepitaxy at low substrate temperatures [265] and later applied to SiGe/Si heteroepitaxy [225]. LEPECVD is characterized by epitaxial growth rates approaching 10 nm s^{-1}, at substrate temperatures around 500 °C. Record hole mobilities in modulation-doped heterostructures proved that defect formation by ion bombardment can be excluded by this technique.

Tuning the morphology of nc-Si material, in terms of nc-Si size, shape and distribution in the a-Si matrix, remains the last challenge for most PECVD processes. The nc-Si size and volume fraction depend in a complex fashion on the plasma density and composition, on the silane/hydrogen ratio and on the chemistry of the surface growth processes.

To bring at least to a preliminary solution the problem, it is common practice to

- carry out the plasma analysis by mass spectroscopy
- model the plasma discharge in a specific reactor, in order to identify the chemical species generated in the plasma phase by the non-elastic electron impact
- model the gas phase and surface kinetics. The chemical species, as ions or radicals, present in the gas phase (outside the plasma phase) undergo a series of reactions, which usually take place both in the gas phase and on the surface of the growing film and that eventually determine the film growth. The kinetic model predicts the evolution of the chemical composition of the gas phase and the surface as a function of the operating parameters.
- evaluate the correlations between the experimental crystallinity fraction and the model parameters
- evaluate the correlation between the experimental morphology and the SiH_4/H_2 ratio.

The modelling of the plasma discharge, of the gas phase and the surface kinetics in a LEPECVD reactor has been carried out by Cavallotti [266]. It is interesting here to note that modelling the full process implied the calculation of

- the kinetic constants for the dissociation and ionization of silane, SiH_3, SiH, (which are the main species present) and H_2 and the ionization of Si.
- the kinetic constants of the ionic reactions occurring in the plasma phase involving the precursors of Si
- the kinetic constants for the formation of radical species in a plasma phase containing SiH_4 and H_2
- the kinetic constants of the surface reactions involving neutral and radical species[13] and dangling bonds σ, which are supposed to be involved in the silicon deposition, of which the most important are

$$SiH_3{}^* + \sigma \rightarrow SiH_2{}^* + H^* \tag{4.83}$$

$$SiH_2{}^* \rightarrow Si_s + H_2{}^* \tag{4.84}$$

$$H + SiH_3{}^* \rightarrow SiH_4 + \sigma \tag{4.85}$$

$$H + SiH_3{}^* \rightarrow SiH_2{}^* + H_2 \tag{4.86}$$

This kind of modelling is unable to predict the microstructure of the deposited film but to support the interpretation of the experimental results. It is also able to show the conditions suitable for optimizing the deposition, as is shown in Figure 4.33 which reports the deposition temperature dependence of the mean surface composition.

Here it is possible to see, as an example, that at 500 °C the fraction of free surface sites is minimized, while the concentration of precursor species $SiH_3{}^*$ takes its maximum value. Unfortunately, this temperature is too high to be profitably used when the deposition is carried out on glass substrates, on which the optimum temperature is around 250 °C.

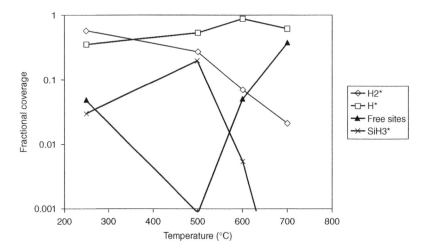

Figure 4.33   *Mean surface composition for the growth of nc-Si in reported as a function of the substrate temperature*

---

[13] Radical species are indicated with a * superscript.

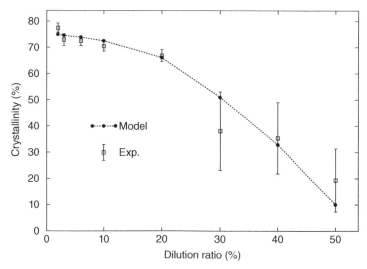

Figure 4.34 *Effect of the dilution ratio on the crystallinity of LEPECVD deposited nc Si. Novikov, et al., 2009, [267]. Reproduced with permission from AIP publishing.*

Another result of modelling, which also fits well with the experimental results, is the evaluation of the role of the hydrogen dilution of SiH_4, $d = \dfrac{\Phi_{SiH_4}}{\Phi_{SiH_4}+\Phi_{H_2}}$, where Φ_{SiH_4} and Φ_{H_2} are the silane and hydrogen fluxes in the reaction chamber, on the crystallinity, reported in Figure 4.34 [267], which represents a trend common to all PECVD processes. One can note that the crystallinity increases with decrease in hydrogen content, leading to a continuous decrease in the a-Si fraction, ending with the formation of a pure μc-Si phase.

As a columnar growth of nanocrystals is common and typical of all PECVD processes, the decrease of the a-Si fraction is associated with a thinner a-Si interphase between the nc-Si columns.

Under low H_2/SiH_4 ratios, see Figure 4.35a,b the microstructure of the film is characterized by columnar nanocrystals separated by very thin intergrain boundaries. Figure 4.35c displays a section of a nc-Si film grown with HW-CVD, with a larger H_2/SiH_4 ratio, leading to crystalline columns embedded in an amorphous silicon matrix, which lead to solar cells with the largest conversion efficiency [261].

The microstructure of nc-Si deposited with LEPECVD, which mimics a distribution of nanowires in an amorphous matrix, is unable to result in efficient light emission, even when the crystallinity is properly tuned [268]. In fact, the emission spectrum is very broad, the intensity is very small even at cryogenic temperatures and the effect of quantum confinement is almost negligible.

The main reason for optoelectronic inefficiency is probably the poor structural quality of the columns (see Figure 4.35a), when compared to that of SiNWs.

4.5.3   Growth of Silicon Nanowires

For more than a decade, SiNWs have been the key subject of multidisciplinary studies aimed at explaining their individual properties and at finding their application where low-dimensionality would be of crucial interest, as is the case for improving the light

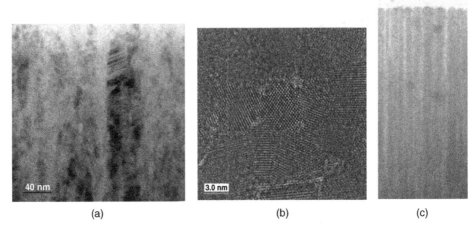

(a)   (b)   (c)

Figure 4.35 *(a) Typical morphology of a vertical section of an a-Si/nc-Si film grown with the LEPECVD process at 210°C; (b) top view of the sample and (c) section of a nc-Si film grown with HW CVD*

emission of silicon [269, 270] or offering *ad hoc* solutions in the field of energy conversion and storage [271, 272].

Two main approaches are used to prepare SiNWs. The traditional one is based on their catalyzed growth from the vapour phase, using gold, copper and aluminium as catalysts and SiH_4 or $SiCl_4$ as precursors. In the case of gold, see Figure 4.36a the presence of a eutectic at 363 °C in the binary Au-Si phase diagram, establishes an equilibrium between a liquid solution containing 18.6% atomic Si, solid silicon and solid gold. In addition, gold catalyzes the thermal decomposition of $SiCl_4$ and SiH_4 to Si. A similar equilibrium occurs, at a slightly higher temperature (577 °C), between an Al-Si liquid solution, solid Al and solid Si (see Figure 4.36b).

Taking Au as a typical example, when a nanometric gold particle is deposited on a silicon single crystal substrate and the substrate is heated slightly above the eutectic temperature, a liquid droplet of Si-Au alloy will form. Under the action of a gradient of temperature between the liquid droplet of the Si-Au alloy and the substrate, a thin silicon single crystal wire, having the diameter of the liquid cap, can grow homoepitaxially as soon as an excess of silicon is delivered by the decomposition of SiH_4 or $SiCl_4$ at the surface of the Au droplet. In the case of an ordered matrix of gold seeds, suitably deposited by sublimation or sputtering, the process proceeds with the formation of a mat of nanowires, as shown in Figure 4.37a.

High temperature, Au-catalyzed growth of SiNWs inevitably leads to Au-contamination of the wire, due to bulk metal in-diffusion ($D_{Au}(cm^2 \ s^{-1}) = 0.28 \ exp(-1.6 \ (eV)/kT)$) with relevant consequences for the electronic properties of the wire, as Au is a deep level not only in bulk silicon but also in SiNWs [273].

Enhanced contamination is expected using copper, which is a fast diffuser in silicon ($D_{Cu}(cm^2 \ s^{-1}) = 0.015 \ exp(-0.86 \ (eV)/kT)$), while Al would be less dangerous $D_{Al}(cm^2 \ s^{-1}) = 8.0 \ exp(-3.47 \ (eV)/kT)$.

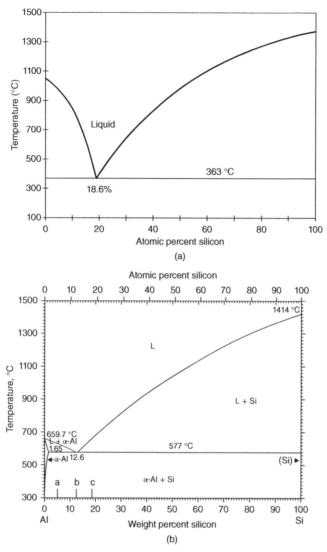

Figure 4.36 *Phase diagrams of (a) Si-Au and (b) Si-Al. Scientific Group Thermodata Europe (SGTE). Reproduced with permission from SGTE*

A low temperature process, therefore, would favour better electronic properties of the individual SiNWs.

This is the case for the metal assisted chemical etching (MACE) process [270, 274], a metal assisted variant of the process used for the formation of porous silicon [275], based on the room temperature, electroless etching of silicon, using Au, Ag or Cu as catalysts.

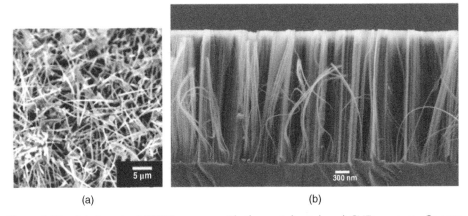

(a)   (b)

Figure 4.37 *(a) A mat of SiNWs grown with the metal-catalyzed CVD process. Courtesy of A. Fukata, Nanostructural Semiconducting Materials Group, Japan. (b) Scanning electron microscopy (SEM) micrograph of an array of SiNWs grown with the MACE process. Courtesy of A.Irrera, Institute of Microelectronics & Microsystems of CNR, Italy*

Electroless processes are electrochemical processes which work without the application of an external dc power and cover a wide range of applications, such as the corrosion protection of metal gas- and oil-pipelines, metal deposition/capping on metallic/insulating substrates and preferential metal or semiconductor etching.

They are eminently metal-catalyzed or self-catalyzed surface processes which occur in the presence of a reducing or oxidizing solution and of properly prepared surfaces.

The electroless growth of ordered Si NWs arrays, see Figure 4.37b, is carried out by reacting an aqueous solution of HF and hydrogen peroxide (H_2O_2) at the surface of a silicon substrate covered with a nanometric (3–10 nm) thin layer of dots of a noble metal, preferably Au, which behaves as the reaction template.

The morphology of nanometrically thin Au- or noble metal-layers depends not only on the actual amount of deposited Au, but on its distribution on the Si surface, which in turn depends on the structure, morphology and chemical configuration of the surface itself and on the process used for its deposition.

For this reason, a homogeneous distribution of pores within a continuous noble metal cover layer would be the preferred configuration, which is, however, not very easy to obtain spontaneously with either a dry or an electroless process.

The first elemental steps of the process are, then, the deposition of an a random distribution or an ordered matrix of Au, Ag or Cu nanometric particles on a silicon substrate,[14] which is dipped in a solution of hydrogen peroxide and HF. Dipping is followed by the cathodic reduction of H_2O_2 at the Au/solution interface

$$H_2O_2 + 2H^+ \rightarrow 2H_2O + 2h^+ \tag{4.87}$$

---

[14] Patterned configurations may be obtained using physical deposition methods (thermal or e-beam sublimation, sputtering etc.) or electroless capping processes, by dipping a silicon sample in a fluoridic solution of a noble metal salt, such as Au Cl_3 or Ag $(NO_3)_3$. The electroless deposition is associated with a non-uniform distribution of the noble metal catalyst at the silicon surface, with dendritic deposits in the case of Ag and Au but with smoother, dense and thicker deposits with Pt and Cu.

(where h^+ is an electron hole), probably associated with the evolution of hydrogen

$$H^+ \rightarrow {}^1/_2 H_2 + h^+ \tag{4.88}$$

which is, in fact, observed experimentally, and at least by two oxidation reactions, of which the first could be the anodic dissolution of silicon as a Si^{4+} ion in solution

$$Si + 2HF + 4F^- + 4h^+ \rightarrow H_2SiF_6 \tag{4.89}$$

and the second is the surface oxidation of silicon

$$Si + 2H_2O + 4h^+ \rightarrow SiO_2 + 4H^+ \tag{4.90}$$

followed by the dissolution of the oxide by HF

$$SiO_2 + 6HF \rightarrow H_2SiF_6 + 2H_2O \tag{4.91}$$

The direct oxidation of silicon to SiO_2 with hydrogen peroxide, followed by reaction (4.91) occurs also on the uncovered silicon surface, but at a much reduced etching rate.

The overall cell processes occurring during MACE are given by the coupling of reactions (4.87), (4.89), (4.87) and (4.90) under electroneutrality conditions

$$Si + 2H_2O_2 + 6HF \rightleftharpoons 4H_2O + H_2SiF_6 \tag{4.92}$$

$$Si + 2H_2O_2 \rightleftharpoons 2H_2O + SiO_2 \tag{4.93}$$

for which the Gibbs free energies of reaction at 298 K are, respectively,

$$\Delta G_{4.92} = -1935.09 \text{ kJ mol}^{-1}$$

$$\Delta G_{4.93} = -615.56 \text{ kJ mol}^{-1}$$

taking for the Gibbs free energies of formation of H_2SiF_6, H_2O_2 and SiO_2 at 298 K the values of 2175.93, 120.42 and 856.4 kJ mol^{-1}, respectively.

Based on these values, reaction (4.93) is thermodynamically favoured, but the rate of both processes is actually ruled by the maximum rate of the slowest electrode reactions occurring at the noble metal cathode and at the silicon anode.

In addition, while theoretical arguments support the hypothesis that the cathodic reaction (4.87) occurs at the noble metal surface with relatively low overvoltage contributions, it is not clear nor experimentally known whether the anodic reaction occurs in relation to noble metal free-silicon surface regions. It is also supposed that silicon diffuses through the nanometrical thin gold layer and that its electrochemical oxidation occurs at the Au/electrolyte interface.

As expected, and different from the case of SiNWs grown with the vapour phase process [273], RBS and deep level transient spectroscopy (DLTS) analysis of SiNWs grown with the MACE process [276] demonstrate an almost negligible metal contamination.

Results on SiNW show how critical is the role of chemistry and physical chemistry for the development of nanostructures of high electronic quality, not only limited to silicon, but also for a variety of compound semiconductors, for which nanowires represent a platform for nanoscience and technology for the years to come [277].

4.5.4   Growth of Films of CdTe and of Copper Indium (Gallium) Selenide (CIGS)

4.5.4.1   *Growth of CdTe Thin Films*

CdTe solar cells are already produced in GW amounts and are very popular, in spite of the use of Cd, whose worse environmental effects are well known.

Different from the liquid phase growth of single crystals of CdTe and ZnCdTe (see Section 4.3.4.3), thin films of CdTe are only prepared by vapour phase processes.

These processes are favoured by the congruent sublimation of CdTe

$$CdTe \rightarrow Cd(v) + \frac{1}{2}Te_2(v) \tag{4.94}$$

and by comparable values of the sublimation pressures of Cd and Te [256] at growth temperatures around 600 °C, which allow the deposition of almost stoichiometric, single phase, polycrystalline films, with micrometric grain sizes when the sublimation is carried out in vacuum at $T > 300\,°C$.

The presence of intragrain defects (native defects and impurity defects, see Table 4.14), with energy levels distributed along the entire band gap, and of a high density of grain boundaries, severely degrades the electrical properties of the as-grown films, and calls for the use of defect- and GB-passivation processes, most of which are still of empirical nature.

As an example, hydrogen and lithium were used as passivating agents [278], which, however, lead only to a temporary passivation.

It is common practice to introduce $CdCl_2$, O_2 and Cu as activation or passivation agents, but only the effect of Cl has been securely ascertained as due to intra-grain and GB passivation [279], as revealed by the increase in the electron beam induced current (EBIC) intensity of grains and GBs of CdTe films treated with $CdCl_2$. This conclusion is supported by photoluminescence (PL) measurements on $CdCl_2$-treated CdTe polycrystalline films, which show no discontinuity of a PL signal when the light beam is scanned across GBs [280]. It could be argued that Cl-passivation is associated with the formation of a shallow state complex between Cl_{Te} and a deep level.

Table 4.14   *Energy levels of native defects and selected impurities in CdTe, making reference to the top of the valence band (VB) or the minimum of the conduction band (CB)*

Native defects	Energy level (eV)	Impurities	Energy levels (eV)
Cd_{Te} (+2)	0.1 (vs CB)	Na_i (+1)	0.01(vs CB)
		Cu_i (+1)	0.01 (vs CB)
Cd_i (+2)	0.33 (vs CB)	Cl_{Te} (+1)	0.35 (vs CB)
$Cd*_i$(+1)	0.46 (vs CB)	—	—
$Cd*_i$(+2)	0.56 (vs CB)		
V_{Te}(+2)	0.71 (vs CB)	—	—
Te_i (−2)	0.74 (vs CB)	—	—
Te_i (−1)	0.67 (vs CB)		
V_{Cd} (−2)	0.21 (vs VB)	Sb_{Te} (−1)	0.23(vs VB)
V_{Cd} (−1)	0.13 (vs VB)	Cu_{Cd} (−1)	0.22 (vs VB)

McCandless and Sites, 2003, [256]. Reproduced with permission from John Wiley & Sons.

4.5.4.2   Growth of Copper Indium (Gallium) Selenide (CIGS) Thin Films

$Cu(In_{1-x}Ga_x)Se_2$ (CIGS)-based solar cells represent today one of the most promising candidates for high efficiency (>20%), low cost thin film solar cells, with the potential to compete in efficiency and cost with single crystal solar cells [258]. CIGS alloys are quaternary solid solutions with the copper indium diselenide $CuInSe_2$ (CIS) structure [281].

It is apparent that the low temperature chalcopyritic α phase is a nearly stoichiometric phase, with a very shallow homogeneity range. It exhibits a solid–solid transition to the high temperature sphalerite δ phase, which is a cation-disordered phase with a wide and temperature dependent extension of its homogeneity range.

It is also apparent that difficulties would be encountered in the growth of the low temperature single phase, stable at the working temperatures of solar cells, as very small deviations from the stoichiometric ratio of the Cu and Te species in the growth atmosphere would lead to biphasic mixtures.

On the other hand, if the growth is performed in the temperature range of the δ phase, the α phase would be obtained by the solid–solid transformation only if the δ phase exhibits the appropriate composition.

Low temperature operation and the absence of thermodynamic constraints favour, therefore, the growth of the α phase, which is generally carried out on a Mo interlayer deposited on glassy, metallic or polymeric substrates

The deposition of CIGS is only a part of a complex process addressed at the manufacture of a solar cell, where CIGS is the adsorbing layer, and is generally carried out by co-sublimation of the metallic precursors and of selenium in the proper composition range. The deposition temperature depends on the substrate nature, and cannot exceed 550 °C in the case of glass substrates and about 450 °C in the case of Kapton. As expected, the quality of the absorber layer depends significantly on the nature of the substrate, on the deposition temperature and on the further steps needed for the completion of the final cell structure [258]. It depends also on the details of the in-line deposition process, which is carried out by solar cell manufacturers on a moving substrate and which could involve a single step, consisting in the simultaneous deposition of all the components, or multiple steps. The distribution of the components in the deposited layer may vary along the in-line manufacturing steps, depending on the temperature profiles involved and on the local variations of the chemical potentials of the alloy components, which are out of any direct control. It is therefore easily understandable that the deposited material could present composition profiles and second phase inclusions (Cu_2Se and In_2Se_3) [282], which would influence the solar cell efficiency. Different from the case of elemental semiconductors, in a quaternary semiconductor, such as CIGS, the defect matrix is extremely complicated and its detailed structure is still partially unknown, in spite of decades of research [258], leading to mostly empirical solutions for their control or passivation.

A significant example is given by Na doping, which is necessary to get high efficiency CIGS devices, independently of the source of Na, and by the role of the Mo buffer layer, which actually seems to work as a filter (or reservoir) for Na and which is systematically used, independently of the nature of the mechanical substrate. The problem here is twofold, and concerns both the role of Mo and the doping effects of Na.

Looking, first, at Mo issues, it is known that if Mo is deposited on a glass substrate a kind of Na-exchange occurs between the glass and the Mo film. The physics of the process, however, is not very clear, although it is shown that the Na-exchange is favoured by the presence of oxygen, that Na is found segregated at Mo-GBs and that the GBs are completely oxidized [283]. GB oxidation is favoured by the high reactivity of oxygen with Mo.[15]

The same beneficial effects are obtained when a sodium halide (NaCl or NaF) is deposited on Mo.

Although it is well known that soda lime-silica glasses exchange Na^+ with molten salts and with ionic solids, and that a Na^+ exchange enhancement is observed at 570 °C [284], physics would impede the exchange of Na^+ ions between the glass or a sodium halide and Mo.

It is possible, however, to argue,[16] even if the details of the physical chemistry of the process are still unknown, that the GBs of Mo are covered by a layer of α-MoO_3, with its anisotropically layered crystalline structure formed by stacking bilayer sheets of MoO_6 octahedra. This layer attributes to the Mo grain boundaries the character of functional oxide interfaces. Due to its intrinsic structural anisotropy and variable oxidation states, α-MoO_3 allows the formation of intercalation compounds with various species, including Na^+ [285] and the occurrence of surface redox reactions needed to allow the Na^+ exchange at the glass/Mo interface and the reduction of Na^+ to Na at the Mo/CIGS interface. The Gibbs energy required for the intercalation is made available by a variation of the oxidation state of Mo, and amounts to a maximum of -83.20 kJ mol^{-1} for the reaction

$$MoO_3 \rightleftharpoons MoO_2 + \frac{1}{2} O_2 \tag{4.95}$$

Instead, the driving force of the Na-doping of CIGS is the gradient of the chemical potential of Na in the Na-source (the MoO_3 surface layer) and the Na-sinks in CIGS.

Turning now to the effects of Na-doping on the overall electrical properties of CIGS, it seems ascertained [283] that sodium passivates the grain boundaries of CIGS, behaves as a surfactant with the formation of Na_2Se_x and enhances the grain growth, while also interacting with point defects.

The physics of these defect interaction processes is still only partially known, although it should be supposed that the formation of a Na_{In} antisite by reaction of sodium with a regular In site is energetically unfavourable (2.5 eV of formation energy) [286–287], contrary to the formation of a Na_{Cu} species by substitution of Na in In_{Cu} antisites [281], which will lead also to a reduction in the density of the (In_{Cu} V_{Cu}) defect complexes, behaving as recombination centres.

It is therefore evident that the development of post-growth treatments addressed at improving the physical properties of a multinary compound is particularly challenging, leaving open the door to empirical solutions, different from those in common use in the case of elemental semiconductors. This will be discussed in Chapter 5.

---

[15] The Gibbs free energy of formation of the MoO_3 phase, $\Delta G_{MoO3} = -366.25$ kJ mol^{-1} at 500 °C.
[16] This is the personal view of this book's author.

References

1. Wilkes, G. (1996) Silicon processing, in *Processing of Semiconductors*, Jackson, K.A. and Schröter, W. Eds Chapter 1, pp. 1–65, Wiley-VCH Verlag GmbH, Weinheim.
2. Mullin, B. (1996) Compound semiconductor processing, in *Processing of Semiconductors*, Jackson, K.A. and Schröter, W. Eds Chapter 2, pp. 67–109, Wiley-VCH Verlag GmbH, Weinheim.
3. (a) Claeys, C. and Simoen, E. (2007) *Germanium Based Technologies: from Materials to Devices*, Elsevier; (b) Holloway, P.H. and McGuire, G.E. (1995) *Handbook of Compound Semiconductors: Growth, Processing, Characterization and Devices*, Noyes Publications, XX.
4. Friedrichs, P., Kimoto, T.T., Ley, L., and Pensl, G. (2009) *Silicon Carbide*, Growth, Defects and Novel Applications, vol. **1**, Wiley-VCH Verlag GmbH, Weinheim.
5. Nicollian, E. and Brews, J.R. (1982) *MOS (Metal Oxide Semiconductors): Physics and Technology*, John Wiley & Sons, Inc., New York.
6. Wolf, S. and Tauber, R.N. (1986) *Silicon Processing for the VLSI Era*, Process Technology, vol. **I**, Lattice Press, Sunset Beach, CA.
7. McGuire, G.E. (1988) *Semiconductor Materials and Process Technology Handbook*, Noyes Publications, Park Ridge, NJ.
8. Tyagi, M.S. (1991) *Introduction to Semiconductor Materials and Devices*, John Wiley & Sons, Inc., New York.
9. Singh, J. (2001) *Semiconductor Devices, Basic Principles*, John Wiley & Sons, Inc., New York.
10. Ruterana, P.P., Albrecht, M., and Neugebauer, J. (2003) *Nitride Semiconductors, Handbook on Materials and Devices*, Wiley-VCH Verlag GmbH, Weinheim.
11. Palmer, D.W. (2006) www.semiconductors.co.uk (accessed 24 December 2014) 2006.02.
12. Henager, C. and Morris, J.R. (2009) Atomistic simulation of CdTe solid–liquid coexistence equilibria. *Phys. Rev. B*, **80**, 245309.
13. Yamaguki, K., Kameda, K., Takeda, Y., and Itagaki, K. (1994) Measurements of high temperature heat content of the II-Vi and IV-VI compounds. *Mater. Trans. JIM*, **35**, 118–124.
14. (a) Pfann, W.G. (1952) Principles of zone refining. *Trans. AIME*, **194**, 747–753; (b) Pfann, W.G. (1958) *Zone Melting*, John Wiley & Sons, Inc., New York.
15. Wang, J. and Isshiki, M. (2007) Wide-Bandgap II-VI semiconductors: growth and properties, in *Springer Handbook of Electronic and Photonic Materials*, Springer-Verlag, XX, p. 325, ISBN: 978-0-387-26059-4.
16. Ceccaroli, B. and Pizzini, S. (2012) Processes in *Advanced Silicon Materials for Photovoltaic Applications* S. Pizzini (Ed). pp. 21–78, John Wiley & Sons, Ltd, Chichester.
17. Dietze, W., Döring, E., Glasow, P. *et al.* (1983) *Technology of Si, Ge, and SiC: Technologie Von Si, Ge und SiC, Landolt-Bornstein*, vol. **17**, Springer.
18. Zulehner, W. (1994) The growth of highly pure silicon crystals. *Metrologia*, **31**, 255–261.
19. Nakajima, K. and Usami, N. (2009) *Crystal Growth of Si for Solar Cells*, Springer-Verlag.

20. Pizzini, S. (ed) (2012) *Advanced Silicon Materials for Photovoltaic Applications*, John Wiley & Sons, Ltd, Chichester.

21. Dhamrin, M., Saitoh, T., Kamisako, K., Yamada, K., Araki, N., Yamaga, I., Sugimoto, H., and Tajima, M. (2009) Technology development of high-quality n-type multicrystalline silicon for next-generation ultra-thin crystalline silicon solar cells. *Sol. Energy Mater. Sol. Cells*, **93**, 1139–1142.

22. Dash, W.C. (1958) Silicon crystals free of dislocations. *J. Appl. Phys.*, **29**, 736–737.

23. Dash, W.C. (1959) Growth of silicon crystals free of dislocations. *J. Appl. Phys.*, **30**, 459–474.

24. Huang, X., Taishi, T., Yonenaga, I., and Hoshikawa, K. (2000) Dash necking in Czochralski method:influence of B concentration. *J. Cryst. Growth*, **213**, 283–287.

25. Sanati, M. and Estreicher, S.K. (2006) Boron–oxygen complexes in Si. *Physica B*, **376–377**, 133–136.

26. Voronkov, V. and Falster, R. (2010) Latent complexes of interstitial boron and oxygen dimers as a reason for degradation of silicon-based solar cells. *J. Appl. Phys.*, **107**, 053509.

27. Barron, A. *Chemistry of Electronic Materials*, (2011) http://archive.org/details/ost-chemistry-col10719 (accessed 23 December 2014) Rice University.

28. Akselrod, M.S. and Bruni, F.J. (2012) Modern trends in crystal growth and new applications of sapphire. *J. Cryst. Growth*, **360**, 134–145.

29. (a) Ceccaroli, B. and Pizzini, S. (2012) Processes, in *Advanced Silicon Materials for Photovoltaic Applications*, John Wiley & Sons, Ltd, Chichester, pp. 54–58; (b) Binetti, S., Acciarri, M., Le Donne, A., Morgano, M., and Jestin, Y. (2013) Key success factors and future perspective of silicon-based solar cells. *Int. J. Photoenergy*. doi: dx.doi.org/10.1155/2013/249502, ID 249502, 6 pp.

30. Stoddard, N., Wu, B., Witting, I., Wagener, M., Park, Y., Rozgonyi, G., and Clark, R. (2008) Casting single crystal silicon: novel defect profiles from BP solar mono wafers. *Solid State Phenom.*, **131**, 1–8.

31. Nakajima, K. (2010) High efficiency solar cells obtained with small size ingots. Proceeding of the 25th EUPVSEC, pp. 1299–1301.

32. Mrcarica, M. (2013) Potential for mono-cast material to achieve high efficiency in mass production. Photovoltaics International, pp. 28–36, www. pv-tech.org (accessed 24 December 2014).

33. Wrick, V.L., Ip, K.T., and Eastman, L.F. (1978) High purity LPE InP. *J. Electron. Mater.*, **7**, 253–261.

34. Li, J.G. and Hausner, H. (1966) Reactive wetting in the liquid-silicon/solid-carbon system. *J. Am. Ceram. Soc.*, **79**, 873–880.

35. Li, J.G. (1994) Wetting of ceramic materials by liquid silicon, aluminium and metallic melts containing titanium and other reactive elements: a review. *Ceram. Int.*, **20**, 391–412.

36. Ciftja, A., Tangstad, M. and Engh, T.A. (2008) Wettability of Silicon with Refractory Materials: A Review, Norwegian University of Science and Technology, Faculty of Natural Science and Technology, Department of Materials Science and Engineering Trondheim, February 2008.

37. Schnurre, S.M., Gröber, J., and Schnid-Fetzer, R. (2004) Thermodynamics and phase stability in the Si-O system. *J. Non-Cryst. Solids*, **336**, 1–25.

38. Okamoto, H. (2007) O-Si(Oxygen-Silicon). *J. Phase Equilib. Diffus.*, **28**, 309–310.
39. Liu, Z. and Carlberg, T. (1992) Reactions between liquid silicon and vitreous silica. *J. Mater. Res.*, **7**, 353–358.
40. Müller, G., Mühe, A., Backofen, R., Tomzig, E., and Ammon, W.v. (1999) Study of oxygen transport in Czochralski growth of silicon. *Microelectron. Eng.*, **45**, 135–147.
41. Gao, B., Chen, X.J., Nakano, S., and Kakimoto, K.(2010) Global growth of high purity multicrystalline silicon using a unidirectional solidification furnace for solar cells. *J. Cryst. Growth*, **312**, 1572–1576.
42. Jolly, W.L. and Latimer, W.M. (1952) The equilibrium Ge(s) +GeO$_2$(s) = 2GeO(g): the heat of formation of germanic oxide. *J. Am. Chem. Soc.*, **74**, 5757–5758.
43. Prabhakaran, K. and Ogino, T. (1995) Oxidation of Ge(100) and Ge(111) surfaces: an UPS and XPS study. *Surf. Sci.*, **325**, 263–271.
44. Haller, E.E., Hansen, W.L., Luke, P., McMurray, R., and Jarret, B. (1982) Carbon in high purity germanium. *IEEE Trans. Nucl. Sci.*, **29**, 745–750.
45. Kelton, K.F., Falster, R., Gambaro, D., Olmo, M., Cornara, M., and Wei, P.F. (1999) Oxygen precipitation in silicon: experimental studies and theoretical investigations within the classical theory of nucleation. *J. Appl. Phys.*, **85**, 8097–8111.
46. Yonenaga, I., Sumino, K., and Hosh, K. (1984) The mechanical strength of silicon crystals as a function of the oxygen concentration. *J. Appl. Phys.*, **56**, 2346–2350.
47. Cochard, J., Yonenaga, I., Gouttebroze, S., M'Hamdi, M., and Zhang, Z.L. (2010) Constitutive modeling of intrinsic and oxygen-contaminated silicon monocrystals in easy glide. *J. Appl. Phys.*, **108**, 103524.
48. Karoni, A., Rozgonyi, G., and Ciszek, T. (2004) Effect of oxygen and nitrogen doping on mechanical properties of silicon using nanoindentation. *MRS Symp. Proc.*, **821**, P8, 36.1.
49. Leroy, C. and Rancoita, P.G. (2007) Particle interaction and displacement damage in silicon devices operated in radiation environments. *Rep. Progr. Phys.*, **70**, 493–625.
50. Pizzini, S., Cagnoni, P., Sandrinelli, A., Anderle, M., and Canteri, R. (1987) Grain boundary segregation of oxygen and carbon in polycrystalline silicon. *Appl. Phys. Lett.*, **51**, 676–678.
51. Pizzini, S., Binetti, S., Acciarri, M., and Acerboni, S. (1993) Interaction of oxygen, carbon and extended defects in silicon. *Phys. Status Solidi A*, **138**, 451–464.
52. Borghesi, A., Pivac, B., Sassella, A., and Stella, A. (1995) Oxygen precipitation in silicon. *Appl. Phys. Res.*, **77**, 4169–4244.
53. Pivac, B., Ilic, S., Borghesi, A., Sassella, A., and Porrini, M. (2003) Gap states produced by oxygen precipitation in Czochralski silicon. *Vacuum*, **71**, 141–145.
54. Gilles, D., Weber, E.R., and Hahn, S.K. (1990) Mechanism of internal gettering of interstitial impurities in Czochralski-grown silicon. *Phys. Rev. Lett.*, **64**, 196.
55. McHugo, S.A., Weber, E.R., Mizuno, M., and Kirscht, F.G. (1995) A study of gettering efficiency and stability in Czochralski silicon. *Appl. Phys. Lett.*, **66**, 2840.
56. Haarahiltunen, A. (2007) Heterogeneous precipitation and internal gettering efficiency of iron in silicon. Doctoral Dissertation. TKK Dissertations 64, Espoo 2007 University of Helsinki.
57. Kim, Y.-H., Lee, K.-S., Chung, H.-Y., Hwang, D.-H., Kim, H.-S., Cho, H.-Y., and Lee, B.-Y. (2001) Internal gettering of Fe, Ni and Cu in silicon wafers. *J. Korea Phys. Soc.*, **39**, S348–S351.

58. Liu, A.Y., Walter, D., Phang, S.P., and Macdonald, D. (2012) Investigating internal gettering of iron at grain boundaries in multicrystalline silicon via photoluminescence imaging. *IEEE J. Photovoltaics*, **2**, 479–484.
59. Li, R., Li, M.-W., Imaishi, N., Akiyama, Y., and Tsukada, T. (2004) Oxygen-transport phenomena in a small silicon Czochralski furnace. *J. Cryst. Growth*, **267**, 466–474.
60. Liu, L., Nakano, S., and Kakimoto, K. (2007) Three-dimensional global modeling of a unidirectional solidification furnace with square crucibles. *J. Cryst. Growth*, **303**, 165–169.
61. Smirnov, A.D. and Kalaev, V.V. (2008) Development of oxygen transport model in Czochralski growth of silicon crystals. *J. Cryst. Growth*, **310**, 2970–2976.
62. Kakimoto, K. and Gao, B. (2015) Modern Aspects of Czochralski and multicrystallinesilicon crystal growth *in Silicon, Germanium and their alloys* Kissinger, G. and Pizzini, S., Edts. Taylor and Francis, Boca Raton, Fl pp. 1–22.
63. Teng, Y.-Y., Chen, J.-C., Lu, C.-W., Chen, H.-I., Hsu, C., and Chen, C.-Y. (2010) Effects of the furnace pressure on oxygen and silicon oxide distributions during the growth of multicrystalline silicon ingots by the directional solidification process. *J. Cryst. Growth*. doi: 10.1016/j.jcrysgro.2010.11.110.
64. Hockett, R.S. (2011) Advanced analytical techniques for solar-grade feedstock, in *Advanced Silicon Materials for Photovoltaic Applications* (ed S. Pizzini), John Wiley & Sons, Inc., pp. 215–234.
65. Wilson, L.O. (1980) The effect of fluctuating growth rates on segregation in crystals grown from the melt: I. No backmelting. *J. Cryst. Growth*, **48**, 435–450.
66. Wilson, L.O. (1980) The effect of fluctuating growth rates on segregation in crystals grown from the melt: II. Backmelting. *J. Cryst. Growth*, **48**, 451–458.
67. Kakimoto, K. and Ozoe, H. (2000) Oxygen distribution at a solid–liquid interface of silicon under transverse magnetic field. *J. Cryst. Growth*, **212**, 429–437.
68. Williams, M.G., Walker, J.S., and Langlois, W.E. (1990) Melt motion in a Czochralski puller with a weak transverse magnetic field. *J. Cryst. Growth*, **100**, 233–263.
69. Walker, J.S. and Williams, M.G. (1994) Centrifugal pumping during Czochralski silicon growth with a strong transverse magnetic field. *J. Cryst. Growth*, **137**, 32–36.
70. Falster, R., Voronkov, V.V., and Quast, F. (2000) On the properties of intrinsic point defects in silicon: a perspective from crystal growth and wafer processing. *Phys. Status. Solidi B*, **222**, 219–244.
71. Kulkarni, M.S. (2008) Lateral incorporation of vacancies in Czochralski silicon crystals. *J. Cryst. Growth*, **310**, 3183–3191.
72. Voronkov, V.V. (1982) The mechanism of swirl defects formation in silicon. *J. Cryst. Growth*, **59**, 625–643.
73. Kulkarni, M.S., Holzer, J.C., and Ferry, L.W. (2005) The agglomeration of self-interstitials in growing Czochralski silicon crystals. *J. Cryst. Growth*, **284**, 353–368.
74. Kulkarni, M.S. (2007) Defect dynamics in the presence of oxygen in growing Czochralski silicon crystals. *J. Cryst. Growth*, **303**, 438–448.
75. Voronkov, V.V. and Falster, R. (1999) Vacancy and self-interstitial concentration incorporated into growing silicon crystals. *J. Appl. Phys.*, **86**, 5975.

76. Kamiyama, E., Sueoka, K., and Vanhellemont, J. (2015) Vacancies in Si and Ge, in *Silicon, Germanium, and Their Alloys: Growth, Defects,Impurities, and Nanocrystals* (eds G. Kissinger and S. Pizzini), CRC Press, p. 119–158.

77. Bloem, J., Haas, C., and Penning, P. (1959) Properties of oxygen in germanium. *J. Phys. Chem. Sol.*, **12**, 22–27.

78. Hansen, W.L., Haller, E.E., and Luke, P.N. (1982) Hydrogen concentration and distribution in high-purity germanium crystals. *IEEE Trans. Nucl. Sci.*, **29**, 738–744.

79. Li, Z., Liu, L., Ma, W., and Kakimoto, K. (2010) Effect of argon flow on heat transfer in a directionally solidification process for silicon solar cells. *J. Cryst. Growth*, **318**, 298–303.

80. Gösele, U. (1986) The role of carbon and point defects in silicon, in *Oxygen, Carbon, Hydrogen and Nitrogen in Crystalline Silicon*, MRS Symposium Proceeding, Cambridge University Press, Cambridge,Vol. **59**, pp. 419–431.

81. Tan, T.Y. (1986) Exigent-accomodation-volume of precipitation and formation of oxygen precipitates in silicon, *Oxygen, Carbon, Hydrogen and Nitrogen in Crystalline Silicon*, MRS Symposium Proceeding, Cambridge University Press, Cambridge, Vol. **59**, pp. 269–279.

82. Taylor, W.J., Tan, T.Y., and Gösele, U. (1993) Carbon precipitation in silicon: why is so difficult? *Appl. Phys. Lett.*, **62**, 3336.

83. Li, J.G. and Hausner, H. (1992) Influence of oxygen partial pressure on the wetting behaviour of silicon nitride by molten silicon. *J. Eur. Ceram. Soc.*, **9**, 101–105.

84. Binetti, S., Acciarri, M., Savigni, C., Brianza, A., Pizzini, S., and Musinu, A. (1996) Effect of nitrogen contamination by crucible encapsulation on polycrystalline silicon material quality. *Mater. Sci. Eng., B*, **36**, 68–72.

85. Swain, S.K., Jones, K.A., Datta, A., and Lynn, K.G. (2011) Study of different cool down schemes during the crystal growth of detector grade CdZnTe. *IEEE Trans. Nucl. Sci.*, **58**, 2341–2345.

86. Fistul, V.I. (2004) *Impurities in Semiconductors: Solubility, Migration and Interactions*, CRC Press, Boca Raton, FL.

87. Kattner, U.R. (1997) The thermodynamic modeling of multicomponent phase equilibria. *J. Mater*, **42**, 14–19.

88. Tang, K., Ovrelid, E. J., Tranell, G., Tangstad, M. (2009) *Thermochemical and Kinetic Databases for the Solar Cell Silicon Materials in Crystal Growth of Si for Solar Cells*, Nakajima, K., Usami, N. Eds. p. 259, Springer.

89. Redlich, O. and Kister, A.T. (1948) Algebraic representation of thermodynamic properties and the classification of solutions. *Ind. Eng. Chem.*, **40**, 345–348.

90. Coletti, G., McDonald, D., Yang, D. (2011) Role of impurities in solar silicon, in *Advanced Silicon Materials for Photovoltaic Applications* (Pizzini, S. Ed) pp. 79–125, John Wiley & Sons, Ltd, Chichester.

91. Kakimoto, K. (2009) Crystallization of silicon by directional solidification, in *Crystal Growth of Si for Solar Cells*, Nakajima, K. and Usami, N. eds, pp. 55–70, Springer.

92. Burton, A., Prim, R.C., and Slichter, W.P. (1953) The distribution of solute in crystals grown from the melt. Part I. Theoretical. *J. Chem. Phys.*, **21**, 1987–1991.

93. Burton, A., Prim, R.C., and Slichter, W.P. (1953) The distribution of solute in crystals grown from the melt. Part II. Experimental. *J. Chem. Phys.*, **21**, 1991–1996.

94. Peaker, A.R. (1989) *Landolt -Börnstein Numerical Data and Funcitainol Relationships in Science and Technologies Group III*, Semiconductors: Impurities and Defects, vol. **22b**, Springer-Verlag, Berlin.

95. Haller, E.E., Hansen, W.H., Luke, P., Murray, R., and Jarret, B. (1981) Carbon in high purity germanium. *IEEE Trans. Nucl. Sci.*, **29**, 745–750.

96. Kobayashi, T. and Osaka, J. (1985) Effective segregation coefficients of carbon in LEC GaAs crystals. *J. Cryst. Growth*, **71**, 240–242.

97. O'Hara, J.W.C., Herring, R.B., and Hunt, L.P. (1990) *Handbook of Semiconductor Silicon Technology*, Noyes Publishers.

98. Böer, K.W. (1992) *Semiconductor Physics*, vol. **2**, Van Nostrand-Reinhold, p. 155.

99. Itoh, Y., Takai, M., Fukushima, H., and Kirita, H. (1990) The study of contamination of carbon, boron, and oxygen in LEC-GaAs. *MRS Proc.*, **163**, 1001. doi: 10.1557/PROC-163-1001.

100. Giannattasio, A., Gianquinta, A., and Porrini, M. (2011) The accuracy of the standard resistivity–concentration conversion practice estimated by measuring the segregation coefficient of boron and phosphorous in Cz-Si. *Phys. Status Solidi A*, **208**, 564–567.

101. Yoshikawa, T. and Morita, K. (2005) Removal of B from Si by solidification refining with Si-Al melts. *Metall. Metal. Trans. B*, **36B**, 731–736.

102. Ostrowski, A.G. and Müller, G. (1992) A model of effective segregation coefficients accounting for convection in the solute layer at growth interface. *J. Cryst. Growth*, **121**, 587–598.

103. Lehovec, K. (1962) Thermodynamics of binary semiconductor–metal alloys. *J. Phys. Chem. Solids*, **23**, 695–709.

104. Trumbore, F.A. (1960) Solid solubilities of impurity elements in silicon and germanium. *Bell Syst. Technol. J.*, **39**, 205–233.

105. Mullin, J.B. (1996) Compound semiconductor processing, in *Processing of Semiconductors* (ed K.A. Jackson), p. 97.

106. Coquille, R., Toudic, Y., Haji, L., Gouneau, M., Moisan, G., and Lecrosnier, D. (1987) Growth of low dislocation semi insulating InP (Fe-Ga). *J. Cryst. Growth*, **83**, 167–173.

107. Wallart, X., Godey, S., Douvry, Y., and Desplanque, L. (2008) Comparative Sb and As segregation at the InP on GaAsSb interface. *Appl. Phys. Lett.*, **93**, 123119.

108. Yatsurugi, Y., Akiyama, N., and Endo, Y. (1973) Concentration, solubility, and equilibrium distribution coefficient of nitrogen and oxygen in semiconductor silicon. *J. Electrochem. Soc.*, **120**, 975–979.

109. Pizzini, S., Acciarri, M., and Binetti, S. (2005) From electronic grade to solar grade silicon: chances and challenges in photovoltaics. *Phys. Status Solidi A*, **15**, 2928–2942.

110. Claeys, C.L., Falster, R., Watanabe, M., Stallhofer, P. (2006) High purity silicon *ECS Trans.*, **3**, 451–468.

111. Labert, B., Toudic, Y., Grandpierre, G., Gauneau, M., and Deveaud, B. (1987) Semi-insulating InP codoped with Ti and Hg. *Semicond. Sci. Technol.*, **2**, 78–82.

112. Kasap, S. and Capper, P. (2006) *Springer Handbook on Electronic and Photonic Materials*, Springer.

113. Fischler, S. (1962) Correlation between maximum solid solubility and distribution coefficient for impurities in Ge and Si. *J. Appl. Phys.*, **33**, 1615.
114. Biloni, H. (1983) Solidification, in *Physical Metallurgy, Part I* (eds R.W. Cahn and P. Haasen), Elsevier Science Publishers BV, pp. 477–579.
115. Gotoh, R., Fujiwara, K., Yang, X., Koizumi, H., Nozawa, J., and Uda, S. (2012) Formation mechanism of cellular structures during unidirectional growth of binary semiconductor Si-rich SiGe materials. *Appl. Phys. Lett.*, **100**, 021903. doi: 10.1063/1.3675860.
116. Tokunaga, T., Nishio, K., Ohtani, H., and Hasebe, M. (2003) Phase equilibria in the Ni–Si–B system. *Mater. Trans.*, **44**, 1651–1654.
117. a118. bArafune, K., Ohishi, E., Kusuoka, F. *et al.* (2006) Growth and characterization of multicrystalline silicon ingots by directional solidification technique. Conference Record of the 2006 IEEE 4th World Conference on Photovoltaic Energy Conversion, 1, pp. 1074–1077.
118. Tiller, W.A., Jackson, A., Rutter, J.W., and Chalmers, B. (1953) The redistribution of solute atoms during the solidification of metals. *Acta Metall.*, **1**, 428–437.
119. Yonenaga, I. (2005) Growth and fundamental properties of SiGe bulk crystals. *J. Cryst. Growth*, **275**, 91–98.
120. Stelian, C. (2012) Numerical analysis of constitutional supercooling during directional solidification of alloys. *J. Mater. Sci.*, **47**, 3454–3462.
121. Gasparini, M., Calligarich, C., Rava, P. *et al.* (1982) Advanced crystallization techniques of "solar grade" silicon. Proceeding of the Sixteent IEEE Photovoltaic Specialist Conference, pp. 74–79.
122. Nepomnyashchikh, A., Presnyakov, R., Eliseev, I., and Sokol'nikova, Y. (2011) Specific features of multicrystalline silicon growth from high-purity commercial silicon. *Tech. Phys. Lett.*, **37**, 739.
123. Taishi, T., Huang, X., Kubota, M., Kajigawa, T., Fukami, T., and Hoshikawa, K. (2000) Heavily boron-doped silicon single crystal growth. Constitutional supercooling. *Jpn. J Appl. Phys.*, **39**, L5–L8.
124. Müller, G. 1987 in *Growth, Characterization, Processing of III-V Materials With Correlations to Device Performances*, ed. Nissim, Y. I. and Glasow, P A. (Les Editions de Physique, Les Ulis), p. 117.
125. Longere, J.Y., Schohe, K., Krawczyk, S.K. *et al.* (1990) Assessment of Fe doped seminsulating InP crystals by scanning photoluminescence measurements. *J. Appl. Phys.*, **68**, 755–759.
126. May, G.S. and Sze, S.M. (2004) *Fundamental of Semiconductor Fabrication*, John Wiley & Sons, Inc., Hoboken, NJ.
127. Yoshikawa, T. and Morita, K. (2005) Removal of B from Si by solidification refining with Si-Al melts. *Metall. Metal Trans.*, **36B**, 731–736.
128. Yoshikawa, T. and Morita, K. (2003) Solid solubilities and thermodynamic properties of Al in solid silicon. *J. Electrochem. Soc.*, **150**, G465–G468.
129. Joshi, S.M., Goesele, U., and Tan, T.Y. (2001) Extended high temperature Al gettering for improvement and homogenization of minority carrier diffusion lengths in multicrystalline Si. *Sol. Energy Mater. Sol. Cells*, **70**, 231–238.
130. Brown, G.F. and Wu, J. (2009) Third generation photovoltaics. *Laser Photonics Rev.*, **3**, 394–405. doi: 10.1002/lpor.200810039.

131. Braunstein, R., Moore, A.R., and Herman, F. (1958) Intrinsic optical absorption in Germanium-silicon alloys. *Phys. Rev.*, **109**, 695–710.

132. ONTI Ioffe Institute NSM Archive www.ioffe.rssi.ru (accessed 23 December 2014).

133. Halberg, L.-I. and Nevin, J.H. (1981) Silicon-germanium alloy growth control and characterization. *J. Electron. Mater.*, **11**, 779–793.

134. Wang, P., Yu, X., Chen, P. *et al.* (2011) Germanium-doped Czochralski silicon for photovoltaic applications. *Sol. Energy Mater. Sol. Cells*, **95**, 2466–2470.

135. Barz, A., Dold, P., Kerat, U., Recha, S., Benz, K.W., Franz, M., and Pressel, K. (1998) Germanium-rich SiGe bulk single crystals grown by the vertical Bridgman method and by zone melting. *J. Vac. Sci. Technol., B*, **16**, 1627–1630.

136. Yonenaga, I., Akashi, T., and Goto, T. (2001) Thermal and electrical properties of Czochralski grown GeSi single crystals. *J. Phys. Chem. Sol.*, **62**, 1313–1317.

137. Chaisakul, P., Marris-Morini, D., Isella, G., Chrastina, D., Rouifed, M.-S., Frigerio, J., and Vivien, L. (2013) Ge quantum well optoelectronic devices for light modulation, detection, and emission. *Solid State Electron.*, **83**, 92–97.

138. Dhanaraj, G., Huang, X.R., Dudley, M., Prasad, V., and Ma, R.-H. (2003) Silicon carbide crystals — part I: growth and characterization, in *Crystal Growth Technology* (eds K. Byrappa, W. Michaeli, H. Waarlimont, and E. Weber), William Andrew Inc.

139. Sevast'yanov, V.G., Nosatenko, P.Y., Gorskib, V.V. *et al* (2010) Experimental and theoretical determination of the saturation vapor pressure of silicon in a wide range of temperatures. *Russ. J. Inorg. Chem.*, **55**, 2073–2088.

140. Halden, F.A. (1960) The growth of silicon carbide from solution, in *Silicon Carbide, A High Temperature Semiconductor* (eds J.R. Connor and J. Smilestens), Pergamon, Oxford, pp. 115–123.

141. Epelbaum, B.M., Hofmann, D., Muller, M., and Winnacker, A. (2000) Top-seeded solution growth of bulk SiC: search for the fast growth regimes. *Mater. Sci. Forum*, **338–342**, 107–110.

142. Yakimova, R. and Janzen, E. (2000) Current status and advances in the growth of SiC. *Diamond Relat. Mater.*, **9**, 432–438.

143. Thurmond, C.D. and Logan, R.A. (1972) The equilibrium pressure of N_2 over GaN. *J. Electrochem. Soc.*, **119**, 622–626.

144. Porowski, S. and Grzegory, I. (1997) Thermodynamic properties of III-V nitrides and crystal growth of GaN at high N_2 pressure. *J. Cryst. Growth*, **178**, 174–188.

145. Grzegory, I., Krukowsi, S., Lesczynski, M., Perlin, P., Suski, T., Porowski, S. (2003) High pressure crystallization of GaN, in *Nitride Semiconductors*, Ruterana, P., Albrecht, M., Neugebauer, J. eds pp. 3–43, Wiley-VCH Verlag GmbH, Weinheim.

146. Karpinski, J. and Porowski, S. (1984) High pressure thermodynamics of GaN. *J. Cryst. Growth*, **66**, 11–20.

147. Karpinski, J., Jun, J., and Porowski, S. (1984) Equilibrium pressure of N_2 over GaN and high pressure solution growth of GaN. *J. Cryst. Growth*, **66**, 1–10.

148. Yamane, H., Shimada, M., Sekiguchi, T., and DiSalvo, F.J. (1998) Morphology and characterization of GaN single crystals grown in a Na flux. *J. Cryst. Growth*, **186**, 8–12.

149. Kawamura, F., Moroshita, M., Omae, K., Yoshimura, M., Mori, Y., and Sazaki, T. (2005) The effect of Na and some additives on nitrogen dissolution in the Ga-Na system: a growth mechanism of GaN in the Na flux method. *J. Mater. Sci.: Mater. Electron.*, **16**, 29–34.

150. Reed, E.L. and Droher, J.J. (1970) Solubility and Diffusivity of Inert Gases in Liquid Sodium, Potassium and NaK. Report LMEC-69-36.

151. Wang, J., Yuan, W., and Li, M. (2007) Thermodynamic modeling of the Ga-N-Na system. *J. Cryst. Growth*, **307**, 59–65.

152. Mori, Y., Imade, M., Takazawa, H., Imbayashi, H., Todoroki, T., Kitamoto, K., Maruyama, M., Yoshimura, M., Kitaoka, Y., and Sasaki, T. (2012) Growth of bulk GaN crystals by Na flux method under various conditions. *J. Cryst. Growth*, **350**, 72–74.

153. Peters, D. (1990) Ammonothermal synthesis of aluminium nitride. *J. Cryst. Growth*, **104**, 411–418.

154. Chen, Q.-S., Yan, J.-Y., Jiang, Y.-N., and Li, W. (2012) Modelling on ammonothermal growth of GaN semiconductor crystals. *Progr. Cryst. Growth Charact. Mater.*, **58**, 61–73.

155. Hashimoto, T., Wu, F., Speck, J.S., and Nakamura, S. (2007) A GaN bulk crystal with improved structural quality grown by the ammonothermal method. *Nat. Mater.*, **6**, 568–571.

156. Hashimoto, T., Wu, F., Speck, J.S., and Nakamura, S. (2008) Ammonothermal growth of bulk GaN. *J. Cryst. Growth*, **310**, 3907–3910.

157. Dwilinski, R., Doradzinski, R., Garczynski, J., Sierzputowski, L.P., Puchalski, A., Kanbara, Y., Yagi, K., Minakuchi, H., and Hayashi, H. (2008) Excellent crystallinity of truly bulk ammonothermal GaN. *J. Cryst. Growth*, **310**, 3911–3916.

158. Yoshikawa, A., Ohshima, E., Fukuda, T., Tsuji, H., and Oshima, K. (2004) Crystal growth of GaN by ammonothermal method. *J. Cryst. Growth*, **260**, 67–72.

159. Nakamura, S., Harada, Y., and Senoh, M. (1991) Novel metalorganic chemical vapor deposition system for GaN growth. *Appl. Phys. Lett.*, **58**, 2021.

160. Nakamura, S. (1991) GaN growth using GaN Buffer layer. *Jpn. J. Appl. Phys.*, **30**, L1705–L1707.

161. Nakamura, S., Mukai, T., and Senoh, M. (1994) Candela-class high-brightness InGaN/AlGaN double-heterostructure blue-light-emitting-diodes. *Appl. Phys. Lett.*, **64**, 1687–1789.

162. Nakamura, S., Senoh, M., Iwasa, N., and Nagahama, S.-I. (1995) High-brightness InGaN blue, green and yellow light-emitting diodes with quantum well structures. *Jpn. J. Appl. Phys.*, **34**, L797–L799.

163. Nakamura, S., Senoh, M., Nagahama, S.-.I. *et al.* (1996) Room-temperature continuous-wave operation of ingan multi-quantum-well structure laser diodes. *Appl. Phys. Lett.*, **69**, 4056.

164. Nakamura, S., Senoh, M., Nagahama, S.-I. *et al.* (1996) InGaN-based multi-quantum-well-structure laser diodes. *Jpn. J. Appl. Phys.*, **35**, L74–L76.

165. Rudolph, P. (2003) Non-stoichiometry related defects at the melt growth of semiconductor compound crystals-a review. *Cryst. Res. Technol.*, **38**, 542–554.

166. Fiechter, S. (2004) Defect formation energies and homogeneity ranges of rock salt-, pyrite-, chalcopyrite- and molybdenite-type compound semiconductors. *Sol. Energy Mater. Sol. Cells*, **83**, 459–477.

167. Berding, M.A., Sher, A., and Cher, A.B. (1990) Vacancy formation and extraction energies in semiconductor compounds and alloys. *J. Appl. Phys.*, **68**, 5064–5076.

168. Puska, M. (1989) Electronic structures of point defects in III-V compound semiconductors. *J. Phys. Condens. Matter*, **1**, 7347–7366.

169. Neumark, G.F. (1997) Defects in wide gap II-VI crystals. *Mater. Sci. Eng.*, **R21**, 1–46.
170. Hegewald, S., Hein, K., Frank, C., John, M., and Burig, E. (1994) Investigation on the equilibrium vapor pressure over a GaAs melt. *Cryst. Res. Technol.*, **29**, 549–554.
171. Blakemore, S. (1987) Multitopic reviews, in *Gallium Arsenide*, Blakemore, S. Ed. pp. 3–93, The American Institute of Physics.
172. Yannakopoulos, P.H., Zardas, G.E., Papaioannou, G.J., Symeonides, C.I., Vesely, M., and Euthymiou, P.C. (2009) Behavior of semiinsulating GaAs energy levels. *Rev. Adv. Mater. Sci.*, **22**, 52–59.
173. King, S.E., Dietrich, H.B., Henry, R.L., Katzer, D.S., Moore, W.J., Phillips, G.W., and Mania, R.C. (1996) Development and characterization of zone melt growth GaAs for gamma-ray detectors. *IEEE Trans. Nucl. Sci.*, **43**, 1376–1380.
174. Northrup, J.E. and Zhang, S.B. (1993) Dopant and defect energetics: Si in GaAs. *Phys. Rev. B*, **47**, 6791–6794.
175. Vázquez-Cortés, D., Cruz-Hernández, E., Méndez-García, V.H., Shimomura, S., and López-López, M. (2012) Optical and electrical properties of Si-doped GaAs films grown on (631)-oriented substrates. *J. Vacuum Sci. Technol. B*, **30**, 02B125-4.
176. Lagowski, J., Lin, D.G., Aoyama, T., and Gatos, H.C. (1984) Identification of oxygen related midgap level in GaAs. *Appl. Phys. Lett.*, **44**, 336.
177. Hulluvarad, S.S., Naddaf, M., and Bhoraskar, S.V. (2001) Detection of oxygen-related defects in GaAs by exo-electron emission spectroscopy. *Nucl. Instrum. Methods Phys. Res., Sect. B*, **183**, 432–438.
178. Laaksonen, K., Komsa, H.-P., Rantala, T.T., and Nieminen, R.M. (2008) Nitrogen interstitial defects in GaAs. *J. Phys. Condens. Matter*, **20**, 235231–235234.
179. Yamada, T., Kudo, T., Tajima, K., Otsuka, A., Narushima, T., Ouchi, C., and Iguchi, Y. (2004) Boron and nitrogen in GaAs and InP melts equilibrated with B_2O_3 flux. *Mater. Trans.*, **45**, 1306–1310.
180. Bachmann, K.J. and Buehler, E. (1974) The growth of InP crystals from the melt. *J. Electron. Mater.*, **3**, 279–301.
181. Müller, G., Schwesig, P., Birkmann, B., Hartwig, J., and Eichler, S. (2005) Types and origin of dislocations in large GaAs and InP bulk crystals with very low dislocation densities. *Phys. Status Solidi A*, **202**, 2870–2879.
182. Schäfer, N., Stierlen, J., and Müller, G. (1991) Growth of InP crystals by the horizontal gradient freeze technique. *Mater. Sci. Eng. B*, **9**, 19–22.
183. Shimizu, A., Nishizawa, J.-I., Oyama, Y., and Suto, K. (2002) InP melts: investigation of wetting between boat materials in Bridgman growth. *J. Cryst. Growth*, **237–239**, 1697–1700.
184. Kubota, E., Katsui, A., and Ohmori, Y. (1987) Growth temperature and phosphorous vapor pressure dependence of silicon incorporation into InP crystals in solution growth process. *J. Cryst. Growth*, **82**, 737–746.
185. Davies, R.H., Dinsdale, A.T., Gisby, J.A. *et al.* (2002) MTDATA – thermodynamic and phase equilibrium software from the national physical laboratory. *Calphad*, **26** (2), 229–271.
186. Lapin, N.V. and Grinko, V.V. (1996) Redistribution coefficients of metal impurities in indium phosphide grown by synthesis-solute diffusion technique. *Ceram. Int.*, **22**, 271–274.

187. Bonner, W.A. (1981) InP synthesis and LEC growth of twin free crystals. *J. Cryst. Growth*, **54**, 21–31.
188. Hirt, G., Hofmann, D., Mosel, F. *et al.* (1991) Annealing and bulk crystal growth of undoped InP under controlled P-pressure: a perspective for the preparation of undoped SI InP. Proceedings of the Third International Conference on Indium Phosphide and Related Materials, pp. 16–19.
189. De Oliveira, C.E.M., De Carvalho, M.M.G., and Miskys, C.R. (1997) Growth of InP by LEC using glassy carbon crucibles. *J. Cryst. Growth*, **173**, 214–217.
190. Kubota, E. and Katsui, A. (1985) Growth of InP single crystals using ceramics AlN crucible. *Jpn. J. Appl. Phys.*, **24**, L69–L71.
191. Allegretti, F., Carrara, A., and Pizzini, S. (1993) Growth and characterization of ZnSe for low temperature calorimetry applications. *J. Cryst. Growth*, **128**, 646–649.
192. Kim, T.S., Shin, Y.J., Jeong, T.S., Choi, C.T., Yu, P.Y., and Hong, K.J. (2001) Growth of zinc selenide single crystal by the sublimation method. *J. Korean Phys. Soc.*, **38**, 47–51.
193. Del Sordo, S., Abbene, L., Caroli, E., Mancini, A.M., Zappettini, A., and Ubertini, P. (2009) Progress in the development of CdTe and CdZmTe semiconductor radiation detectors for Astrophysical and medical applications. *Sensors*, **9**, 3491–3527.
194. Jung, I., Perkins, J., Krawczynski, H., Sobotka, L., and Komarov, S. (2004) CZT detectors as MeV gamma-ray calorimeters with excellent position and energy resolution. *IEEE Nucl. Sci. Symp. Conf. Rec.*, **7**, 4383–4387.
195. James, R.B. (2009) Novel Method for Growing Te Inclusion Free CdZnTe Crystals, June 2–4 2009 unclassified web reference.
196. Chu, M. (2004) Role of zinc in CdZnTe radiation detectors. *IEEE Trans. Nucl. Sci.*, **51**, 2405–2411.
197. Zappettini, A., Zha, M., Marchini, L., Calestani, D., Mosca, R., Gombia, E., Zanotti, L., Zanichelli, M., Pavesi, M., Auricchio, N., and Carol, E. (2009) Boron oxide encapsulated vertical bridgman grown CdZnTe crystals as X-ray detector material encapsulant. *IEEE Trans. Nucl. Sci.*, **56**, 1743–1746.
198. Rudolph, P. (1994) Fundamental studies on Bridgman growth of CdTe. *Progr. Cryst. Growth. Charact. Mater.*, **29**, 275–381.
199. Okamoto, H. (2012) Cd-Te (Cadmium-Tellurium). *J. Phase Equilb. Diffus.*, **33**, 414.
200. Szeles, C., Cameron, S.E., Ndap, J.O., and Chalmers, W.C. (2002) Advances in the crystal growth of semi-insulating CdZnTe for radiation detector applications. *IEEE Trans. Nucl. Sci.*, **49**, 2535–2543.
201. Pavesi, M., Marchini, L., Zha, M., Zappettini, A., Zanichelli, M., and Manfredi, M. (2011) On the role of Boron in CdTe and CdZnTe crystals. *J. Electron. Mater.*, **40**, 2043–2050.
202. Grzegory, I., Luczink, B., Bockowski, M., Pastuska, B., Krysko, M., Kamler, G., Novak, G., and Porowski, S. (2006) Growth of bulk GaN by HVPE on pressure grown seeds. *Proc. SPIE*, **6121**, 61207–1.
203. Franzosi, A., Giarda, L., Pelosini, L., and Pizzini, S. (1978) Growth of single crystal silicon rods from the vapour phase, in *Proceedings of the 1st E.C. Photovoltaic Solar Energy Conference*, D. Reidel Publishers, pp. 153–163.
204. Bauerle, D., Leyendecker, G., Wagner, D., Bause, E., and Lu, Y.C. (1983) Laser grown single crystals of silicon. *Appl. Phys. A*, 30147–149.

205. Mendelson, S. (1964) Stacking fault nucleation in epitaxial silicon on variously oriented silicon substrates. *J. Appl. Phys.*, **35**, 1570–1580.

206. Chen, Q-S., Prasad, V., Zhang, H., Dudley, M. (2003) Silicon carbide crystals part II: process physics and modeling in *Crystal Growth for Modern Technology*, K. Byrappa, T. Ohachi, (Ed) Chapter 7, pp. 233–268, William Andrew Publishing, Norwich, NY.

207. Yakimova, R., Syvjarvi, M., Tuominen, M., Yakimov, T., Råback, R., Vehanen, A., and Janzen, E. (1999) Seeded sublimation growth of 6H and 4H-SIC crystals. *Mater. Sci. Eng., B*, **61–62**, 54–57.

208. Hartman, J.S., Berno, B., Hazendonk, P., Kirby, C.W., Ye, E., Zwanziger, J., and Bain, A.D. (2009) NMR studies of nitrogen doping in the 4H polytype of silicon carbide: site assignments and spin–lattice relaxation. *J. Phys. Chem. C*, **113**, 15024–15036.

209. Tymicki, E., Grasza, K., Racka, K. *et al.* (2001) Effect of Nitrogen Doping on the Growth of 4H Polytype on the 6H-SiC Seed. SICMAT Project Contract No. UDA-POIG.01.03.01-14-155/09.

210. Chen, J., Lien, S.C., Shin, Y.C., Feng, Z.C., Kuan, C.H., Zhao, J.H., and Lu, W.J. (2009) Occurrence of polytype transformation during nitrogen doping of SiC bulk wafer. *Mater. Sci. Forum*, **600–603**, 39–42.

211. Straubinger, T.L., Bickermann, M., Weingartner, R., Wellmann, P.J., and Winnacker, A. (2002) Aluminum p-type doping of silicon carbide crystals using a modified physical vapor transport growth method. *J. Cryst. Growth*, **240**, 117–123.

212. Trassodaine, A., Cadoret, R., Aujol, E. (2003) Growth of gallium nitride by hydride vapour phase epitaxy, in *Nitride Semiconductors, Handbook on Materials and Devices*, P. Ruterana, M. Albrecht, J. Neugebauer Ed., pp. 193–219, Wiley-VCH Verlag GmbHWeinheim.

213. Shibata, T., Sone, H., Yahashi, K., Yamaguchi, M., Hiramatsu, K., Sawaki, N., and Itoh, N. (1998) Hydride vapor-phase epitaxy growth of high-quality GaN bulk single crystal by epitaxial lateral overgrowth. *J. Cryst. Growth*, **189/190**, 67–71.

214. Kumagai, Y., Murakami, H., Seki, H., and Koukitu, A. (2002) Thick and high-quality GaN growth on GaAs substrates for preparation of freestanding GaN. *J. Cryst. Growth*, **246**, 215–222.

215. Kim, S.T., Lee, Y.J., Moon, D.C., Hong, C.H., and Yoo, T.K. (1998) Preparation and properties of free-standing HVPE grown GaN substrates. *J. Cryst. Growth*, **194**, 37–42.

216. (2013) *Compound Semiconductors*. Vol. **19** (6).

217. (2013) *Compound Semiconductors*. Vol. **19** (8).

218. O'Donnell, K.P., Mosselmans, J.F.W., Martin, R.W., Pereira, S., and White, M.E. (2001) Structural analysis of InGaN epilayers. *J. Phys. Condens. Matter*, **13**, 6977–6991.

219. Fitzgerald, E. A., Xie, Y.-H., Green, M. L. *et al.* (1991) Totally relaxed Ge_xSi_{1-x} layers with low threading dislocation densities grown on Si substrates *Appl. Phys. Lett.* **59** 811.

220. Tahini, H., Chroneos, A., Grimes, R.W., Schwingenschlogl, U., and Dimoulas, A. (2012) Strain-induced changes to the electronic structure of germanium. *J. Phys. Condens. Matter*, **24**, 195802 (4 pp).

221. Jain, S.C. and Hayes, W. (1991) Structure, properties and applications of Ge_xSi_{1-x} strained layers and superlattices. *Semicond. Sci. Technol.*, **6**, 547–576.

222. Nanishi, Y., Araki, T., Yamaguchi, T. (2010) Molecular Beam epitaxy of InN *in Indium Nitride and Related Alloys* Veal, T.D., McConville, C.F., Shaff, W.J. Ed, CRC Press, Boca Raton, FL, pp. 1–50.
223. Bolkhovityanov, Y.B., Pchelyakov, O.P., Sokolov, L.V., and Chikichev, S.I. (2003) Artificial GeSi substrates for heteroepitaxy: achievements and problem*s*. *Semiconductors*, **37**, 493–518.
224. Chrastina, D., Isella, G., Bollani, M., Roessner, B., Mueller, E., Hackbarth, T., Wintersberger, E., Zhong, Z., Stangl, J., and von Kaenel, H. (2005) Thin relaxed SiGe virtual substrates grown by low-energy plasma enhanced chemical vapor deposition. *J. Cryst. Growth*, **281**, 281–289.
225. Kummer, M., Rosenblad, C., Dommann, A., Hackbarth, T., Höck, G., Zeuner, M., Müller, E., and von Känel, H. (2002) Low energy plasma enhanced chemical vapor deposition. *Mater. Sci. Eng., B*, **89**, 288–295.
226. Falub, C.V., von Känel, H., Isa, F., Bergamaschini, R., Marzegalli, A., Chrastina, D., Isella, G., Müller, E., Niedermann, P., and Miglio, L. (2012) Scaling hetero-epitaxy from layers to three-dimensional. *Cryst. Sci.*, **335**, 1330–1334.
227. Pezzoli, F., Isa, F., Isella, G., Falub, C.V., Kreiliger, T., Salvalaglio, M., Bergamaschini, R., Grilli, E., Guzzi, M., and von Kaenel, H. (2014) Leo Miglio, Germanium crystals on silicon show their light. *Phys. Rev. Appl.*, **1**, 044005.
228. Ruterana, P. (2010) Transmission electron microscopy and XRD investigations of InAlN(GaN) thin heterostructures for HEMT applications. *Proc. SPIE*, **7602**, 6020K.
229. Ruterana, P., Albrecht, M., and Neugebauer, J. (eds) (2003) *Nitride Semiconductors, Handbook on Materials and Devices*, Wiley-VCH Verlag GmbH, pp. 379–438.
230. Ruterana, P., Chauvat, M.P., ArroyoDasilva, Y., Lei, H., Lahourcade, L., and Monroy, E. (2010) Extended defects in nitride layers, influence on the quantum wells and quantum dots. *Proc. SPIE*, **7602**, 760210.
231. Monroy, E., Kandaswamy, P.K., Machhadani, H., Wirthmüller, A., Sakr, S., Lahourcade, L., Das, A., Tchernycheva, M., Ruterana, P., and Julien, F.H. (2010) Polar and semipolar III-nitrides for long wavelength intersubband devices. *Proc. SPIE*, **7608**, 76081G.
232. Arroyo-Rojas Dasilva, Y., Ruterana, P., Lahourcade, L., Monroy, E., and Nataf, G. (2010) Extended defects in semipolar (1122) gallium nitride. *Proc. SPIE*, **7602**, 76020G.
233. Ruterana, P., Sanchez, A.M., and Nouet, G. (2003) Extended defects in wurtzite GaN layers. Atomic structure,formation and interaction mechanism, in *Nitride Semiconductors, Handbook on Materials and Devices* (eds P. Ruterana, M. Albrecht, and J. Neugebauer), Wiley-VCH Verlag GmbH, Weinheim, pp. 379–438.
234. Ambacher, O. (1998) Growth and applications of Group III-nitrides. *J. Phys. D: Appl. Phys.*, **31**, 2653–2710.
235. Kung, P., Sun, C.J., Saxler, A., Ohsato, H., and Razeghi, M. (1994) Crystallography of epitaxial growth of wurtzite-type thin films on sapphire substrates. *J. Appl. Phys.*, **75**, 4515–4519.
236. Yao, T. and Hong, S.-K. (2009) *Oxide and Ntride Semiconductors, Processing, Properties and Applications*, Springer.
237. Kukushkin, S.A., Osipov, A.V., Bessolov, V.N., Medvedev, B.K., Nevolin, V.K., and Tcarik, K.A. (2008) Substrates for epitaxy of gallium nitride: new materials and techniques. *Rev. Adv. Mater. Sci.*, **17**, 1–32.

238. Leszczynski, M., Suski, T., Teisseyre, H., Perlin, P., Grzegory, I., Jun, J., Porowski, S., and Moustakas, T.D. (1994) Thermal expansion of gallium nitride. *J. Appl. Phys.*, **76**, 4909–4911.

239. Dadgar, A., Hums, C., Diez, A., Schulze, F., Bläsing, J., and Krost, A. (2006) Epitaxy of GaN LEDs on large substrates: Si or sapphire? *Proc. SPIE*, **6355**, 63550R.

240. Leszczynski, M., Teisseyre, H., Suski, T., Grzegory, I., Bockowski, M., Jun, J., Porowski, S., Pakula, K., Baranowski, J.M., Foxon, C.T., and Cheng, T.S. (1996) Lattice parameters of gallium nitride. *Appl. Phys. Lett.*, **69**, 73.

241. Yoshida, S., Misawa, S., and Itoh, A. (1975) Epitaxial growth of aluminum nitride films on sapphire by reactive evaporation. *Appl. Phys. Lett.*, **26**, 461.

242. Ambacher, O., Brandt, M.S., Dimitrov, R., Metzger, T., Stutzmann, M., Fisher, R.A., Miehr, A., Bergmaier, A., and Dollinger, G. (1996) Thermal stability and desorption of group III nitrides prepared by metal organic chemical vapor deposition. *J. Vac. Sci. Technol.*,. *B*, **14**, 3532–3542.

243. Sienz, S., Gerlach, J.W., Höche, T., Sidorenko, A., Mayerhöfer, T.G., Benndorf, G., and Rauschenbach, B. (2004) Comparison of ion-beam-assisted molecular beam epitaxy with conventional molecular beam epitaxy of thin hexagonal gallium nitride films. *J. Cryst. Growth*, **264**, 184–191.

244. Parikh, R.P. and Adomaitis, R.A. (2006) An overview of gallium nitride growth chemistry and its effect on reactor design: application to a planetary radial-flow CVD system. *J. Cryst. Growth*, **286**, 259–278.

245. Jacko, M.G. and Prince, S.J.W. (1963) The Pyrolysis of trimethylgallium. *Can. J. Chem.*, **41**, 1560–1567.

246. Hirako, A., Yashitani, M., Nishibayashi, M., Nishikawa, Y., and Ohkawa, K. (2002) GaN-MOVPEgrowth and its microscopic chemistry of gaseous phase by computational thermodynamic analysis. *J. Cryst. Growth*, **237–239**, 931–935.

247. Konnov, A.A. and DeRuyck, J. (2000) Kinetic modeling of the thermal decomposition of ammonia. *Combust. Sci. Technol.*, **152**, 23–37.

248. Srinivasan, S., Geng, L., Liu, R., Ponce, F.A., Narukawa, Y., and Tanaka, S. (2003) Slip systems and misfit dislocations in InGaN epilayers. *Appl. Phys. Lett.*, **83**, 5187.

249. Arslan, I. and Browning, N.D. (2002) Intrinsic electronic structure of threading dislocations in GaN. *Phys. Rev. B*, **65**, 075310.

250. Blumenau, A.T., Elsner, J., Jones, R., Heggie, M.I., Öberg, S., Frauenheim, T., and Briddon, P.R. (2000) Dislocations in Hexagonal and Cubic GaN. *J. Phys. Condens. Matter*, **12**, 10223–10234.

251. Dumiszewska, E., Strupinski, W., and Zdunek, K. (2007) Interaction between dislocations density and carrier concentration of gallium nitride layers. *J. Superhard Mater.*, **29**, 174–176.

252. Saritas, M. and McKell, H.D. (1988) Comparison of minority carrier diffusion length measurements in silicon by the photoconductive decay and surface photovoltage methods. *J. Appl. Phys.*, **63**, 4561–4567.

253. Deng, X. and Schiff, E.A. (2003) Amorphous silicon–based solar cells, in *Handbook of Photovoltaic Science and Engineering* (eds A. Luque and S. Hegedus), John Wiley & Sons, Ltd, Chichester, pp. 505–565.

254. Gaugash, P. and Milnes, A.G. (1981) Minority carriers diffusion length in Cadmium Telluride. *J. Electrochem. Soc.*, **128**, 921–924.

255. Brown, G., Faifer, V., Pudov, A., Anikeev, S., Bykov, E., Contreras, M., and Wu, J. (2010) Determination of the minority carrier diffusion length in compositionally graded Cu InGaSe$_2$ solar cells using electron beam induced current. *Appl. Phys. Lett.*, **96**, 022104.

256. McCandless, B. E. and Sites, J. R. (2003) Cadmium telluride solar cells, in *Handbook of Photovoltaic Science and Engineering* Eds Luque, A. and Hegedus, S., pp. 617–662 John Wiley & Sons, Ltd, Chichester.

257. Shah, A., Torres, P., Tscharner, R., Wyrsch, N., and Keppner, H. (1999) Photovoltaic technology: the case for thin-film solar cells. *Science*, **285**, 692–698.

258. Niki, S., Contreras, M., Repins, I., Powalla, M., Kushiya, K., Ishizuka, S., and Matsubara, K. (2010) CIGS absorbers and processes. *Prog. Photovoltaics Res. Appl.*, **18**, 453–466.

259. Kroll, U., Shah, A., Keppner, H., Meier, J., Torres, P., and Fischer, D. (1997) Potential of VHF-plasmas for low-cost production of a-Si: H solar cells. *Sol. Energy Mater. Sol. Cells*, **48**, 343–350.

260. Saito, K., Sano, M., Okabe, S., Sugiyama, S., and Ogawa, K. (2005) Microcrystalline silicon solar cells fabricated by VHF plasma CVD method. *Sol. Energy Mater. Sol. Cells*, **86**, 565–575.

261. Rath, J. (2012) Thin film deposition processes, in *Advanced Silicon Materials for Photovoltaic Applications* (ed S. Pizzini), John Wiley & Sons, Ltd, Chichester, pp. 235–286.

262. Binetti, S., Acciarri, M., Bollani, M., Fumagalli, L., von Känel, H., and Pizzini, S. (2005) Nanocrystalline silicon film grown by LEPECVD for optoelectronic applications. *Thin Solid Films*, **487**, 19–25.

263. Pizzini, S., Acciarri, M., Binetti, S., Cavalcoli, D., and Cavallini, A. (2006) Nanocrystalline silicon films as multifunctional material for optoelectronic and photovoltaic applications. *Mater. Sci. Eng., B*, **134**, 118–124.

264. Cavalcoli, D., Cavallini, A., Rossi, M., and Pizzini, S. (2007) Micro-and nano structures in silicon studied by DLTS and scanning probe methods. *Semiconductors*, **41**, 421–426.

265. Rosenblad, C., Deller, H.R., Dommann, A., Meyer, T., Schröter, P., and von Känel, H. (1998) Silicon epitaxy by low-energy plasma enhanced chemical vapor deposition. *J. Vac. Sci. Technol., A*, **16**, 2785–2791.

266. Cavallotti, C. (2012) Modeling of thin film deposition processes, in *Advanced Silicon Materials for Photovoltaic Applications* (ed S. Pizzini), John Wiley & Sons, Inc., pp. 287–310.

267. Novikov, P.L., Le Donne, A., Cereda, S., Miglio, L., Pizzini, S., Binetti, S., Rondanini, M., Cavallotti, C., Chrastina, D., Moiseev, T., von Känel, H., Isella, G., and Montalenti, F. (2009) Crystallinity and microstructure in Si films grown by plasma-enhanced chemical vapor deposition: a simple atomic-scale model validated by experiments. *Appl. Phys. Lett.*, **94**, 051904.

268. Poliani, E., Somaschini, C., Sanguinetti, S., Grilli, E., Guzzi, M., Le Donne, A., Binetti, S., Pizzini, S., Chrastina, D., and Isella, G. (2009) Tuning by means of laser annealing of electronic and structural properties of nc-Si/a-Si:H. *Mater. Sci. Eng., B*, **159–160**, 31–33.

269. Pavesi, L., Dal Negro, L., Mazzoleni, C., Franzo, G., and Priolo, F. (2000) Optical gain in silicon nanocrystals. *Nature*, **408**, 440–444.
270. Huang, Z., Geyer, N., Wener, P., de Boor, J., and Gösele, U. (2011) Metal-assisted chemical etching of silicon: a review. *Adv. Mater.*, **23**, 285–308.
271. Peng, K-Q., Wang, X., Li, L., Hu, Y., Lee, S.-T. (2013) Silicon nanowires for advanced energy conversion and storage *Nano Today* **8**, 75–97.
272. Hasan, M., Huq, M.F., and Mahmood, Z.H. (2013) A review on electronic and optical properties of silicon nanowire and its different growth techniques. *SpringerPlus*, **2**, 151–9.
273. Sato, K., Castaldini, A., Fukata, N., and Cavallini, A. (2012) Electronic level scheme in boron- and phosphorus-doped silicon nanowires. *Nanoletters*, **12**, 3012–3017.
274. Nassiopoulou, A.G., Gianneta, V., and Katsogridakis, C. (2011) Si nanowires by a single-step metal-assisted chemical etching process on lithographically defined areas: formation kinetics. *Nanoscale Res. Lett.*, **6**, 597.
275. Lehmann, V. and Gösele, U. (1991) Porous silicon formation: a quantum wire effect. *Appl. Phys. Lett.*, **58**, 856–858.
276. Irrera, A., Artoni, P., Iacona, F., Pecora, E.F., Franzo, G., Galli, M., Fazio, B., Boninelli, S., and Priolo, F. (2012) Quantum confinement and electroluminescence in ultrathin silicon nanowires fabricated by a maskless etching technique. *Nanotechnology*, **23**, 075204.
277. Lieber, C.M. (2011) Semiconductor nanowires: a platform for nanoscience and nanotechnology. *MRS Bull.*, **36**, 1052–1063.
278. Thorpe, T.P. Jr.,, Fahrenbruch, A.L., and Bube, R.H. (1986) Electrical and optical characterization of grain boundaries in polycrystalline cadmium telluride. *J. Appl. Phys.*, **60**, 3622–3630.
279. Zywitzki, O., Modes, T., Morgner, H., Metzner, C., Siepchen, B., Späth, B., Drost, C., Krishnakumar, V., and Frauenstein, S. (2013) Effect of chlorine activation treatment on electron beam induced current signal distribution of cadmium telluride thin film solar cells. *J. Appl. Phys.*, **114**, 163518.
280. Alberi, K., Fluegel, B., Moutinho, H., Dhere, R.G., Li, J.V., and Mascarenhas, A. (2013) Measuring long-range carrier diffusion across multiple grains in polycrystalline semiconductors by photoluminescence imaging. *Nat. Commun.*, **4**. doi: 10.1038/ncomms3699, Article No. 2966.
281. Amin, N. (2011) Promises of $Cu(In, Ga) Se_2$ thin film solar cells from the perspective of material properties, fabrication methods and current research challenges. *J. Appl. Sci.*, **11**, 401–410.
282. Romero, E., Calderon, C., Bartolo-Perez, P., Mesa, F., and Gordillo, G. (2006) Phase identification and AES depth profile analysis of $Cu(In, Ga)Se_2$ thin films. *Braz. J. Phys.*, **36**, 1050–1053.
283. Rockett, A., Granath, K., Asher, S., Al Jassim, M.M., Hasoon, F., Matson, R., Basol, B., Kapur, V., Britt, J.S., Gillespie, T., and Marshall, C. (1999) Na incorporation in Mo and $CuInSe_2$ from production processes. *Sol. Energy Mater. Sol. Cells*, **59**, 255–264.
284. Schäffer, H.A. (2012) Transport phenomena and diffusion anomalies in glass. *Ceram. Mater.*, **64**, 156–161.

285. Tarascon, J.M. and Hull, G.W. (1986) Sodium intercalation into the layer oxides $Na_xMo_2O_4$. *Solid State Ionics*, **22**, 85–96.

286. Wei, S.H., Zhang, S.B., and Zunger, A.J. (1999) Effects of Na on the electrical and structural properties of $CuInSe_2$. *J. Appl. Phys.*, **85**, 7214–7218.

287. Niki, S., Contreras, M., Repins, I., Powalla, M., Kushiya, K., Ishizuka, S., and Matsubara, K. (2010) CIGS absorbers and processes. *Prog. Photovolt: Res. Appl.* **18**: 453–466.

5

Physical Chemistry of Semiconductor Materials Processing

5.1 Introduction

It was shown in previous chapters that impurities are unavoidably introduced and defects are generated in semiconductor materials during growth and device manufacturing processes, which often, if not always, operate in conditions which enhance the chemical and thermal aggressiveness of the environment. The associated, also unavoidable, consequence is the electrical degradation of the semiconductor quality due to impurity contamination and point- and extended-defects generation.

Impurities and defects in semiconductors can, however, be manipulated, that is engineered, in the sense that it is possible to use thermal, chemical and/or mechanical processes to stimulate inter-defect interaction, aggregation or segregation at extended defects or at internal and external surfaces in order to electrically deactivate defects and impurities.

If the reacted impurities remain inactive under further thermal treatments or under photon or particle irradiations,[1] impurities are stability gettered or passivated.

Some kinds of manipulation are purely physico-chemical in character, as is the case for hydrogenation of impurities and dangling bonds at extended defects. Other defect engineering tools imply the application of strain fields or concentration gradients to drive impurities to suitable sinks, where they are immobilized and electrically deactivated.

Thermal annealing is another type of defect engineering which relaxes mechanical stresses generated by previous treatments to which a particular sample has been submitted, induces decomposition, condensation or segregation of defects and impurities, with the objective to bring the material properties to the quality needed for further process steps. Recovery of irradiation damage by thermal annealing of ion implanted layers or np junction formation using high temperature dopant diffusion are typical examples of thermal processes systematically applied in semiconductor device manufacturing.

---

[1] Semiconductor radiation detectors, such as X-ray sensors for medicine or security applications, should withstand high radiation fields without detection degradation.

Thermal annealing can be applied to alloy a heterogeneous multicomponent matrix preliminarily deposited on a convenient substrate by a sequential sublimation process, as is the case for copper indium selenide-based solar cells, as discussed in Chapter 4.

Gettering and hydrogen passivation processes are the most straightforward application of defect/impurity chemical interaction schemes and are in common use in a variety of processes for manufacturing silicon-based devices. Examples are the passivation of grain boundaries (GBs) and dislocations in mc-Si (multicrystalline silicon) solar cells, the intrinsic gettering at oxide precipitates in ultra large scale integration (ULSI), the extrinsic gettering processes at P-rich surface layers or at Al-sinks in the silicon solar cell manufacturing processes.

Gettering processes are applied also in compound semiconductor (CS) technology, though with different modalities or aims, as CS devices consist generally of micrometres-thin single crystalline or polycrystalline layers deposited on a substrate of different nature which gives the necessary mechanical resistance to the active layers.

An examples is the use of reactive impurities to getter oxygen in III–V CS [1], or to reduce the dislocation multiplication in CS [2], or to passivate the electrical activity of GBs in CS [3].

The aim of this chapter is, therefore, to discuss the conceptual physico-chemical background of defect/impurities manipulation procedures in semiconductors, limiting, however, most of the attention to silicon to which the main efforts have been devoted so far, with the perspective to address this approach to the micro- and the nano-scale.

5.2   Thermal Annealing Processes

Thermal annealing is currently applied in a number of processes involving semiconductors. However, it is not new, being already used by our ancestors in the iron age for hardening iron, a dual stage process starting with the mechanical damage of hot iron followed by a thermal quench, and is systematically applied in metallurgy till today.

An annealing process consists in the supply of work, in the form of an amount $Q = \int_0^t q \, dt$ of heat, to a physical system in order to drive or enhance the rate of a transformation.

The transformation could be a change of state, the formation of a solution or compound or the dissolution of a precipitate in a solid system, each strictly obeying thermodynamic rules. Heat can be supplied isothermally by an extended heating source (a conventional furnace) or delivered in the requireded amount by a local and fast irradiation of the sample, carried out by using a flash lamp, a laser or a microwave (MW) source.

Microwave irradiation could be particularly suitable when the system involved presents *in toto* or in a specific region a selective absorbance in the MW range, which could allow very localized heating.

Depending on the kind of transformation involved, on the specific region of the sample where the transformation should occur[2] and on the thermodynamic and kinetic parameters of the process, isothermal, rapid or ultrarapid thermal annealings could be preferred.

As an example, high activation energies for a transformation dominated by a true reaction limited process

$$k(\mathrm{s}^{-1}) = \omega \exp -(E_\mathrm{R}/kT) \qquad (5.1)$$

---

[2] A typical case is that of shallow or ultra-shallow junction formation.

or for the diffusion of the species generated by a decomposition process

$$D = D^o \exp -\frac{E_B}{kT} \tag{5.2}$$

would favour the use of fast annealing processes to enhance the direct transformation and to suppress or retard the inverse one.

Slow reaction rates, such as that of the nucleation and precipitation of oxygen in Czochralski (CZ) silicon occurring at $T > 700\,^{\circ}\text{C}$ or the decomposition of thermal donors nucleated in large diameter CZ silicon ingots[3] during the post-growth ingot cooling would favour, instead, the use of conventional annealing processes.

Intermediate conditions are satisfied by low thermal capacity furnaces and high speed radiation sources which allow fast heating and cooling cycles, of which the heating and cooling rates are determined uniquely by the thermal capacity and by the heat transfer coefficients of the material to be annealed.

Doping processes carried out by dopant diffusion from a heterogeneous source deposited at the surface of a semiconductor material or by ion implantation are a common practice in semiconductor technology to form pn and np junctions, but the formation of ultra-shallow junctions requires authoritative experience and technological excellence.

Several problems arise, however, even when the doping process is carried out in a conventional furnace, and the sample[4] is submitted to a thermal cycle consisting of a heating ramp to bring the sample to the doping temperature, followed by a constant temperature segment during the dopant diffusion and, finally, the cooling stage which could be carried out inside or outside the furnace.

The doping impurity concentration profile, which at the end of a constant temperature stage depends only on the diffusion length $L_D = (D_i t)^{1/2}$ of the dopant, would, in fact, depend on the thermal inertia of the furnace when the sample is cooled inside the furnace, or on the thermal inertia of the sample itself and of its container, if extracted from the furnace and suitably quenched to room temperature.

Additional problems arise with high efficiency solar cells, where the junction depth should be as shallow as possible to limit the loss of short diffusion length carriers generated by blue light at the surface of the emitter region [4]. Rapid thermal annealing (RTA) [5], flash lamp annealing and ultrafast annealing provide the necessary conditions to avoid the undesired thermal diffusion of the dopant [6], as well as the thermal ionization of metallic impurities, as a millisecond annealing (0.2–20 ms) allows one to heat only the surface of the sample, using the substrate as the heat sink.

The ion implantation doping processes require, instead, a post-implantation annealing to recover the implantation damage associated with the generation of defects (point and extended defects) which would cause severe device degradation unless suitably suppressed.

The success of thermal annealing processes of ion-implanted materials depends strongly on the nature of the material, the characteristics of the implantation process and the depth of the junction.

---

[3] The large thermal capacity of these ingots favours a long stage at temperatures around 450–500 °C during the post-growth cooling. In this temperature range the formation rate of thermal donors is particularly high.

[4] The sample is generally placed inside a quartz tube and maintained under a protective atmosphere or under vacuum.

As an example, to recover the radiation damage and the partial amorphization of Al^+-implanted 6H-SiC, the implantation process should be better carried out at 500 °C rather than at room temperature (RT) to reduce the damage. An annealing at 1600 °C allows, in this case, a complete recovery of the damage and recrystallization of the amorphized regions of the sample, which is more difficult in an RT implanted material [7].

It has also been observed that a conventional thermal annealing of B^+-implanted silicon is unable to remove the implantation damage and, additionally, induces precipitation and redistribution of the dopant. Instead, pulsed laser annealing is able to recover the damage by operating inside the implanted region, with negligible, if any, degradation of the minority carriers' lifetime [8].

The annealing problem is particularly severe in the case of the ion-implanted ultra-shallow junctions in MOS and MOSFET (metal oxide semiconductor field-effect transistor) devices approaching a depth of 10 nm, fabricated with the source/drain (S/D) technology, since the annealing time and temperatures must be precisely controlled in order to minimize changes in the junction depth due to thermal diffusion. Another problem occurs when low energy implantation of As is used to form a shallow n-type junction, due to the presence of oxygen and vacancy-defects generated by irradiation. Oxygen interacts with vacancies and dopants, leading to the formation of As–O and O–As–V complexes, which cause at least partial dopant deactivation [9]. Eventually, oxygen interaction with vacancies favours the segregation of SiO_x precipitates inside the junction, with a risk of local impurity gettering during annealing.

In these cases, the annealing temperature should be kept as low as possible and MW annealing could be a potential solution [10–13].

Microwave heating in semiconductors depends on dielectric losses associated with dipole and electronic polarization [12] that cause the onset of conditions of only partial coupling of the sample with the applied electromagnetic field.

In turn, the damped vibration of a polarized dipole induces a change in its dielectric constant ε^*

$$\varepsilon^* = \varepsilon_0(\varepsilon' - i\varepsilon'') \tag{5.3}$$

(where ε_0 is the dielectric constant of vacuum, ε' is the real part of the dielectric constant and ε'' is the imaginary part of the dielectric constant which accounts for the dielectric losses) and, thus, a fraction of the energy is converted to thermal energy within the material itself.

The MW power dissipated by a dielectric loss as thermal energy dP_{th} in a volume fraction dV of the sample is given by the following equation

$$dP_d = \omega\varepsilon_0\varepsilon''|E|^2 dV \tag{5.4}$$

where ω is the MW excitation frequency and $|E|$ is the electromagnetic field energy [11].

As not all the excitation frequencies of charged particles and dipoles can lie in the MW frequency range, MW annealing is, however, strongly dependent on the nature of the material.

Efficient MW annealing is obtained in materials with a high density of dipolar species, as is the case for amorphous silicon with its abundant content of interstitial-vacancy pairs, double-vacancies and other point defects pairs, which behave as polarized dipoles.

As MW heating was also shown to operate well at temperatures lower than 300 °C, on CZ silicon but not on FZ-Si (float zone), one could suppose that in CZ silicon dopant-oxygen dipolar species, such as B-O and As-O will interact properly with the MW field, although[5] only MW spectroscopy could give experimental proof.

Also, on ion implanted 6H-SiC, which is a typical good MW absorbing material [14], MW heating was found to be very effective, as it allows the annealing temperature to be reached in a matter of minutes, with a rate of 200 °C min^{-1} and to cool at a rate of 400 °C min^{-1} [15].

It is, therefore, apparent that a straightforward application of selective MW annealing would imply preliminary knowledge of the defect species or the lattice dipoles capable of activating dielectric losses and then the definition of the MW energy range in which to work, thus requiring preliminary MW-spectroscopy information.

It is, however, possible to conclude that a number of heating sources are available for semiconductor annealing, but few are also potentially able to bring to success processes otherwise probably unfeasible, as is the case with MW heating.

5.2.1 Thermal Decomposition of Non-stoichiometric Amorphous Phases for Nanofabrication Processes

The preparation of disordered arrays of silicon nanodots in an amorphous silicon matrix was discussed in Chapter 4, where it was shown that the amorphous matrix does not represent a suitable quantum mechanical (qm) confining medium [16, 17].[6] In addition, the nanocrystal (nc) size is generally far from that required to get qm-confinement (see Figure 5.1), with serious negative outcomes concerning their optoelectronic application.

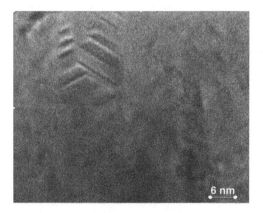
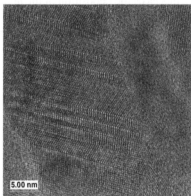

6 nm   5.00 nm

Figure 5.1   *HRTEM images of silicon nanocrystallites embedded in an a-Si matrix*

---

[5] To the author's knowledge MW spectroscopy is mostly carried out on organic molecules on which rotational spectroscopy could be applied. The fingerprints of excited dipoles are also observed in the cosmic microwave radiation background.

[6] The physics of quantum-mechanical confinement will not be discussed in this book but is left to the specialized literature.

Better results could be obtained using as confining hosts high band-gap materials, among which SiO_2 and SiC represent today the best alternatives, as they can be used in their non-stoichiometric configuration as the precursors of Si nanodots.

Non-stoichiometric SiO_x decomposes by heating in stoichiometric SiO_2 and Si [18–22]

$$SiO_x \rightarrow \frac{x}{2}SiO_2 + \left(1 - \frac{x}{2}\right)Si \tag{5.5}$$

like the non-stoichiometric $Si_{1+x}C$

$$Si_{1+x}C \rightarrow xSi + SiC \tag{5.6}$$

The preparation of Si nanodots in a SiO_2 matrix is carried out by reactive sublimation of SiO in oxygen atmosphere on a Si wafer. This process leads to an amorphous SiO_x material which is then submitted to an annealing stage at 1100 °C for phase decomposition, which automatically results in amorphous grains of Si embedded in a SiO_2 matrix and further crystallization of the decomposition products.

By taking x very close to 1, the nc size may be selectively tuned by depositing a SiO_x layer of appropriate thickness.

Adopting the configuration of a multilayer structure (see Figure 5.2), the sequential deposition of SiO_x and stoichiometric SiO_2 results in ordered linear arrays of almost spherical Si-nc embedded in SiO_2. This material exhibits a strong photoluminescence emission, with a remarkable blue-shift (up to 0.4 eV from the regular band to band emission of single crystal silicon, depending on the Si-nc size) induced by quantum confinement [18].

The limitation of structures based on Si-nc embedded in an insulating matrix with respect to direct charge injection applications could be overcome by the use of SiC as the embedding matrix, which presents good electronic properties.

In this case, the deposition of the non-stoichiometric amorphous SiC is carried out using a plasma enhanced chemical vapour deposition (PECVD) process with SiH_4, CH_4 and H_2 as the source gases, whose composition ratio determines the stoichiometry of the amorphous a-SiC layer deposited. The deposition is carried out at 325–350 °C, that is at a

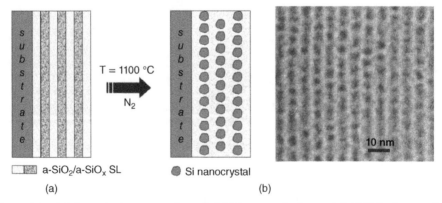

Substrate

T = 1100 °C
N_2

a-SiO$_2$/a-SiO$_x$ SL
(a)

Si nanocrystal
(b)

10 nm

Figure 5.2 *(a) Growth of an amorphous SiO/SiO$_2$ superlattice and (b) TEM micrograph of sample after thermally induced phase separation. Zacharias et al., 2003, [19]. Reproduced with permission from Trans Tech Publications*

temperature low enough to get the amorphous phase. Also, for nc-SiC a post-deposition annealing is carried out to decompose the homogeneous amorphous phase and crystallize the SiC nanodots, but here a delicate problem occurs with the annealing process, whose preliminary step is carried out at 600 °C to strip out the residual hydrogen followed by a second step at 1100 °C.

The superlattice structure (see Figure 5.3), consisting of a sequence of layers of stoichiometric SiC and non-stoichiometric SiC_x survives only for high values of the thickness d_{SCR} (>2.5–3 nm) of the silicon carbide recrystallized layer (SCR). The crystalline fraction, determined by Raman spectroscopy, increases with the SCR layer thickness and the total silicon volume fraction V_{Si}.

Finally, the Si-nc size remains within that of the SRC layer only for large values of the sum $d_{RSC} + d_{SiC}$, where d_{SiC} is the thickness of the SiC layer. These complex features could be qualitatively explained [22] by considering that the Si/SiC system is a highly mismatched system, with a large difference in lattice parameters (19% at RT) and in thermal expansion coefficients ($\approx 23\%$ at high deposition temperatures and 8% at RT), making the Si/SiC interface a high energy surface [23]. Due to this high energy barrier, the crystalline layer inside the amorphous region is prevented to reach the SiC layer, resulting in a non-negligible fraction of a-SiC left in the annealed material.

5.2.2 Other Problems of a Thermodynamic or Kinetic Nature

A different aspect of challenges inherent to annealing processes concerns environmental problems. This is the case for the surface degradation of germanium devices when annealed at 773–873 K without a protective capping layer, even when the process is carried out under a protective atmosphere of nitrogen or argon. It is observed that annealing causes severe

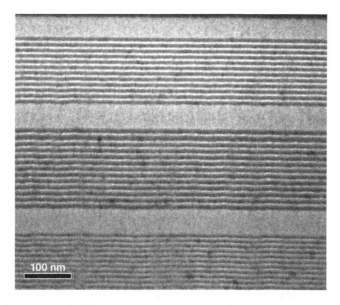

Figure 5.3 *TEM image of a Si/SiC superlattice. Courtesy of C. Summonte, CNR-IMM, Bologna*

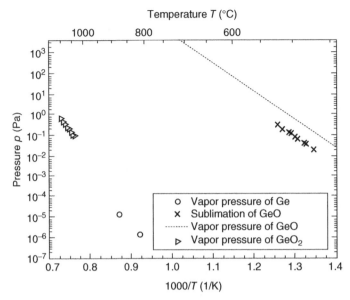

Figure 5.4   *Sublimation pressures of Ge, GeO$_2$ and GeO. Kaiser et al., 2011, [24]. Reproduced with permission from Elsevier*

surface degradation with consistent loss of Ge bulk material [24], amounting to 3 nm min^{-1} at 873 K. Direct sublimation of Ge can be excluded due to its low vapour pressure at these temperatures (see Figure 5.4). Instead, the most probable cause is associated with the oxidation of Ge to GeO$_2$ due to residual oxidants in the protective atmosphere, thanks to a very negative value of the Gibbs energy of GeO$_2$ formation ($\Delta G_{GeO_2} = -410.875$ kJ mol^{-1} at 873 K) and the successive sublimation of GeO formed by reaction of GeO$_2$ with Ge, GeO being a volatile component. The equilibrium vapour pressure of GeO in equilibrium with solid Ge and GeO$_2$ at 859 K is $7.7 \cdot 10^{-6}$ atm (0.77 Pa) [25], in good agreement with the data reported in Figure 5.4 [24].

Challenges are also encountered when doping a semiconductor well above the solubility limits of the dopant. Using conventional doping processes, including ion implantation, spontaneous segregation of the excess dopant as a second phase component occurs, either during the doping process itself or after a subsequent annealing. As expected, the second phase consists of the dopant itself or of a compound, depending on the phase diagram of the system. The segregation process occurs spontaneously, driven by the excess Gibbs energy of supersaturation

$$-\Delta G_{ss} = RT \ln \frac{c}{c^*} \tag{5.7}$$

unless metastability conditions can set-up. These conditions could be achieved either by operating the process at temperatures sufficiently low, such as to prevent the second phase segregation thanks to the low mobility of the dopants, or doping the sample with a high speed pulsed laser at room temperature, using a layer of the solid dopant deposited at the surface of the sample to be doped as the source.

This process was successfully used to hyperdope silicon with selenium [26, 27], which is only slightly soluble in silicon (see Figure 5.5), using femtosecond laser pulses and a

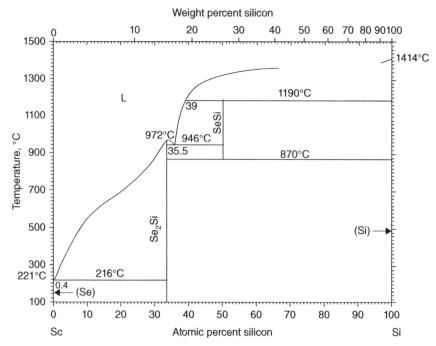

Figure 5.5 *Phase diagram of the Se–Si system. Odin and Ivanov, 1991, [27]. Reproduced with permission from ASM International*

65 nm thick solid selenium target sublimed on top of a silicon wafer as the source. Doping Si with chalcogens leads to near-unity absorption of radiation from UV to below band gap IR wavelengths, with potential impact for silicon use in optoelectronic applications. The near-unity absorbance is attributed to the formation of a supersaturated solution of trapped chalcogen dopants and point defects in the silicon lattice, which modify the electronic structure of the surface layer.

After laser irradiation, a 300 nm thick surface layer was hyperdoped up to a selenium content of 10^{20}–10^{21} at cm^{-3} compared with a solubility of only $5 \cdot 10^{15}$ at cm^{-3}. Extended X-ray absorption fine structure (EXAFS) measurements on hyperdoped, un-annealed samples demonstrate that Se occupies substitutional and dimeric substitutional sites in the silicon lattice, while annealing in the 600–1225 K interval induces the segregation of $SiSe_2$, which is the stable intermediate compound in this temperature interval.

Flash lamp heating is used for the suppression of bulk microdefects (BD) in CZ silicon, when used for the fabrication of power devices, instead of the more expensive low oxygen FZ material,[7] which is the material currently employed for these applications.

Suppression of BDs is mandatory as they behave as nuclei for oxygen segregation as SiO_2 during the high temperature processes used in power device fabrication and also reduce the lifetime of the minority carriers.

Flash lamp annealing in the ms range has been successfully used [28] and the annealing effect was studied by submitting un-annealed and annealed samples to a standard oxygen

---

[7] In FZ silicon the interstitial oxygen content is sufficiently low as to have a negligible density of BDs.

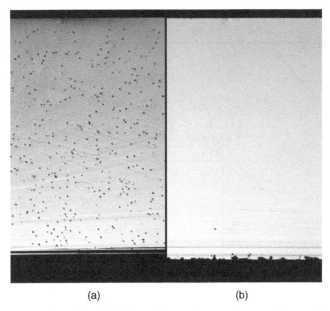

Figure 5.6 *Micrographs of the etched surfaces of an as grown CZ sample (a) and of a flash-lamp annealed sample (b), after an oxygen precipitation test (see text). Kissinger et al., 2014, [28]. Reproduced with permission from AIP*

precipitation test,[8] followed by Secco etching (H_2O 67.6 mol%, HF 32.2 mol%, $K_2Cr_2O_7$ 0.17 mol%), which evidences the selectively etched microprecipitates at the sample surface.

The results of this procedure are shown in Figure 5.6 [28] on a CZ sample directly submitted to the oxygen precipitation test (a) and on a sample previously flash lamp annealed (b). The virtual absence of etch pits in the flash annealed CZ sample indicates that the BDs are efficiently dissolved by a 20 ms light flash, with a normalized irradiance very close to that needed to melt the sample surface.

We can conclude that rapid and ultrarapid thermal annealings are very effective tools, and that their use will be greatly improved by a more theoretical approach to their design.

5.3   Hydrogen Passivation Processes

Atomic hydrogen from various dry or wet sources is known to passivate surface states as well as bulk shallow- and deep-levels arising from dopants, impurities, point and extended defects, in silicon and other semiconductors, as shown in Sections 2.3.4.1 and 3.5.2.

Here, passivation means the electrical deactivation of a defect or an impurity as the consequence of a bonding interaction with hydrogen, that is of a true chemical reaction, which shifts their energy levels out of the gap [29].

A typical wet hydrogen passivation process is that occurring when silicon wafers are dipped in an HF solution. Surface passivation of dangling bonds then sets-up, associated

---

[8] Consisting of an annealing at 780 °C for 3 hours in oxygen atmosphere followed by a stage at 1000 °C for 16 hours to stabilize the nuclei and grow them to a size compatible with the detection sensitivity.

with the formation of covalent Si-H bonds, which lead to fully hydrogenated and remark-ably hydrophobic surfaces. Although HF can be an effective, but temporary, surface pas-sivating agent, permanent surface and bulk defects passivation is usually obtained using plasma hydrogenation.

The effective passivation depth[9] and yield comes, however, from a delicate balance of inward and outward hydrogen diffusion and of trapping and de-trapping processes, all hav-ing a strong temperature dependence.

The development of efficient hydrogen passivation processes was, therefore, possible only with deep knowledge of the physical properties of hydrogen in semiconductors and the features of its interaction with defects and impurities that became available as the result of more than 30 years of basic research devoted to hydrogen physics [30–37].

Apart from its technological interest, hydrogen plays a key role in semiconductor physics because of its properties of an amphoteric deep donor/acceptor centre and the strong depen-dence of its physical properties on the nature of the host in which it is dissolved, typically shown by the different behaviour of hydrogen in silicon and germanium [37].

The shift from the neutral H^0 charge state to negative H^- to positive H^+ depends on the Fermi level and it has been demonstrated [31] that in p-type Si the donor H^+ state is the stable state, while the acceptor H^- is the stable one in n-type Si (see Figure 5.7) [31].

This hydrogen property provides the condition for an at least partial compensation of silicon dopants by electrostatic and partially covalent bonding

$$H_i^+ + B_{Si}^- \rightleftharpoons H - B \tag{5.8}$$

and

$$H_i^- + P_{Si}^+ \rightleftharpoons H - P \tag{5.9}$$

depending on the temperature and on the bonding enthalpy of the hydrogen complexes, whose calculated values are 0.6–0.9 eV for B-H and around 0.35–0.65 eV for H-P [32]. This kind of reversible reaction with ionized hydrogen holds for every ionized shallow or

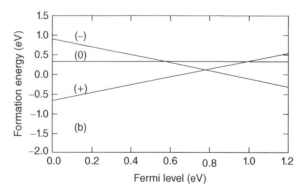

Figure 5.7 *Relative formation energies of H in the different charge states in silicon. Van de Walle et al., 1989, [31]. Reproduced with permission from American Physical Society*

---

[9] In several cases the passivation should be extended to the entire thickness of the device, that in mc-Si solar cells is around 200 μm.

deep impurity, while the passivation of dangling bonds of dislocations and GBs would, instead, involve hydrogen in its neutral state, demanding attention to the doping conditions of the substrate to be passivated, as will be seen later in this section.

The dependence of the hydrogen properties on the nature of the matrix in which it is dissolved can be demonstrated by the calculated alignment of the hydrogen level in different hosts [35, 36], including hydrogen in aqueous electrolytes. This conclusion results straightforwardly from the calculation of the position of the Fermi level $\varepsilon(+/-)$ for which the formation energies of the donor and acceptor charge states are equal (see Figure 5.8) [36] in an absolute scale energy, taking as the reference the energy of the H_2 molecule at 0 K.

One can see, as an example, that the calculated $\varepsilon(+/-)$ level for Ge lies around the valence maximum, as in the case of InSb and GaSb, while the opposite is true for InN and ZnO. This will cause H to behave as a shallow acceptor in Ge, InSb and GaSb because their donor states, see Figures 5.7 and 5.8 are inside the valence band, and as a donor in InN and ZnO. Therefore, the presence of hydrogen in Ge, which is actually grown in a H_2 atmosphere, will lead to the formation of stable complexes with donor dopants such as P

$$H_i^- + P_{Ge}^+ \rightleftharpoons H - P \tag{5.10}$$

but not with acceptors.

Hydrogenation of a partially ionic compound semiconductor of formal composition AB leads to particularly interesting consequences, as here strong ionic bonds can be formed also between positively charged H^+ and an anionic species B^-, resulting in the creation of a cationic dangling bond db_A. Conversely, negatively charged hydrogen H^- may form a strong bond with cations A^+, with the creation of an anionic dangling bond db_B.

In this case hydrogen does not behave as a passivating agent, as dangling bonds may behave as recombination centres or as donor and acceptor species, depending on the position of their energy levels in the gap.

A particularly interesting case of a hydrogen passivation process of major technological importance is that adopted for the passivation of extended defects (dislocations and GBs) and deep level impurities in mc-Si used for the cost-competitive manufacture of solar cells.

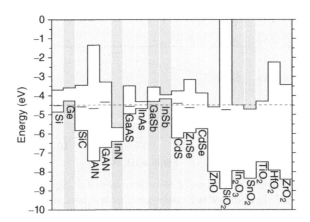

Figure 5.8 *Position of the $\varepsilon(\pm)$ level for a number of semiconductors and insulators. van de Walle, 2006, [36]. Reproduced with permission from Elsevier*

Although plasma hydrogenation or catalytically dissociated hydrogen may be successfully used [38, 39], the current state of the art process employs as the hydrogen source the antireflection (AR) layer. This consists of hydrogenated silicon nitride $SiNH_x$: H deposited on the solar cell front using PECVD with silane and ammonia mixtures [40–42], a process similar to that used for the deposition of nc-Si (see Chapter 4).

The chemistry of the mc-Si hydrogenation process is based on the decomposition of the surface nitride $SiNH_x$: H, which releases atomic hydrogen available for passivation of defects during the firing and annealing steps in the cell manufacturing process.

The process itself is, however, not free of problems and the passivation efficiency is limited by a number of drawbacks which are intrinsic to the nature of hydrogen and to the technology of the solar cell manufacturing process.

The first is that the passivation of extended defects requires the availability of neutral interstitial hydrogen H_i^0 as reactant, while it ionizes to H^+ or H^- when it diffuses through the highly doped n-type or p-type layer[10] that forms the n/p or p/n junction on the front side of the cell, with the consequent generation of the complex species H–B and H–P, respectively. The diffusion of hydrogen is, therefore, inhibited by a true trapping process (see Section 2.2.3) and only a fraction of the hydrogen potentially delivered by the source is available for passivation. The concentration of neutral hydrogen arriving in the volume of the solar cell, that is the active part of the device, is also influenced by a partial compensation of the bulk dopants. The situation is not improved if hydrogen is supplied by an external source, consisting of a hydrogen plasma or a forming gas atmosphere.

The second problem concerns the temperature that is adopted at the end of the solar cell manufacturing process to form the front metal contacts, a process that implies the firing of a metal past a screen printed on the cell front, in order to let it diffuse through the AR layer.

If the process is carried out at temperatures above 600 °C, the thermal release of hydrogen from the passivated defects in the bulk is favoured, followed by degradation of the quality of the solar cell, that can be monitored by the sudden decrease in lifetime of the minority carriers (see Figure 5.9a) [43] and by the decrease in the absorption associated with Si-H (right panel) with increasing the annealing temperature [43].

The third problem is associated with the configuration of the solar cell on the rear, where Al is used as the contact and also as a gettering agent (see next section). Al dissolved in silicon behaves as an acceptor and works as an extended sink for hydrogen properly diffused in the volume of the cell, thus reducing its concentration.

A compromise should be found, therefore, between the optimum firing temperature for contact formation and the hydrogenation properties, but the solution, in most cases, is purely empirical.

An advanced solution is proposed in a recent patent [44], which foresees the use of a strong light source to inject excess carriers into the front junction during the hydrogen diffusion process

$$nil \xrightarrow{h\nu} e + h \qquad (5.11)$$

(where, by convention, nil means a couple of electrons and holes in the conduction and valence bands, respectively) with $h\nu \geq E_G$, in order to compensate the majority carriers in

---

[10] Si solar cells are typically manufactured on slightly p-type (B-doped) substrates, and the front side n/p junction is made by diffusing P. Today high efficiency solar cells are manufactured using n-type substrates, and the p/n junction is made by diffusing B. The advantage of n-type Si relates to the lowest recombination efficiency of several impurities, including Fe, in this type of material.

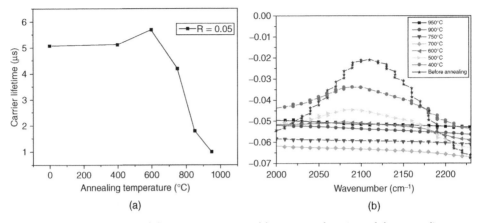

Figure 5.9 *(a) Evolution of the minority carriers lifetime as a function of the annealing temperature. (b) Evolution of the Si-H IR absorption as a function of the annealing temperature. Bousbih et al., 2012, [43]. Reproduced with permission from John Wiley & Sons*

the front junction and to allow neutral hydrogen to diffuse through the doped layer without complex formation. Illumination should also be beneficial for suppressing the hydrogen sinks present in the parasitic junction on the rear due to Al doping. The solution proposed is scientifically sound, but no information is available on its technological success.

5.4   Gettering and Defect Engineering

5.4.1   Introduction

Trapping and deactivation of lifetime killer impurities at internal or external gettering structures has been a research [45] and engineering challenge since the discovery of the internal gettering (IG) effect in silicon wafers, at the end of the 1970s [46, 47].

Gettering implies the localization and inactivation of impurities in a predetermined region of the semiconductor sample, far from or inside[11] the active region of the device, leaving the active region or the entire bulk of the sample with an impurities content possibly well below their solubility.

IG is provided by localized bulk defects (BDs), suitably generated by preliminary thermal treatments. BDs behave as sinks where impurities diffuse under the driving force of a chemical potential gradient or/and of a stress field [47] and interact irreversibly. IG is a typical attribute of CZ-silicon, since it is promoted by the presence of BDs consisting of oxygen precipitates induced by a proper thermal treatment of the oxygen-rich sample. External gettering (EG) processes should be, instead, used to recondition mc-Si, where GBs, intragrain defects and dislocations behave as deep traps of metallic impurities, that lead to their electrical activation [48].

---

[11] Depending on the device configuration: In microelectronic devices, where the active region is at the front surface of a wafer, gettering is carried out in the bulk of the wafer. In Si solar cells, where the active region is the bulk of the wafer, gettering at the external surfaces and extended defect passivation should be carried out.

EG is promoted by the presence of surface defects (dislocations, precipitates) generated by mechanical or chemical means, including ion implantation. Phosphorus, boron and aluminium gettering are other traditional forms of EG, although operating with different mechanisms [48].

The physics and the engineering of gettering processes in silicon have been discussed in thousands of papers and in dozens of excellent books, of which a few are referenced here for their general relevance [49–51], and a series of dedicated symposia.[12]

The aim of the next sections is, therefore, not to give a critical review of the work done on this subject but to reconsider some critical physico-chemical aspects of a subject of extreme technological relevance, in order to propose an appropriate model of understanding.

5.4.2   Thermodynamics of Gettering

Most of the thermodynamic and kinetic aspects of gettering processes have been indirectly introduced in the previous chapters. A gettering process can be treated as the heterogeneous repartition of single or multiple chemical species between a solid solution and a localized sink, which behaves as a deep trap. The trap itself may consist of an array of extended defects (dislocations or GBs), an assembly of 3D defects or a different (solid or liquid) phase. Gettering occurs either with localized chemical bond formation, with the nucleation and segregation of a second phase, that in the case of silicon could be a silicide,[13] or with the dissolution of the impurity, respectively.

In either case, the equilibrium distribution of an impurity i between the bulk phase and the gettering 'phase' could be formally expressed following the concepts discussed in Sections 1.6.2 and 3.4.3 using the following equation (Eq. (3.15))

$$\mu_i^{0,\alpha} + RT \ln \gamma_i^{\alpha}(x) x_i^{\alpha} = \mu_i^{0,t} + RT \ln \gamma_i^{t}(x) x_i^{t} \qquad (5.12)$$

where the superscripts t and α mean trap and bulk phase, $\gamma_i^{\alpha}(x)$ and $\gamma_i^{t}(x)$ are the activity coefficients of i, x_i^{α} and x_i^{t} are the concentrations of i and $\mu_i^{0,t}$ and $\mu_i^{0,\alpha}$ are the standard chemical potentials of the impurity i in the trap and in the bulk phase.

In most cases, the activity coefficient of i in the bulk phase can be approximated to unity, as the solutions of impurities in semiconductors are normally dilute solutions, but interaction with other impurities would bring $\gamma_i^{\alpha}(x)$ far from unity, as is typical for Fe in the presence of B, that forms Fe–B complexes [52].

The distribution coefficient of i between the bulk and the trap is given by the following equation

$$k = \frac{x_i^{t}}{x_i^{\alpha}} = \Gamma(x_i^{t}, x_i^{\alpha}) \exp -\frac{\mu_i^{0,t} - \mu_i^{0,\alpha}}{kT} = A\Gamma(x_i^{t}, x_i^{\alpha}) \exp -\frac{\Delta H_B}{kT} \qquad (5.13)$$

where the difference $\mu_i^{0,\alpha} - \mu_i^{0,t}$ is the Gibbs free energy of the bonding reaction of i at a trap site, ΔH_B is the bonding enthalpy, the pre-exponential A has an entropic character and $\Gamma(x_i^{t}, x_i^{\alpha})$ is the ratio of the activity coefficients of i which accounts for all the interactions occurring between the impurity atoms and the atoms of the matrix in both the bulk and trap 'phases'.

---

[12] See Gettering and Defect Engineering in Semiconductor Technology I–XIV, Trans Tech Publ. TTP Inc.
[13] In CZ silicon the impurity could also be trapped as an oxide.

The driving force for this process is the interaction or solution[14] Gibbs energy ΔG_B whose enthalpic term is given in Eq. (5.13). In the case of a true trapping process, the absolute value of the binding enthalpy ΔH_B also allows a distinction between traps which can be emptied and those which remain filled at a predetermined temperature, as discussed in Chapter 3 and illustrated in Figure 3.25 [53].

A different case occurs when the impurity solution in the immediate vicinity of the gettering site is brought to supersaturation conditions by a proper thermal treatment and the segregation of a secondary phase (a metal or a compound, depending on the phase diagram of the system under consideration), therefore, becomes thermodynamically possible.

In this case the supersaturation ratio is the thermodynamic driving force $-\Delta G_{ss} = RT \ln \frac{x_i}{x_i^0}$, where x_i^0 is the solubility of the impurity, of the segregation process and the trap function is to promote the nucleation and further growth of the compound by reducing the activation barrier, favouring, therefore, the segregation of the impurity in a predetermined region of the sample.

These kinds of procedure have been applied to model the thermodynamics and kinetics of internal [54] and external gettering [55] in silicon, by assuming a Fickian diffusion of the impurity towards the gettering region where they react or dissolve.

It is of immediate understanding that the absolute value of the diffusion coefficients of the impurities to be gettered has a substantial role in the yield of the gettering process, fast diffusers being the best candidates for high process yields.

The topical details of the gettering processes, the role of defects on the process rate, the specific configuration of the gettering site and the nature of the species to be gettered are the issues considered in the next two sections. No direct mention will be made of the physics and chemistry of gettering ability of dislocations, as this was the subject of Sections 3.4 and 3.5, in which the interaction of dislocations with impurities was discussed in detail.

5.4.3   Physics and Chemistry of Internal Gettering

The main interest in IG has been addressed to Fe, Ni and Cu,[15] which are the critical metallic contaminants of electronic grade silicon, arising from different sources during ingot growth and further device processing.[16] As most of the common transition metal impurities, they have small solid solubilities and, for this reason, their precipitation as silicides occurs even during fast quenching processes [56]. Fe, Ni and Cu are fast diffusers also at room temperature, form deep levels in the silicon gap, which behave as lifetime killers and cause oxide breakdown when their silicides segregate at the MOS surface [57].

Metallic contaminants of solar grade silicon include Ti, Cr and other 3d metals [58] but in this case external gettering (EG) processes are mandatorily followed [50], as will be discussed in the next section.

IG[17] in silicon is a process associated with the segregation of metallic impurities in a region of a silicon wafer where oxygen has been precipitated in the form of small oxygen

---

[14] Depending on whether the process consists of the binding of an impurity species on a localized trap (a dislocation, for example) or the dissolution of the impurity in a liquid or solid phase.

[15] But also to Au, used in metal interconnects.

[16] The polycrystalline electronic grade silicon used as the starting material for ingot growth is the purest synthetic material available today, with an impurity content at the ppt level or lower.

[17] Internal or intrinsic gettering are used in the literature with the same meaning.

clusters, called bulk defects, by a two-step process, of which the first is carried out at around 1100 °C to diffuse oxygen out of the wafer, leaving an oxygen-denuded zone[18] and the second, generally around 700–800 °C,[19] to precipitate oxygen.

In spite of the most recent advances in ULSI technology, that demands low-oxygen silicon for better gate oxide quality and applies lower thermal budgets that would make the oxygen precipitation process more troublesome and the out-diffusion of oxygen to form a defect-free subsurface region [57], IG remains of unique value for gettering Fe, Ni and Cu. IG gettering, at least for Cu, Pt and Ni, proceeds before an appreciable precipitation of oxygen [59].

IG gettering was originally attributed by Tan *et al.* [46] to the interaction of impurities with dislocations punched-off from oxygen precipitates. This was, and remains, a well-founded hypothesis based on transmission electron microscopy (TEM) and scanning infrared microscopy SIRM) investigations [59], reminiscent of the earliest results of Dash [60], who used copper to decorate dislocations, Schwuttke [61] and the Tiller model [62], that requires a dislocation density of at least 10^3 cm^{-2} to get copper segregated in a single crystal material.

It fits well with the properties of dislocations as heterogeneous sinks for impurities, discussed in Chapter 3, and also with the common experience that segregation of impurities at extended defects (dislocations, GBs and stacking faults) is a typical feature of impurity gettering in mc-Si, as will be discussed in the next section.

As gettering of Ni also occurs in the absence of a measurable TEM density of punched-off dislocations, Ourmazd and Schröter [45] suggested that IG of Ni arises as a consequence of a local supersaturation of self-interstitials, generated as a side-product of the silicon oxide precipitation (see Sections 3.5.3 and 3.7.2).

Self-interstitial supersaturation, in turn, was supposed to favour the epitaxial segregation of NiSi$_2$ by delivering the required amount of Si interstitials to form a NiSi$_2$ precipitate

$$Ni_{Si} + 2I \xrightarrow{\Delta V \approx 0} NiSi_2 \qquad (5.14)$$

From the phase diagram of the Si–Ni systems it can be seen that NiSi$_2$ crystallizes from Ni-saturated Si solutions, at temperatures below 981 °C, with a cubic CaF$_2$ type of structure,[20] where Si occupies the interstitial sites occupied by fluorine in the CaF$_2$ lattice.

The NiSi$_2$ structure and it molar volume fit well with that of silicon [63] and also makes feasible the heteroepitaxial deposition of NiSi$_2$ on silicon [64] and its faceted growth in the Si bulk [65].

It can, however, be seen that the Ourmazd and Schröter model [45] could be straightforwardly applied only to Ni.

In fact, a saturated solution of Fe in Si is in thermodynamic equilibrium with the slightly non-stoichiometric β-FeSi$_2$ silicide phase at temperatures below 1107 °C and with the orthorhombic stoichiometric α-SiFe$_2$ phase at $T < 938$ °C [66, 67], whose cell constants ($a = 0.9863$ nm, $b = 0.7791$ nm and c $= 0.7833$ nm) largely deviate from that of silicon, making difficult its heteroepitaxial segregation.

---

[18] The denuded zone is the top region of the wafer where the microelectronic device is processed.

[19] This temperature results to be the optimal temperature to segregate metal silicide phases, as will be seen later in this section.

[20] The cell constant of the NiSi$_2$ phase is 0.5406 nm, very close to the lattice constant of Si ($a = 0.54306$ nm) and the volume change associated with the NiSi$_2$ precipitation is very small.

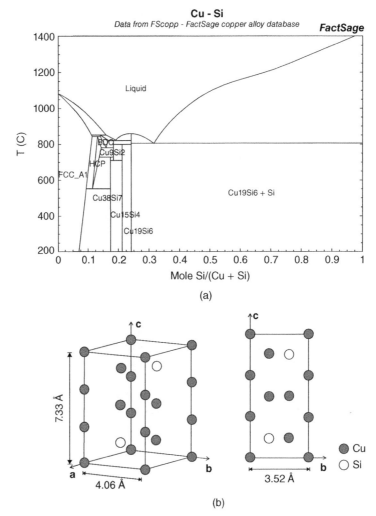

Figure 5.10 *(a) Phase diagram of Cu–Si system. Reprinted with permission of the Chairman of the Scientific Group Thermodata Europe. (b) Structure of Cu_3Si. Wen and Spaepen, 2007, [70]. Reprinted with permission of Taylor & Francis*

Concerning the copper case, at temperatures lower than 802 °C, corresponding to the eutectic (see Figure 5.10a) [69] the phase in equilibrium with Cu-saturated Si is the slightly non stoichiometric hexagonal η-Cu_3Si phase (see Figure 5.10b) [70], whose segregation[21] leads to a volume expansion of 150% [63].

In both cases, the segregation of the silicide phase is associated with a large volume expansion and crystal strain, and induces the ejection of self-interstitials or the adsorption of vacancies [49], not the adsorption of self-interstitials as suggested by Eq. (5.1).

---

[21] The segregation of copper silicide was also treated with similar arguments in Section 3.7.3.

Ni segregation in an IG process is, therefore, an exceptional case and, in general, gettering at oxide particles occurs by a complex mechanism, which starts with the segregation of a metal precipitate or a colony of metal precipitates at dislocations punched-off from an oxide precipitate (see Figure 3.19) and continues with the dislocation multiplication and propagation, associated with self-interstitial generation by a further segregation of silicides [59].

When gettering occurs with the segregation of silicide phases, the impurity concentration in solid solution after gettering will, necessarily, remain at its thermodynamic solubility limit, decreasing with decrease in the gettering temperature, as the segregation of the silicides occurs for impurity saturation or supersaturation conditions.

If gettering instead occurs with a trapping process at dislocations or at heterogeneous trapping sites, the final impurity concentration could go well below the solubility, depending only on the trapping efficiency and on the density of the heterogeneous sinks.

It should, therefore, be expected that the segregation yield of metallic impurities at BDs, and then the IG efficiency, will depend on several factors, among which the BD density, the chemical nature of the impurity to be gettered and the IC process temperature will play major roles, making IG a multi-variables process and each impurity a unique case.

The arguments reported in what follows do not represent the synthesis of the vast amount of work addressed at giving an answer to the aforementioned questions, but just give an indication of the problems encountered when looking for a conclusion.

Concerning the effect of IG process temperatures and BD density, it was demonstrated that the overall IG efficiency depends on the temperature of the treatments carried out for the oxygen precipitation, as high temperatures lead to the (partial)dissolution of the oxygen clusters responsible for the gettering. It depends also on the effective impurity supersaturation developed in the cooling stage of the thermal cycle [54]. Since the supersaturation is inversely proportional to the solubility of the impurity, which is higher for Fe than other metallic impurities [71] and increases with B-doping, consequent to the formation of Fe–B complexes [52], the IC gettering of Fe is difficult and the experience is that its final concentration lies above its solubility and depends on the BD density (see Figure 5.11) [54].

Ni gettering requires a higher BD density than Fe and its external gettering is a very efficient process, associated with the $NiSi_2$ segregation on Si wafer surfaces [72].

The favourable Ni trapping efficiency of surfaces is qualitatively confirmed by the segregation of Ni at the surface of voids generated by ion implantation, which is a more efficient

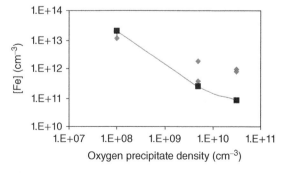

Figure 5.11   *Dependence of the residual Fe concentration on the oxygen precipitates density. Geranzani et al., 2002, [54]. Reproduced with permission from Trans Tech Publications*

process than that at oxygen defects [73]. The proposed background mechanism is based on the assumption that the binding energies of Ni with its silicide, with oxygen clusters and void surfaces are similar, an argument, however, rather difficult to be physically sensible.

Copper gettering efficiency has also been shown to depend on the BD density, reaching a maximum at a defect density, measured by preferential etching, above 10^9 cm^{-3}, but it is supposed that IG depends also on small BDs which are difficult to identify by these preferential etching procedures [74]. It is also known from studies on heterogeneous copper precipitation at dislocations that Cu segregates as copper silicide particles bound by extrinsic dislocation loops [75].

In an attempt to rationalize the main features of IG, it is, first, possible to remark that the precipitation of oxygen in silicon is a very complex physico-chemical process, whose precise and reliable control *during* the processing of wafers into integrated circuits remains unsatisfactory. This is mainly due to the uneven distribution of oxygen in a silicon ingot and to the amount of microdefects, capable of influencing the process, present in uncontrolled density. A temporary solution was found with the development of the 'perfect silicon' (see Section 4.2.2.2) but the defect engineering of conventional silicon wafers can still be considered a sophisticated empirical exercise.

Further, self-interstitials play a key role, as they either favour the homogeneous nucleation of a silicide phase (the case of NiSi$_2$) or are directly involved in the dislocation multiplication.

Thus, the IG efficiency depends on the average BD density, but BDs are not directly responsible for gettering, as dislocations and self-interstitials seem to have the active role.

5.4.4   Physics and Chemistry of External Gettering

External gettering arises as a consequence of the irreversible transfer of impurities to heterogeneous (mostly consisting of dislocation networks) or homogeneous surface sinks, where impurities are trapped or preferentially dissolve.

Therefore, an EG process requires the initial creation of a highly damaged region or the deposition of a suitable metal alloy at the sample surface, followed by a proper annealing process to drive-in the impurities.

5.4.4.1   Gettering at Mechanically Damaged Surfaces

Dislocation arrays can be generated by different techniques, of which that relying on the application of a mechanical damage followed by plastic deformation was the first applied in microelectronic technology, using abrasives to scratch the sample surface under a predetermined contact pressure [76].

In correspondence to the scratches, after suitable thermal annealing, the local elastic stress is relieved by plastic deformation and dislocation emission (see Figure 5.12) and for this reason mechanical damage followed by plastic deformation has also been used for years as a convenient means to generate arrays of dislocations used to study the dislocation properties.

In spite of being highly efficient, relatively cheap and affordable in conventional production lines, this technique was very soon abandoned, due to several intrinsic drawbacks. The main problem is the difficulty in avoiding the uncontrolled penetration of the microdefects deep in the volume of the wafer, in spite of a careful choice of the abrasives used to scratch

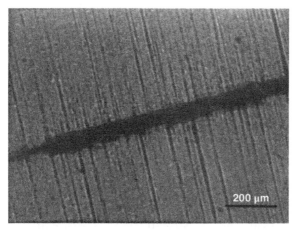

Figure 5.12 *Optical micrograph of an array of parallel 60° segments, generated by scratching with a diamond tool and thermal annealing a CZ (100) sample. The dislocations and the trace of the scratch are revealed by a selective chemical etch*

the sample surface, and the generation of surface microcracks. These side problems lead to limited reproducibility and impurity pollution due to traces of abrasives left in the material.

Dislocation networks acting as efficient impurity traps can, however, be generated by non-mechanical means, as will be seen in the next section.

5.4.4.2 *Phosphorus Diffusion Gettering*

Physico-chemical processes of external gettering allow milder, but effective and reliable conditions of operation, as demonstrated already 30 years ago by Ourmazd and Schröter [45], who showed that Ni is efficiently gettered at the surface of a silicon wafer doped with phosphorus from P_2O_5 vapours in a N_2 flux at 900 °C.[22]

The argument given to rationalize this process is that silicon monophosphide (SiP) particles segregated from a supersaturated P_xSi_{1-x} solid solution ($c_P(T) > c_p^0(T)$, where $c_p^0(T)$ is the P-solubility at the process temperature) at the surface of the sample behave as a source of self-interstitials, generated to relieve the lattice misfit[23] [77] due to the difference in the molar volumes of silicon and SiP

$$P_{Si} + (1 + x)Si_i \rightleftharpoons SiP + xSi_i \qquad (5.15)$$

In turn, the availability of a local self-interstitial excess favours the occurrence of a very effective Ni gettering, assumed to occur according to Eq. (5.14).

$$Ni_{Si} + 2I \overset{\Delta\Omega\approx0}{\rightleftharpoons} NiSi_2 \qquad (5.14)$$

Microstructural evidence of processes occurring in supersaturated P solutions is provided by X-ray and electron diffraction [78] and high resolution transmission electron microscopy

---

[22] Today phosphorus diffusion gettering (PDG) is carried out either using $POCl_3$ as the source gas or ion implantation.

[23] $\delta = +0.084$ in the <111> direction of silicon (*c*-axis of orthorhombic SiP) and -0.086 in the <110> direction (*a*-axis of SiP, with δ positive when the spacing of the SiP planes is larger than the corresponding planes of the matrix).

Table 5.1   *Solid solubilities of B and P in silicon in the temperature range of P gettering processes*

T (°C)	B N_{sl} (cm^{-3})	P N_{sl} (cm^{-3})
900	3.7×10^{20}	6.0×10^{20}
950	3.9×10^{20}	7.8×10^{20}
1000	4.1×10^{20}	1.0×10^{21}
1050	4.3×10^{20}	1.2×10^{21}
1100	4.5×10^{20}	1.4×10^{21}
1150	4.8×10^{20}	1.5×10^{21}
1200	5.0×10^{20}	1.5×10^{21}
1250	5.2×10^{20}	1.4×10^{21}
1300	5.4×10^{20}	1.1×10^{21}
1350	5.7×10^{20}	7.1×10^{20}

(HRTEM) measurements, that show the presence at the surface of the P-diffused sample of a network of flat, rod-like SiP precipitates [77].

Since it was demonstrated that self-interstitial injection dissolves existing precipitates or inhibits their formation [79], all silicide formation reactions involving a net volume change ($\Delta\Omega \neq 0$), like the following one involving dissolved iron[24] taken as typical

$$\text{Fe}_i + (2+x)\text{Si}_{Si} \underset{}{\overset{\Delta\Omega \neq 0}{\rightleftharpoons}} \text{FeSi}_2 + x\text{Si}_i \qquad (5.16)$$

may be treated as true equilibrium reactions.

PDG is, however, founded on a such a complex pattern of defect interaction processes, depending on the process temperature and regime, that it may occur also in the absence of SiP precipitates when $c_P(T) < c_P^0(T)$, or in the simultaneous presence of SiP precipitates and dislocations [79] or in the sole presence of dense dislocation networks.

In addition, the surface preoxidation could be a mean to improve the overall gettering efficiency, due to the presence of the surface oxide [80] which works as an independent gettering system.

As an example, the absence of SiP precipitates is typical of a standard PDG carried out at 1050–1150 °C for 1–7 hours using POCl$_3$ as the P-source [51, pp. 141–146]. Here, the P concentration in the diffused layer is close to 10^{21} cm^{-3} [81] (see also the P-solubility values reported in Table 5.1[25]), but below saturation conditions. Here, the surface of P-diffused layer is a concentrated solution ($\approx 3\%$ molar) of P in Si, in which metallic impurities could be dissolved and confined forming a multinary alloy.

This process regime is, however, also ideal for developing a network of misfit dislocations running up to a depth of 2 µm or more, induced by the elevated concentration of dissolved phosphorus [79, 82] and the consequent mechanical stress.

---

[24] Most 3d impurities sit preferentially in interstitial positions, with the exception of Au and Pt which sit preferentially in substitutional positions.

[25] Also the solubility of B is reported in this table, as B is used as a dopant and a diffusion getter.

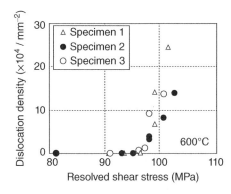
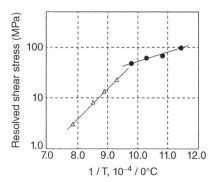

Figure 5.13   *Resolved shear stress for silicon and its temperature dependence. Ohta et al., 2001, [85]. Reproduced with permission from John Wiley & Sons*

Dissolution in Si of substitutional dopant impurities like P (and also B) with their atomic radii[26] very different from that of silicon, creates, in fact, the conditions for the set-up of local mechanical stress σ [83, 84]

$$\sigma = \frac{E}{1-\nu}\gamma C_s \qquad (5.17)$$

where E is Young's modulus, ν the Poisson coefficient, $\gamma = \frac{r_{imp}-r_{Si}}{r_{Si}}$ and C_s is the dopant concentration at the surface of the sample at the beginning of the diffusion process.

When σ exceeds the critical stress necessary for the generation of dislocation [84] (see Figure 5.13) [85] the system relaxes with the emission of self-interstitials and the set-up of a network of extrinsic misfit dislocations, which acts as a very efficient impurity gettering system. Similar conditions occur with boron diffusion gettering, although the critical stress is lower due to the lower solubility of B in Si.

An additional condition for dislocation generation occurs when the supersaturated dopant segregates as an intermediate compound (SiP in the case of PDG), leading to the local onset of elastic stress due to molar volume misfits between the silicon host and SiP, which relaxes with self-interstitials emission and extrinsic dislocation generation.

As impurity-decorated dislocations behave as recombination centres for minority carriers [86] (see Section 3.2.5), their presence should however be avoided in correspondence of the front junction of solar cells manufactured with both p-type or n-type silicon [83], requiring dedicated attention to the management of the phosphorus gettering process during solar cell manufacturing processes.

Additional effect associated to self-interstitial emission in P-gettering processes is not only the enhanced diffusion toward the P-rich surface of transition metal impurities, which, as shown in Chapter 2, mainly diffuse with a kick-off mechanism, but also the self-enhancement of the P-diffusion process.

This anomalous behaviour was explained by Mulvaney and Richardson [87] assuming that in a strongly n-type material P diffuses paired with silicon self-interstitials [88] as a

[26] $r_{B-} = 0.098\,\text{nm}, r_{p+} = 0.107\,\text{nm}, r_{Si} = 0.117\,\text{nm}.$

neutral PI species

$$I + e \rightleftharpoons I^-$$ (5.18)

$$P_{Si}^+ + I^- \rightleftharpoons PI$$ (5.19)

When the pair dissociates and the dopants return to substitutional positions (reverse of Eq. (5.19)) becoming electrically active [89] self-interstitials are injected in the bulk which becomes supersaturated.

Even if subsequent work [90] suggests the involvement of P-V pairs in the P-diffusion process, the conclusions of Morehead and Lever[89] remain fully valid, since the contribution of P–V pairs is orders of magnitude lower.

Thus, PDG occurs by self-interstitials enhanced diffusion of impurities towards the P-rich surface layer, where they are trapped at dislocations or segregate as solutes in the $S_{i1-x}P_x$ solution with a segregation coefficient less than one

$$K_i = \frac{x_i^b}{x_i^{Si_{1-x}P_x}} < 1$$ (5.20)

if the solubility of the impurity in the P-diffused surface layer is larger than in the bulk.

Several hypotheses have been proposed to explain the enhanced solubility of metal impurities in P-diffused regions, which increases with increase in the P-doping [55]. The formation of ternary complexes $M_x P_{1-x} Si$ [91] or of electrostatically stabilized pairs $M_{Si}^- P_{Si}^+$ [55] which are generated with metallic impurities sitting in substitutional positions, seem the most physically sensible explanation of the enhanced solubility, considering that metal impurities in silicon are mostly amphoteric species.

Taking as an example the case of $Au^- - P^+$ pairs, their formation energy could be calculated as the sum of the electrostatic stabilization energy ΔE^*, holding around 0.4 eV, and of a $\Delta\xi$, term, which derives from the difference between the ionization energy of donor phosphorus (0.044 eV) and the electron affinity of acceptor gold (0.54 eV), that is 0.496 eV [55].[27] Therefore, the formation energy of the pair is 0.896 eV or -86.56 kJ mol^{-1}, leading to an enhanced solubility of Au in the $S_{i1-x}P_x$ alloy.

The presence of a local self-interstitial excess leads, however, to a decrease in the concentration of substitutional impurities

$$I + M_s \rightleftharpoons M_i$$ (5.21)

which is the case for Au, Pt and Zn, which sit preferentially in substitutional positions, and also that of the substitutional concentration of most 3d impurities that sit preferentially in interstitial positions, see Section 2.3.

This is, however, only one of the problems which make difficult the predictive simulations of the PDG process [92] and, therefore, also the achievement of a fully satisfactory yield of the technological process itself.

Recent progress has been made by applying ion implantation [93] with a shallower getter region, showing however that P-gettering is case sensitive, since it depends on the amount of impurity present.

---

[27] Baldi *et al.* assume that P diffuses as a P-V species [55].

5.4.4.3   Aluminium Gettering

Aluminium gettering is the simplest gettering process among those discussed in this section as it consists of the equilibrium segregation of impurities from a bulk silicon phase to a diluted Al–Si phase, which is obtained by depositing a thin Al layer on Si [90]. The process is carried out with excellent gettering yield at 700–900 °C, that is above the eutectic temperature of the binary Al–Si system (577 °C).

As already shown in Sections 1.7.5 and 4.2.7, see also Figure 1.36, Al has a relatively low solubility in solid silicon at any temperature (the maximum solubility is $4.3 \cdot 10^{-4}$ in molar fraction or $2.9 \cdot 10^{19}$ at cm^{-3} at 1450 K) [94]. In addition, the solubility of impurities in Al-Si alloys, irrespective of being liquid or solid [95] is higher than in silicon. For example, at 1093 K the segregation coefficient of cobalt in Al, $K_{Co} = x_{Co(s)}/x_{Co(l)}$, is 10^4 times smaller than that of Co in silicon [96].

The segregation coefficients of Fe and Ti are even smaller ($K_{Fe}(1073 \text{ K}) = 1.7 \cdot 10^{-11}$, $K_{Fe}(1273 \text{ K}) = 5.9 \cdot 10^{-9}$: $K_{Ti}(1073 \text{ K}) = 3.8 \cdot 10^{-9}$; $K_{Ti}(1273 \text{ K}) = 1.6 \cdot 10^{-7}$), compared with those of B and P ($K_B(1073 \text{ K}) = 7.6 \cdot 10^{-2}$; $K_B(1273 \text{ K}) = 2.2 \cdot 10^{-1}$; $K_P(1073 \text{ K}) = 4 \cdot 10^{-2}$; $K_P(1273 \text{ K}) = 8.5 \cdot 10^{-2}$) [97].

It should be noted here that the solubility values used to calculate the segregation coefficient of a specific impurity M should refer to the ternary system involved, not to the corresponding binary Al–M system, on which more information is normally available from the literature.

In ternary systems involving 3d elements as the third component, see Figure 5.14[28] [99–101] the composition range of the Al-rich liquid phase depends strongly on the nature of the 3d impurity. In addition, the 3d impurity solubility depends on the composition of the liquid phase, since the nature of the solid phase in equilibrium with the liquid, see again Figure 5.14, depends on the alloy composition.

Therefore, the predictive simulation of the Al gettering is even more uncertain than that predicted by Abdelbarey *et al.* [102], if dedicated to a system containing, simultaneously, several 3d impurities.

In fact, this is a problem involving a multinary solution and each impurity cannot be considered an independent variable of the system.

In the approximation that each impurity dissolved in a common host consisting of a binary Al-Si alloy will actually behave independently of the other ones, one could argue that the thermodynamic stability of these ternary solutions and, therefore, their Gibbs energy G_s could be described, see Section 1.7.11.2, by

$$G^s = G^{o,s}_{AC} y_{AC} + G^{o,s}_{BC} x_{BC} + RT(y_{AC} \ln y_{AC} + y_{BC} \ln x_{BC}) + L^s_{AC-BC} y_{AC} y_{BC} \qquad (5.22)$$

where A is Al, B is the 3d impurity and C is Si, where the interaction effects are dominated by the term $L^s_{AC-BC} y_{AC} y_{BC}$ involving the Al–3d and Si–3d impurities interactions.

While the L^s_{BC} values could be deduced from the formation enthalpies of the silicides, $-20.4 \text{ kJ mol}^{-1}$) for Fe, $-78.62 \text{ kJ mol}^{-1}$ for Ni and (-148 kJ mol^{-1}) for cobalt disilicide [98, 103, 104] the L^s_{AC} values could be evaluated from the respective values of the aluminides [105] which are -25 kJ mol^{-1} for FeAl, 55.2 kJ mol^{-1} for CoAl

---

[28] It should be noted that the phase diagram of the system Al–Co–Si is relative to a temperature of 600 °C, and that only a partial diagram of this system is available for a temperature of 900 °C [98].

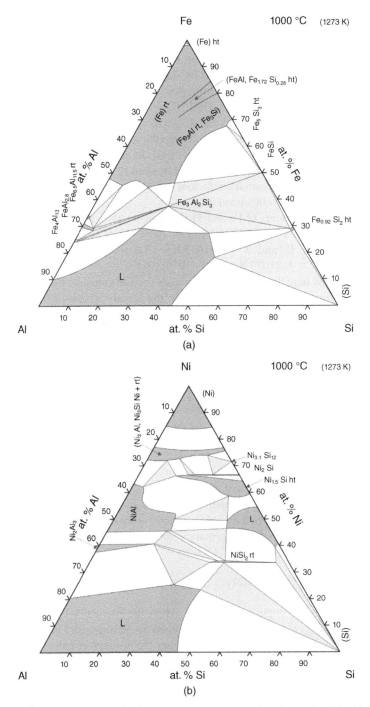

Figure 5.14 *Phase diagrams of the ternary systems (a) Al–Fe–Si, (b) Al–Ni–Si and (c) Al–Co–Si. Ghosh, 1992, [99]; Raghavan, 2005, [100] and Schmid Fetzer, 1991, [101]. Reproduced with permission from ASM International*

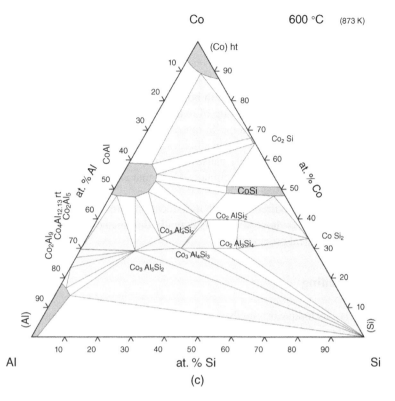

Figure 5.14   *(continued)*

and 58.99 kJ mol^{-1} for NiAl. On this basis, the L^s_{AC-BC} term is -4.6 kJ mol^{-1} for Fe, -19.63 kJ mol^{-1} for Ni and -92.8 kJ mol^{-1} for Co, showing a progressive deviation from ideality passing from Fe to Ni and Co.

5.4.4.4   Cleaning of Extended Defects in mc-Si

One of the main problems in the use of mc-Si for high efficiency solar cells is the minority carrier lifetime degradation induced by electrically active GBs and dislocations. Electrical activity of dislocations is mostly associated, as already seen in Section 3.2.5, with impurity decoration, caused by impurity segregation in their strain field, by formation of chemical bonds at the dislocation core or/and by the segregation of a solid phase consisting of a silicide or an oxide in the case of simultaneous oxygen gettering.

Also in the case of GBs, as already seen in Sections 3.6.3 and 3.6.4, the electrical activity is mostly due to impurity segregation and there is experimental evidence that segregation of impurities occurs via formation of bonds at the GB core or via the segregation of silicide microprecipitates at or close to the GB. It has also been seen that the segregation configuration depends on the nature of the impurity. In both dislocations and GBs, also oxygen segregates extended defects, with the formation of oxide phases or even silicate phases.

To clean dislocations and GBs, therefore, it is necessary to use a process for which the Gibbs energy increase (in absolute values) ΔG_{proc} is higher than that associated with dislocation gettering ΔG_{dislo}, such that

$$\Delta G_{proc} \ll \Delta G_{dislo} \tag{5.23}$$

P-diffusion gettering of Fe, Ni and Cu in both single crystal and mc-Si is demonstrated to be capable of a very efficient clean-up of extended defects, including dislocations and BDs [106]. It is also known that in a mc-Si material contaminated with iron and nitrogen, exhibiting the presence of silicon and iron silicides, detected by TEM and electron diffraction experiments, P-gettering not only causes a dramatic increase in the lifetime but also the dissolution of the iron silicides precipitates [107] that are additional causes of lifetime degradation.

There is, therefore, the indirect proof that dislocations are not the most active species in the P-gettering process whose activity is actually dominated by a complex pattern of different physico-chemical processes.

5.5 Wafer Bonding

Wafer bonding is a technique originally conceived around the 1980s as a means to fabricate multilayer structures consisting of a thin layer of a crystalline and/or amorphous semiconductor, integrated in a hybrid functional structure [108]. Today it is used in a number of applications, of which typical examples are given by the integration of a micro-electro mechanical system (MEMS) on a silicon wafer, the fabrication of silicon on insulator (SOI) structures and the fabrication of multi-junction structures for high concentration, high efficiency solar cells [109]. There are, however, three basic problems to be considered, in order to transform the technique into a viable industrial process. The first is to design a bonding process which would have a negligible effect on the mechanical or electronic properties of the layers to be joined. The second is to prepare *perfectly* flat surfaces capable of working in a stick and bond process. The third is to remove matter from one of the components layers, which should work as the substrate, in order to transform it into a thin layer of uniform thickness.

Depending on the specific device configuration which should be developed and on the temperature sensitivity of the active layers, different types of bonding should be chosen, among which direct bonding, fusion bonding, alloy bonding and adhesive bonding processes are widely employed.

While fusion bonding, alloy bonding and adhesive bonding processes lead to the formation of an intermediate layer, which should be (thermally and/or electrically) insulating or conductive, direct bonding establishes a physical or a chemical bond between two physically flat surfaces, with obvious advantages with respect to electronic device integration.

In addition, direct bonding between hydrophilic and hydrophobic surfaces has the advantage of being a low temperature process and of using well known chemical processes. Limiting the interest to the case of silicon, hydrophilic or hydrophobic surfaces could be obtained with simple thermal or chemical etching processes. A hydrophilic surface is a surface covered by its native oxide layer or by an oxide layer obtained by a mild chemical oxidation, using a hydrogen peroxide solution[29] as the oxidant. A common chemical

---

[29] A $1:1:5$ $NH_3:H_2O_2:H_2O$ solution.

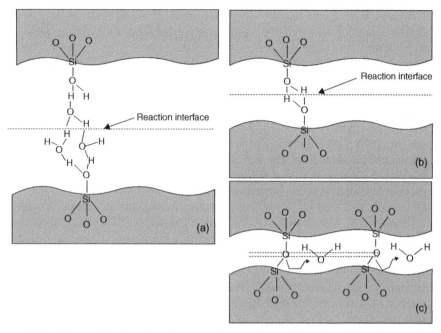

Figure 5.15 *Hydrophilic bonding of two silicon wafers (a) bonding with water molecules, (b) bonding via silanolic bonds and (c) bonding with oxygen bonds. Suni, 2006, [110]. Reproduced with permission from VTT Technical Research Centre of Finland*

process used for getting, instead, a temporary hydrophobic surface is room temperature hydrofluorination, which removes the native oxide layer and leads to the formation of surface fluorine or hydrogen bonds. The re-oxidation process of these surfaces is very fast, and the material should be stored under vacuum to preserve the hydrophobicity, unless immediately processed.

Both processes are routinely employed in microelectronics, making wafer bonding easily integrated in microelectronic processes.

A schematic flowsheet of the hydrophilic bonding during the thermal annealing of two flat silicon surfaces is displayed in Figure 5.15 [110], which shows that in the initial step bonding is due to water molecule bridges. Once the water molecules are released, silanolic bonding occurs, while at the end of water removal the two layers are bonded by strong oxygen bonds.

References

1. Baratte, H., deSouza, J.P. and Sadana, D.K. (1993) Internal gettering of oxygen in III-V compound semiconductors. US Patent 5,272,373, Dec. 21, 1993.
2. Sumino, K. (1992) Role of impurities in reducing grown-in dislocations in compound semiconductor crystals. *MRS Symp. Proc.*, **262**, 87.
3. Greuter, F. and Blatter, G. (1990) Electrical properties of grain boundaries in polycrystalline compound semiconductors. *Semicond. Sci. Technol.*, **5**, 111–137. doi: 10.1088/0268-1242/5/2/001.

4. Normann, H.B., Svensson, B.G., and Monakhov, E. (2012) Formation of shallow front emitters for solar cells by rapid thermal processing. *Phys. Status Solidi C*, **9**, 2138–2140.

5. Serra, J.M. (2012) Silicon processing for photovoltaics based on incoherent radiation power. *Phys. Status Solidi C*, **9**, 2169–2172.

6. Reichel, D. and Skorupa, W. (2012) Precise millisecond annealing for advanced material processing. *Phys. Status Solidi C*, **9**, 2045–2049.

7. Rawski, M., Zuk, J., Kulik, M., Drozdziel, A., Lin, L., Prucnal, S., Pyszniak, K., and Turek, M. (2011) Influence of hot implantation on residual radiation damage in silicon carbide. *Acta Phys. Pol. A*, **120**, 192–195.

8. Narayan, J., Young, R.T., and White, C.W. (1978) A comparative study of laser and thermal annealing of boron implanted silicon. *J. Appl. Phys.*, **49**, 3912–3917.

9. Oberemok, O., Klad'ko, V., Litovchenko, V., Romanyuk, B., Popov, V., Melnik, V., Sarikov, A., Gudymenko, O., and Vanhellemont, J. (2014) Stimulated oxygen impurity gettering under ultra-shallow junction formation in silicon. *Semicond. Sci. Technol.* **29**, 05508–055015.

10. Kowalski, M., Kowalski, J.E. and Lojek, B. (2007) Microwave annealing for low temperature activation of As in Si. Proceedings of the 15th IEEE International Conference on Advanced Thermal Processing of Semiconductors – RTP2007, Catania, Italy, pp. 51–56.

11. Lojek, B. (2008) Low temperature microwave annealing of S/D. Proceedings of the 16th IEEE International Conference on Advanced Thermal Processing of Semiconductors – RTP2008 978-1.

12. Sameshima, T., Hayasaka, H., and Haba, T. (2009) Analysis of microwave absorption caused by free carriers in silicon. *Jpn. J. Appl. Phys.*, **48**, 0212041–6.

13. Xu, P., Fu, C., Hu, C., Zhang, D.W., Wu, D., Luo, J., Zhao, C., Zhang, Z.-B., and Zhang, S.-L. (2013) Ultra-shallow junctions formed using microwave annealing junctions formed using microwave annealing. *Appl. Phys. Lett.*, **102**, 122114.

14. Yang, H.-J., Yuan, J., Li, Y., Hou, Z.-L., Jin, H.-B., Fang, X.-Y., and Cao, M.-S. (2013) Silicon carbide powders: temperature-dependent dielectric properties and enhanced microwave absorption at gigahertz range. *Solid State Commun.*, **163**, 1–6.

15. Gardner, J.A., Rao, M.V., Tian, Y.L., Holland, O.W., Roth, E.G., Chi, P.H., and Ahmad, I. (1997) Rapid thermal annealing of ion implanted 6H-SiC with microwave processing. *J. Electron. Mater.*, **26**, 144–149.

16. Acciarri, M., Binetti, S., Bollani, M., Comotti, A., Fumagalli, L., Pizzini, S., and von Kaenel, H. (2005) Nanocrystalline silicon grown by LEPECVD for photovoltaic applications. *Sol. Energy Mater. Sol. Cells*, **87**, 11–24.

17. Binetti, S., Acciarri, M., Bollani, M., Fumagalli, L., von Kanel, H., and Pizzini, S. (2005) Nanocrystalline silicon film grown by LEPECVS for optoelectronic applications. *Thin Solid Films*, **487**, 19–25.

18. Zacharias, M., Heitmann, J., Scholz, R., Kahler, U., Schmidt, M., and Bläsing, J. (2002) Size-controlled highly luminescent silicon nanocrystals: a SiO/SiO_2 superlattice approach. *Appl. Phys. Lett.*, **80**, 661.

19. Zacharias, M., Yi, L.X., Heitmann, J., Scholta, R., Reiche, M., and Gösele, U. (2003) Size-controlled Si nanocrystals for photonic and electronic applications. *Solid State Phenom.*, **94**, 95–104.

20. Shukla, R., Summonte, C., Canino, M., Allegrezza, M., Bellettato, M., Desalvo, A., Nobili, D., Mirabella, S., Sharma, N., Jangir, M., and Jain, I.P. (2012) Optical and electrical properties of Si nanocrystals embedded in SiC matrix. *Adv. Mater. Lett.*, **3**, 297–304.

21. Löper, P., Canino, M., Qazzazie, D., Schnabel, M., Allegrezza, M., Summonte, C., Glunz, S.W., Janz, S., and Zacharias, M. (2013) Silicon nanocrystals embedded in silicon carbide: investigation of charge carrier transport and recombination. *Appl. Phys. Lett.*, **102**, 033507.

22. Allegrezza, M., Bellettato, M., Liscio, F., Canino, M., Desalvo, A., López-Vidrier, J., Hernández, S., López-Conesa, L., Estradé, S., Peiró, F., Garrido, B., Löper, P., Schnabel, M., Janz, S., Guerra, R., and Ossicini, S. (2014) Silicon nanocrystals in carbide matrix. *Sol. Energy Mater. Sol. Cells*, **128**, 138–149.

23. Severino, A. (2012) 3C-SiC epitaxial growth on large area silicon: thin films, in *Silicon Carbide Epitaxy* (ed F. La Via), Research Signpost, Trivandrum, pp. 145–191.

24. Kaiser, R.J., Koffel, S., Pichler, P., Bauer, A.J., Amon, B., Frey, L., and Ryssel, H. (2011) Germanium substrate loss during thermal processing. *Microelectron. Eng.*, **88**, 499–502.

25. Sheehy, M.A., Tull, B.R., Friend, C.M., and Mazur, E. (2007) Chalcogen doping of silicon via intense femtosecond-laser irradiation. *Mater. Sci. Eng., B*, **137**, 289–294.

26. Newman, B.K., Ertekin, E., Sullivan, J.T., Winkler, M.T., Marcus, M.A., Fakra, S.C., Sher, M.-J., Mazur, E., Grossman, J.C., and Buonassisi, T. (2013) Extended X-ray absorption fine structure spectroscopy of selenium-hyperdoped silicon. *J. Appl. Phys.*, **114**, 133507–133514.

27. N. Odin, V.A. Ivanov, Phase equilibria in the Cd-Si-Se system and properties of the compound Cd4SiSe6, *Russ. J. Inorg. Chem.* **36**, 1650–1653, (1991) see also ASM Alloy Phase Diagrams, (1991).

28. Kissinger, G., Kot, D., and Sattler, A. (2014) The influence of flash lamp annealing on the minority carrier lifetime of Czochralski silicon wafers. *AIP Proc.*, **1583**, 94–99.

29. Pearton, S.J., Corbett, J.W., and Stavola, M. (1992) *Hydrogen in Crystalline Semiconductors*, Springer, Berlin.

30. Pankove, J.I., Carlson, D.E., Berkeyheiser, J.E., and Wance, H.O. (1983) Neutralization of shallow acceptor levels in silicon by atomic hydrogen. *Phys. Rev. Lett.*, **51**, 2224.

31. Van de Walle, C.G., Denteneer, P.J.H., Bar-Yam, Y., and Pantelides, S.T. (1989) Theory of hydrogen diffusion and reactions in crystalline silicon. *Phys. Rev. B*, **39**, 10791–10808.

32. Johnson, N.M., Doland, C., Ponce, F., Walker, J., and Anderson, G. (1991) Hydrogen in crystalline semiconductors. *Physica B*, **170**, 3–20.

33. Myers, S.M., Baskes, M.I., Corbett, J.W., DeLeo, G.G., Estreicher, S.K., Haller, E.E., Johnson, N.M., and Pearton, S.J. (1992) Hydrogen interactions with defects in crystalline solids. *Rev. Mod. Phys.*, **64**, 559–617.

34. Estreicher, S.K. (1995) Hydrogen-related defects in crystalline semiconductors: a theorist's perspective. *Mater. Sci. Eng., R*, **14**, 319–412.

35. Van de Walle, G. and Neugebauer, J. (2003) Universal alignment of hydrogen levels in semiconductors, insulators and solutions. *Nature*, **423**, 626–628.

36. Van de Walle, C.G. (2006) Universal alignment of hydrogen levels in semiconductors and insulators. *Physica B*, **376-377**, 1–6.
37. Estreicher, S.K., Stavola, M., and Weber, J. (2015) Hydrogen in Si and Ge, in *Silicon, Germanium and Silicon-Germanium Alloys: Growth, Defects, Impurities and Nanocrystals* (eds G. Kissinger and S. Pizzini), CRC Press pp. 217–254.
38. Pizzini, S., Acciarri, M., Binetti, S., Narducci, D., and Savigni, C. (1997) Recent achievements in semiconductor defect passivation. *Mater. Sci. Eng., B*, **45**, 126–133.
39. Binetti, S., Basu, S., Savigni, C., Acciarri, M., and Pizzini, S. (1997) Passivation of extended defects in silicon by catalytically dissociated molecular hydrogen. *J. Phys. III (France)*, **7**, 1487–1493.
40. Aberle, A.G. (2001) Overview on SiN surface passivation of crystalline silicon solar cells. *Sol. Energy Mater. Sol. Cells*, **65**, 239–248.
41. Duerinckx, F. and Szlufcik, J. (2002) Defect passivation of industrial multicrystalline solar cells based on PECVD silicon nitride. *Sol. Energy Mater. Sol. Cells*, **72**, 231–246.
42. Duttagupta, S., Ma, F., Hoex, B., Mueller, T., and Aberle, A.G. (2012) Optimised antireflection coatings using silicon nitride on textured silicon surfaces based on measurements and multidimensional modelling. *Energy Procedia*, **15**, 78–83.
43. Bousbih, R., Dimassi, W., Haddadi, I., and Ezzaouia, H. (2012) The effect of thermal annealing on the properties of PECVD hydrogenated silicon nitride. *Phys. Status Solidi C*, **9**, 2189–2193.
44. Wenham, S.R. (2013) Advanced hydrogenation of silicon solar cells. WO Patent 2013173867, PCT/AU2013/000528 May 20, 2013.
45. Ourmazd, A. and Schröter, W. (1984) Phosphorus gettering and intrinsic gettering of nickel in silicon. *Appl. Phys. Lett.*, **45**, 781.
46. Tan, T.Y., Gardner, E.E., and Tice, W.K. (1977) Intrinsic gettering by oxide precipitate induced dislocations in Czochralski Si. *Appl. Phys. Lett.*, **30**, 175.
47. Gilles, D., Weber, E.R., and Hahn, S.-K. (1990) Mechanism of internal gettering of interstitial impurities in czochralski-grown silicon. *Phys. Rev. Lett.*, **64**, 196.
48. Macdonald, D., Liu, A.Y., and Phang, S.P. (2014) External and internal gettering of interstitial iron in silicon for solar cells. *Solid State Phenom.*, **205-206**, 26–33.
49. Schröter, W., Seibt, M., and Gilles, D. (2000) High temperature properties of transition elements in silicon, in *Handbook of Semiconductors Technology: Electronic Structure and Properties of Semiconductors*, vol. **1** (eds K.A. Jackson and W. Schröter), Wiley-VCH Verlag GmbH. doi: 10.1002/9783527621842.ch10.
50. Seibt, M. and Kveder, V. (2012) Gettering processes and the role of extended defects, in *Advanced Silicon Materials for Photovoltaic Applications* (ed S. Pizzini), John Wiley & Sons, Ltd, pp. 127–188.
51. Perevostchikov, V.A. and Skoupov, V.D. (2005) *Gettering Defects in Semiconductors*, Springer.
52. Istratov, A.A., Hieslmair, H., and Weber, E.R. (1999) Iron and its complexes in silicon. *Appl. Phys. A*, **69**, 13–44.
53. Sumino, K. (1999) Impurity reaction with dislocations in semiconductors. *Phys. Status Solidi A*, **171**, 111–122.

54. Geranzani, P., Pagani, M., Pello, C., and Borionetti, G. (2002) Internal gettering in silicon: experimental studies based on fast and slow diffusing metals. *Solid State Phenom.*, **82-84**, 381–386.
55. Baldi, L., Cerofolini, G.F., Ferla, G., and Frigerio, G. (1978) Gold solubility in silicon and gettering by phosphorus. *Phys. Status Solidi A*, **48**, 523–532.
56. Seibt, M. (1992) Relaxation-induced gettering of metal impurities in silicon: microscopic properties of effective gettering sites. *MRS Symp. Proc.*, **262**, 957–962.
57. Kim, Y.-H., Lee, K.-S., Chung, H.-Y., Hwang, D.-H., Kim, H.-S., Cho, H.-Y., and Lee, B.-Y. (2001) Internal gettering of Fe, Ni and Cu in silicon wafers. *J. Korean. Phys. Soc.*, **39**, S348–S351.
58. Ceccaroli, B. and Pizzini, S. (2012) Processes, in *Advanced Silicon Materials for Photovoltaic Applications* (ed S. Pizzini), John Wiley & Sons, Ltd, pp. 21–78.
59. Falster, R., Laczik, Z., Booker, G.R., Bhatti, A.R., and Török, P. (1992) Gettering and gettering stability of metals at oxide particles in silicon. *MRS Symp. Proc.*, **262**, 945–956.
60. Dash, W.C. (1956) Copper precipitation on dislocations in silicon. *J. Appl. Phys.*, **27**, 1193–1195.
61. Schwuttke, G.H. (1961) Study of copper precipitation behaviour in silicon single crystals. *J. Electrochem. Soc.*, **108**, 163–167.
62. Tiller, W.A. (1958) Production of dislocations during growth from the melt. *J. Appl. Phys.*, **29**, 611–617.
63. Seibt, M., Hedemann, H., Istratov, A.A., Riedel, F., Sattler, A., and Schröter, W. (1999) Structural and electrical properties of metal silicide precipitates in silicon. *Phys. Status Solidi A*, **171**, 301–310.
64. Feng, Y.Z. and Wu, Z.Q. (1996) High resolution electron microscopy study of nickel silicide-silicon interface by molecular beam epitaxy. *J. Mater. Sci. Lett.*, **15**, 2000–2001.
65. Nash, P. and Nash, A. (1987) The Ni-Si (Nickel-Silicon) system. *Bull. Alloy Phase Diagr.*, **8**, 1–14.
66. Fe-Si (Iron-silicon), in *Landolt-Börnstein New Series IV*, vol. **5**, Springer-Verlag (1991).
67. Tang, K. and Tangstad, M. (2012) A thermodynamic description of the Si-rich Si-Fe system. *Acta Metall. Sinica (Engl. Lett.)*, **25**, 249–255.
68. Sumino, K. (1989) Interaction of impurities with dislocations in semiconductors, in *Point and Extended Defects in Semiconductors*, NATO ASI Series, Series B: Physics, vol. **202**, Plenum Press, pp. 77–94.
69. Olesinski, R.W. and Abbaschian, G.J. (1986) The Cu-Si (copper-silicon) system. *Bull. Alloy Phase Diagr.*, **7**, 702–781.
70. Wen, C.-Y. and Spaepen, F. (2007) *In situ* electron microscopy of the phases of Cu_3Si. *Philos. Mag.*, **87**, 5581–5599.
71. Weber, E. and Riotte, H.G. (1980) The solution of iron in silicon. *J. Appl. Phys.*, **51**, 1484–1488.
72. Hirsch, F., Orschel, B., Rouvinov, S., and Shabani, M. (2002) Comparison of nickel and iron gettering in CZ silicon wafers. *Solid State Phenom.*, **82-84**, 367–372.

73. Regula, G., El Bouayadi, R., Pichaud, B., and Ntsoenzok, E. (2002) Nickel gettering in silicon: the role of oxygen. *Solid State Phenom.*, **82-84**, 355–360.
74. Isomae, S., Ishida, H., Itoga, T., and Hozawa, K. (2002) Efficiency of intrinsic gettering for copper in silicon. *Solid State Phenom.*, **82-84**, 349–354.
75. Seibt, M., Griess, M., Istratov, A.A., Hedemann, H., Sattler, A., and Schröter, W. (1998) Formation and properties of copper silicide precipitates in silicon. *Phys. Status Solidi A*, **166**, 171–182.
76. Vanhellemont, J., Claeys, C., and Van Landuyt, J. (1995) In situ HVEM study of dislocation generation in patterned stress fields at silicon surfaces. *Phys. Status Solidi A*, **150**, 497–506.
77. Servidori, M. and Armigliato, A. (1975) Electron microscopy of silicon monophosphide precipitates in P-diffused silicon. *J. Mater. Sci.*, **10**, 306–313.
78. Schmidt, P.F. and Stickler, R. (1964) SiP precipitates in P diffused silicon. *J. Electrochem. Soc.*, **111**, 1188–1189.
79. Polignano, M.L., Cerofolini, G.F., Bender, H., and Claeys, C. (1988) Gettering mechanisms in silicon. *J. Appl. Phys.*, **64**, 869–876.
80. Savigni, C., Acciarri, M., and Binetti, S. (1996) About a novel gettering procedure for multicrystalline silicon samples. *Solid State Phenom.*, **51-52**, 485–490.
81. Trumbore, F.A. (1976) Solid solubilities of impurity elements in germanium and silicon. *Bell Syst. Tech. J.*, **19**, 911, 38-43.
82. Rozgony, G.A., Petroff, P.M., and Read, M.H. (1975) Elimination of oxidation-induced stacking faults by peroxidation gettering of silicon wafers I. Phosphorous diffusion-induced misfit dislocations. *J. Electrochem. Soc.*, **122**, 1725–1729.
83. Queisser, H.J. (1961) Slip patterns on boron doped silicon surfaces. *J. Appl. Phys.*, **32**, 1776–1780.
84. Prussin, S. (1961) Generation and distribution of dislocations by solute diffusion. *J. Appl. Phys.*, **32**, 1876–1881.
85. Ohta, H., Miura, H., and Kitano, M. (2001) A ball-indentation method to evaluate the critical stress for dislocation generation in a silicon substrate. *Fatigue Fract. Eng. Mater. Struct.*, **24**, 877–884.
86. Radzimski, Z.J., Zhou, T.Q., Rozgonyi, G.A., Finn, D., Hellwing, L.G., and Ross, J.A. (1992) Recombination at clean and decorated misfit dislocations. *Appl. Phys. Lett.*, **60**, 1096.
87. Mulvaney, B.J. and Richardson, W.B. (1987) Model for defect-impurity pair diffusion in silicon model for defect-impurity pair diffusion in silicon. *Appl. Phys. Lett.*, **51**, 1439.
88. Cowern, N.E.B. (1989) Ion pairing effects on substitutional impurity diffusion in silicon. *Appl. Phys. Lett.*, **54**, 703.
89. Morehead, F.F. and Lever, R.F. (1989) The steady-state model for coupled defect-impurity diffusion in silicon. *J. Appl. Phys.*, **66**, 5349–5352.
90. Kveder, V., Schroter, W., Sattler, A., and Seibt, M. (2000) Simulation of Al and P diffusion gettering in Si. *Mater. Sci. Eng., B*, **71**, 175–181.
91. Wilcox, R.W., La Chapelle, T.J., and Forbes, D.H. (1964) Gold in silicon: effect on resistivity and diffusion in heavily-doped layers. *J. Electrochem. Soc.*, **111**, 1377–1380.
92. Spiecker, E., Seibt, M., and Schröter, W. (1997) Phosphorous-diffusion gettering in the presence of a nonequilibrium concentration of silicon interstitials: a quantitative model. *Phys. Rev. B*, **55**, 9577–9583.

93. Vähänissi, V., Haarahiltunen, A., Yli-Koski, M., and Savin, H. (2014) Gettering of iron in silicon solar cells with implanted emitters. *IEEE J. Photovoltaics*, **4**, 142–147.

94. Yoshikawa, T. and Morita, K. (2003) Solid solubilities and thermodynamic properties of Al in solid silicon. *J. Electrochem. Soc.*, **150**, G465–G468.

95. Joshi, S.M., Goesele, U., and Tan, T.Y. (2001) Extended high temperature Al gettering for improvement and homogenization of minority carrier diffusion lengths in multicrystalline Si. *Sol. Energy Mater. Sol. Cells*, **70**, 231–238.

96. Apel, M., Hanke, I., Schindler, R., and Schröter, W. (1994) Aluminium gettering of cobalt in silicon. *J. Appl. Phys.*, **76**, 4432–4433.

97. Yoshikawa, T. and Morita, K. (2005) Refining of Si by the solidification of Si–Al melt with electromagnetic force. *ISIJ Int.*, **45**, 967–971.

98. Richter, K.W. and Gutierrez, D.T. (2005) Phase equilibria in the system Al-Co-Si. *Intermetallics*, **13**, 848–856.

99. Ghosh, G. (1992) Aluminium-iron-silicon, in *Ternary Alloys*, vol. **5**, VCH Publishers, pp. 394–438.

100. Raghavan, V. (2005) Al-Ni-Si (aluminum-nickel-silicon). *J. Phase Equilib. Diffus.*, **26**, 262–267.

101. Schmid Fetzer, R. (1991) Aluminium-cobalt-silicon, in *Ternary Alloys*, vol. **4**, VCH Publishers, pp. 254–257.

102. Abdelbarey, D., Kveder, V., Schröter, W., and Seibt, M. (2009) Aluminum gettering of iron in silicon as a problem of the ternary phase diagram. *Appl. Phys. Lett.*, **94**, 061912.

103. Acker, J., Bohmhammel, K., van den Berg, G. J. K., van Miltenburg, J. C. (1999) Thermodynamic properties of iron silicides FeSi and α-FeSi2. *J. Chem. Thermodyn.*, **31**, 1523–1536

104. Van Bockstael, Ch. (2009-2010) In situ study of the formation and properties of Nickel silicides. Thesis, Universiteit Ghent.

105. Kubaschewski, O. and Heymer, G. (1960) Heats of formation of transition-metal aluminides. *Trans. Faraday Soc.*, **56**, 473–478.

106. Shabani, M., Yamashita, T., and Morita, E. (2008) Study of gettering mechanism in silicon: competitive gettering between phosphorous diffusion gettering and other gettering sites. *Solid State Phenom.*, **131-133**, 399–402.

107. Binetti, S., Acciarri, M., Savigni, C., Brianza, A., Pizzini, S., and Musinu, A. (1996) Effect of nitrogen contamination by crucible encapsulation on polycrystalline silicon material quality. *Mater. Sci. Eng., B*, **36**, 68–72.

108. Alexe, M. and Gösele, U. (eds) (2004) *Wafer Bonding, Application and Technology*, Springer.

109. Krause, R., Ghyselen, B. and Dimroth, F. (2013) Wafer Bonding Creates a Record- Breaking Four-Junction Cell, November/December 2013, pp. 36–39, www.compoundsemiconductor.net (accessed 7 April 2015).

110. Suni, T. (2006) Direct wafer bonding for MEMS and Microelectronics. Dissertation, Helsinki University of Technology, VTT Publication 609, ISBN: 951.38.6851.6, July 2006.

Index

Printed and bound by CPI Group (UK) Ltd, Croydon, CR0 4YY

27/10/2024

14580307-0004